Principles of Bioinorganic Chemistry

PRINCIPLES OF
Bioinorganic Chemistry

Stephen J. Lippard

MASSACHUSETTS INSTITUTE OF TECHNOLOGY

Jeremy M. Berg

JOHNS HOPKINS SCHOOL OF MEDICINE

University Science Books
Mill Valley, California

University Science Books
20 Edgehill Road
Mill Valley, CA 94941
Fax: (415) 383-3167

Associate publisher: *Jane Ellis*
Production manager: *Susanna Tadlock*
Manuscript editor: *Aidan Kelly*
Designer: *Robert Ishi*
Illustrators: *Georg Klatt, Audre Newman, John and Judy Waller*
Compositor: *ASCO Trade Typesetting*
Color separator: *Color Response*
Printer and binder: *Edwards Brothers*

This book is printed on acid-free paper.

Library of Congress Cataloging-in-Publication Data

Lippard, Stephen J.
   Principles of bioinorganic chemistry / Stephen J. Lippard, Jeremy
M. Berg.
      p.   cm.
   Includes bibliographic references and index.
   ISBN 0-935702-72-5 (cloth) :   ISBN 0-935702-73-3 (paper)
   1. Bioinorganic chemistry.   I. Berg, Jeremy Mark.   II. Title.
QP531.L55   1994
574.19'214 —dc20   91-67871
                                                          CIP

Printed in the United States of America
10 9 8 7 6 5 4 3

*This book is dedicated to
Judy, Josh, Alex, and in memory of Andrew
and to
Wendie, Alexander, and Corey
whose love and patience with us were
essential to its completion.*

Contents

CHAPTER 3   Properties of Biological Molecules 43

CHAPTER 4   Physical Methods in Bioinorganic Chemistry 75

CHAPTER 5   Choice, Uptake, and Assembly of Metal Containing Units in Biology 103

CHAPTER 6   Control and Utilization of Metal-Ion Concentration in Cells 139

CHAPTER 12   Protein Tuning of Metal Properties to Achieve Specific Functions 349

Preface

Bioinorganic chemistry is a leading discipline at the interface of chemistry and biology. Many critical processes require metal ions, including respiration, much of metabolism, nitrogen fixation, photosynthesis, development, nerve transmission, muscle contraction, signal transduction, and protection against toxic and mutagenic agents. Unnatural metals have been introduced into human biology as diagnostic probes and drugs. As more systems requiring metal ions have been discovered, the number of research papers in bioinorganic chemistry has multiplied to the point where what was once an expanding frontier is now a maturing one. From the many beautiful studies of bioinorganic systems have emerged not only deep insight into individual cases, but also principles that tie together seemingly unrelated facts.

This book attempts to unify the field of bioinorganic chemistry by identifying the principles that have emerged and arranging them in a logical and consistent order. We first consider which metal ions are used in living organisms, why nature might have chosen them, how they get into cells, and how their concentrations are regulated. We next discuss how metals bind to biopolymers, how metal binding can fold biopolymers, leading to function, and how they are inserted into their active centers. We finally treat the major roles of metal ions in biological systems, as electron carriers, centers for binding and activating substrates, agents for transferring atoms and groups, and as core units, or "bioinorganic chips," the functions of which can be tuned in a given biological environment to perform a critical function. For each of these topics we first identify the fundamentals and then illustrate them with selected examples for which, in general, a consensus has emerged about the pertinent structural and mechanistic themes. Attention is paid not only to naturally occurring metal ions, but also to those that are introduced in chemotherapy, are employed as radiopharmaceutical or magnetic resonance imaging agents, or whose concentrations must be limited to avoid toxic or environmentally harmful effects. Consequently, the book should appeal to members of the medical as well as the chemical and biological communities.

The level of detail employed in the book was chosen to facilitate exposition of principles, with only limited detailed examples being provided. By elucidating principles, the book will enable faculty who already teach bioinorganic chemistry courses to use their existing notes, with our text serving as an organizational framework. For material not previously covered in existing courses, or for new courses or students who wish to pursue the subject independently, we refer the reader to the bibliographic sources listed at the ends of the individual chapters.

The book itself has grown out of courses in bioinorganic chemistry offered at Columbia University and the Massachusetts Institute of Technology by SJL, and at The Johns Hopkins University by JMB. In addition, several of the chapters have been used in courses at a variety of other institutions during the preparation of the final manuscript, and all of the chapters have had the benefit of external review by one or more experts in the field. Attempts have been made to modify or clarify the language to meet the needs of a diverse body of readers and to produce a book that is as free of factual errors as possible.

The following individuals read and offered critical comments on the entire book: Walther Ellis, Robert Scott, and Thomas Sorrell. Individual chapters were critiqued by Ronald Breslow, John Caradonna, David Christianson, James Collman, Robert Hausinger, Kenneth Karlin, David Lavalee, George McLendon, Thomas O'Halloran, William Orme-Johnson, Lawrence Que, Jr., Wesley Sundquist, and William Tolman. We are grateful to these and other members of the bioinorganic chemistry community who have helped us in this enterprise, and stress that the remaining errors are solely the responsibility of the two of us.

We also thank the many MIT students who have proofread and commented on the final draft as well as those who have devoted countless hours to chasing down figures and references, especially Linda Doerrer, Petra Turowski and David Wright, the teaching assistants in Chem. 5.067, the course on bioinorganic chemistry. Many typographical errors were eliminated as a result of the careful proofreading of the text by these individuals, by students in the course, and, especially, by members of the Lippard group to whom we are most grateful.

The ribbon diagrams of proteins were generated with the aid of MOLSCRIPT (P. Kraulis, *J. Appl. Crystallogr.*, 24, 946, (1991)), and we thank Jeffrey Buchsbaum for assistance in getting the program running at Johns Hopkins. We owe a special debt of gratitude to Julie Croston, Kathleen Kolish, Colette Laurencot, Marilyn Mason, and Nancy Williams for invaluable assistance in preparing the manuscript. Finally, we thank Bruce

Armbruster, our publisher, for ensuring the production of a handsome book at an affordable price, as well as Audre Newman and Georg Klatt for working many long and creative hours to provide illustrations that not only illuminated our intended message, but also improved it.

Stephen J. Lippard
Jeremy M. Berg

Principles of Bioinorganic Chemistry

CHAPTER 1

Overview of Bioinorganic Chemistry

1.1. What Is Bioinorganic Chemistry?

Bioinorganic chemistry constitutes the discipline at the interface of the more classical areas of inorganic chemistry and biology. Although biology is generally associated with organic chemistry, inorganic elements are also essential to life processes. Table 1.1 lists the essential inorganic elements together with some of their known roles in biology. Bioinorganic chemists study these inorganic species, with special emphasis on how they function *in vivo*.

Inorganic elements have also been artificially introduced into biological systems as probes of structure and function. Heavy metals such as mercury and platinum are used by X-ray crystallographers and electron microscopists to help elucidate the structures of macromolecules. Paramagnetic metal ions have been valuable in magnetic-resonance applications. Metal-containing compounds have been used not only as biological probes, but also as diagnostic and therapeutic pharmaceuticals. The mechanisms of action of platinum anticancer drugs, gold antiarthritic agents, and technetium radiopharmaceuticals are some currently active topics of investigation in bioinorganic chemistry. The elements of central interest to bioinorganic chemists are presented in Figure 1.1.

Bioinorganic chemistry thus has two major components: the study of naturally occurring inorganic elements in biology and the introduction of metals into biological systems as probes and drugs. Peripheral but essential aspects of the discipline include investigations of inorganic elements in nutrition, of the toxicity of inorganic species, including the ways in which such toxicities are overcome both by the natural systems and by human intervention, and of metal-ion transport and storage in biology. Even these added topics do not exhaust all aspects of the field, however. Like any subject, bioinorganic chemistry is defined by what the most creative people are doing.

Table 1.1
Biological functions of selected metal ions

Metal	Function
Sodium	Charge carrier; osmotic balance
Potassium	Charge carrier; osmotic balance
Magnesium	Structure; hydrolase; isomerase
Calcium	Structure; trigger; charge carrier
Vanadium	Nitrogen fixation; oxidase
Chromium	Unknown, possible involvement in glucose tolerance
Molybdenum	Nitrogen fixation; oxidase; oxo transfer
Tungsten	Dehydrogenase
Manganese	Photosynthesis; oxidase; structure
Iron	Oxidase; dioxygen transport and storage; electron transfer; nitrogen fixation
Cobalt	Oxidase; alkyl group transfer
Nickel	Hydrogenase; hydrolase
Copper	Oxidase; dioxygen transport; electron transfer
Zinc	Structure; hydrolase

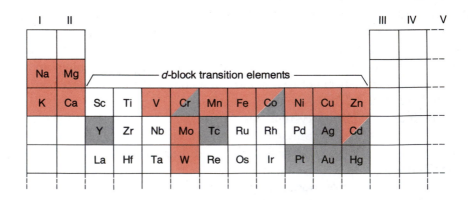

Figure 1.1
Selected elements important in bioinorganic chemistry. Naturally occurring elements are shown in color; those used as probes or drugs are depicted in gray.

These activities will therefore change with time as the population of bio-inorganic chemists changes.

Bioinorganic chemistry is more than a collection of facts with no unifying principles. A major purpose of this book is to systematize our knowledge of the subject, extracting and organizing important principles from the many research areas currently being pursued by members of the bioinorganic chemistry community. The remainder of this chapter surveys most of the currently studied topics in the field, organized according to the functions performed by the metal centers, in order to provide a background setting for the principles and descriptive material to be discussed later.

1.2. Metal Functions in Metalloproteins

Metals are commonly found as natural constituents of proteins. Nature has learned to use the special properties of metal ions to perform a wide variety of specific functions associated with life processes. In this respect, it is interesting to inquire whether the metal centers in metalloproteins are endowed with functional capabilities that, owing to the complexities of the protein environment, bioinorganic chemists may not be able to duplicate. We shall return to this point later. Metalloproteins that perform a catalytic function are called *metalloenzymes*; they constitute a special class treated separately in Section 1.3.

1.2.a. Dioxygen Transport. One function uniquely performed by metalloproteins is respiration. There are three known classes of dioxygen-transport proteins: the hemoglobin-myoglobin family, hemocyanins, and hemerythrins. The functional metallic cores of these proteins are illustrated in Figure 1.2. In these proteins, a delicate balance is achieved whereby the O_2 molecule binds to an iron or copper center without undergoing an irreversible electron transfer, or redox, reaction leading to O–O bond cleavage and concomitant oxidation chemistry. In hemoglobin (Hb) and myoglobin (Mb), the dioxygen-binding site is an iron-porphyrin complex that undergoes structural changes upon O_2 binding. For hemoglobin, these structural changes trigger subtle movements of the protein chains that lead to cooperativity in the uptake of dioxygen. The structures of the porphyrin ring and some other organic cofactors are shown in Figure 1.3.

The other two respiratory proteins employ a pair of metal ions in the dioxygen-binding reaction (Figure 1.2). In hemocyanin (Hc), found in

4

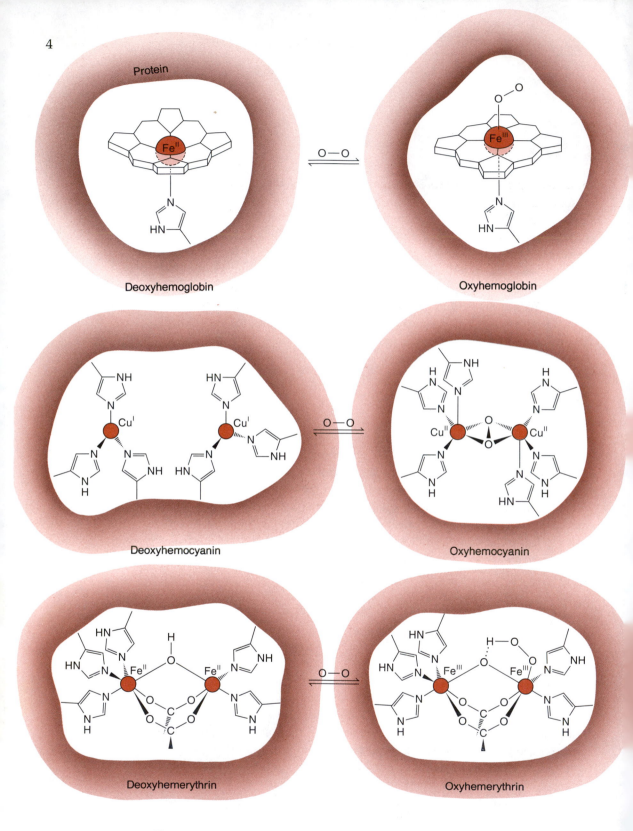

Figure 1.2
Functional units found in three classes of dioxygen-binding proteins.
Protoporphyrin IX is depicted as the heme.

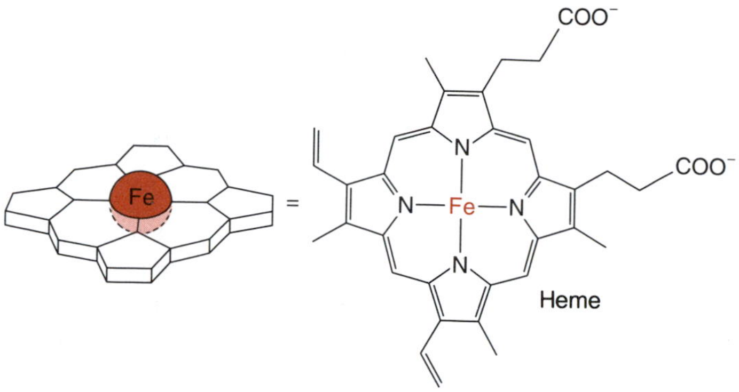

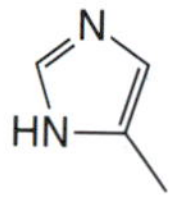

Heme

Imidazole,
the side chain
of histidine

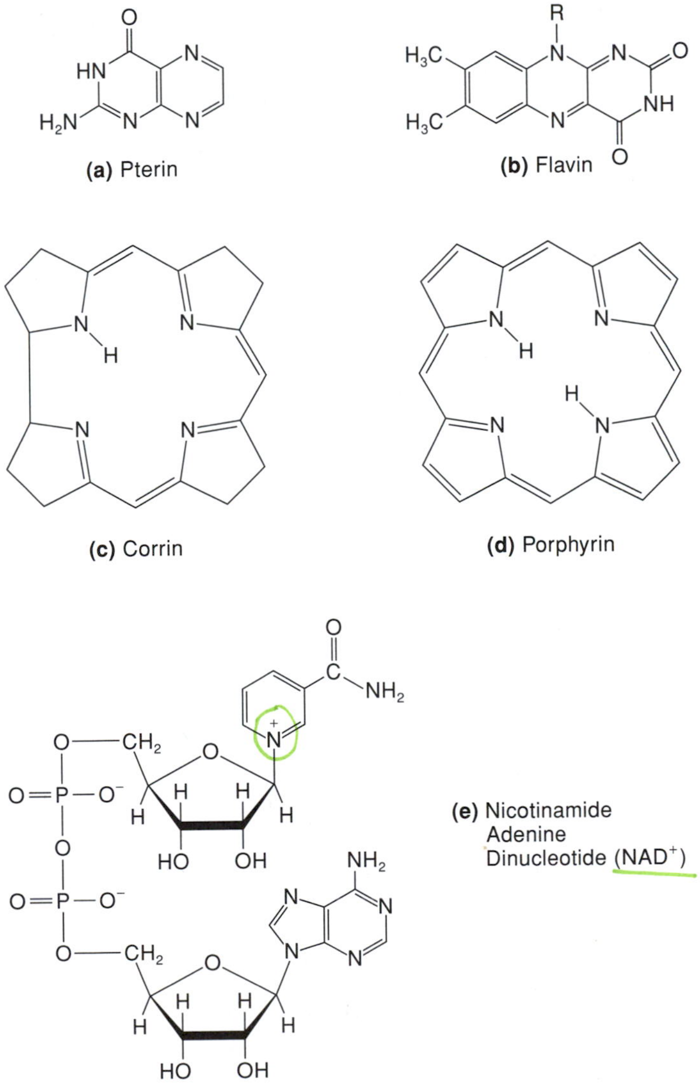

(a) Pterin

(b) Flavin

(c) Corrin

(d) Porphyrin

(e) Nicotinamide
Adenine
Dinucleotide (NAD$^+$)

Figure 1.3
Selected organic ligands and cofactors.

mollusks and arthropods, dioxygen binds between two copper atoms, whereas in the marine invertebrate protein hemerythrin (Hr), the O_2 molecule coordinates at a terminal site in the Fe_2 unit. In both cases, the O_2-binding reaction involves oxidative addition to the reduced, or deoxy, form of the dimetallic center, generating the peroxide or hydroperoxide derivative of the oxidized form. Hc and Hr are but two of several examples in bioinorganic chemistry in which a function (here, reversible O_2 binding) performed by a pair of metal ions is identical to that achieved by a metalloporphyrin unit, Hb or Mb, in a different protein class. Moreover, the chemical strategy employed in all three reactions is very similar. Thus nature uses different transition metals in different organisms to carry out an identical function. These realizations, and others like them, form the basis behind the principles of bioinorganic chemistry that constitute the core of this book.

1.2.b. Electron Transfer.   The dioxygen-binding metal center in hemoglobin illustrates how bioinorganic chemistry uses one of the major classes of chemical reactions, acid-base chemistry. A Lewis base, O_2, binds reversibly to a Lewis acid, the porphyrin-bound iron. Another main class of chemical reactions, those involving net electron transfer, occurs in proteins that undergo redox transformations without themselves catalyzing an overall chemical change in a substrate molecule. Such bioinorganic current carriers usually pass their electrons to or from enzymes that require redox chemistry in order to perform a specific function, for example, nitrogen fixation. Sometimes the redox-active center is directly incorporated into the metalloenzyme. Two electron-transfer centers widely encountered in bioinorganic chemistry are iron-sulfur clusters and cytochromes. The iron-porphyrin groups in the latter resemble the one, illustrated in Figure 1.2, used for dioxygen transport. The structures of several Fe_nS_n units and their most common bioinorganic redox states are shown in Figure 1.4. Table 1.2 lists redox potentials for several metal ions encountered in biology as well as corresponding values for O_2, a substrate that reacts frequently with metalloproteins. The metal potentials quoted are for standard conditions in water, and will differ significantly when the metal ion is coordinated to a biological ligand and embedded in a protein matrix.

1.2.c. Structural Roles for Metal Ions.   Several families of proteins that regulate the expression of genes have recently been found to contain Zn^{2+} ion. The first example of this type of metalloprotein to be discovered was transcription factor IIIA (TFIIIA) from the African clawed toad *Xenopus*, a protein required for the accurate transcription of 5S ribosomal RNA genes. In

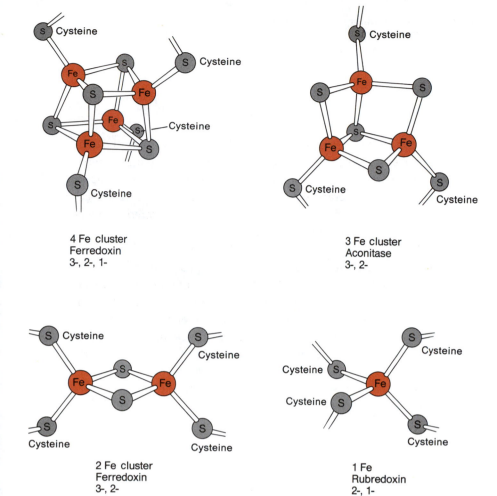

4 Fe cluster
Ferredoxin
3-, 2-, 1-

3 Fe cluster
Aconitase
3-, 2-

2 Fe cluster
Ferredoxin
3-, 2-

1 Fe
Rubredoxin
2-, 1-

Figure 1.4
Structures of biologically occurring iron-sulfur clusters. Also depicted are
representative proteins that house these clusters and the overall cluster charges.

Table 1.2
Selected reduction potentials for important redox-active metal ions and for oxygen species[a]
in biology

Acid solutions of metal ions or substrates					
Species		$E^0/(V)$		Species	$E^0/(V)$
$Cu^{2+} + e^-$	$= Cu^+$	$+0.153$	$O_2(g) + e^-$	$= O_2^-$	-0.33
$Fe^{3+} + e^-$	$= Fe^{2+}$	$+0.771$	$O_2(g) + H^+ + e^-$	$= HO_2$	-0.13
$Mn^{3+} + e^-$	$= Mn^{2+}$	$+1.51$	$O_2(g) + 2H^+ + 2e^-$	$= H_2O_2(aq)$	$+0.281$
$Co^{3+} + e^-$	$= Co^{2+}$	$+1.842$	$O_2(g) + 4H^+ + 4e^-$	$= 2H_2O$	$+0.815$
			$O_2^- + e^- + 2H^+$	$= H_2O_2$	$+0.89$
			$OH + H^+ + e^-$	$= H_2O$	$+2.31$

[a] V. vs. NHE @ pH 7, 25°C

these proteins, the metal ions play a structural role, forming the central core of small nucleic-acid-binding domains referred to as "zinc fingers." In the class of proteins typified by TFIIIA, the metal-binding domains appear to be directly involved in interactions with DNA. Other classes of Zn^{2+}-containing proteins involved in gene regulation have been discovered in which metal-stabilized domains may play other roles. For example, Zn^{2+} ions are present in most DNA and RNA polymerases, and the functions of the metal ions in these important enzymes remain to be elucidated in detail. Finally, several proteins involved in metalloregulation of gene expression have now been characterized. In such metalloregulatory systems, the activity of a protein depends on the presence or absence of bound metal ions. Examples include the bacterial MerR protein, which regulates the expression of a mercury-detoxification system at the level of transcription from DNA into RNA, and the iron-responsive system in mammals, which controls the expression of ferritin and other iron-storage and iron-transport proteins at the level of translation of RNA into protein.

1.3. Metalloenzyme Functions

Metalloenzymes are a subclass of metalloproteins that perform specific catalytic functions. A net chemical transformation occurs in the molecule, termed a substrate, being acted upon by the metalloenzyme. Table 1.3 summarizes many of the reactions catalyzed by metalloenzymes, grouped according to function. Some remarkable transformations for which no simple analogs exist in small-molecule chemistry under comparable conditions are listed. Included are the catalytic reduction of N_2 to NH_3 (nitrogen fixation), the oxidation of water to O_2, and the reduction of gem-diols to monoalcohols (reduction of ribonucleotides). As indicated in the table, metalloenzymes can be classified according to function. Within each category there are usually several kinds of metal centers that can catalyze the required chemical transformation, a situation analogous to that already encountered for respiratory proteins. The reasons for this diversity are shrouded in evolutionary history, but most likely include bioavailability of a given element at the geosphere/biosphere interface during the initial development of a metalloenzyme, as well as pressure to evolve multiple biochemical pathways to secure the viability of critical cellular functions.

1.3.a. Hydrolytic Enzymes. These proteins catalyze addition or removal of the elements of water in a substrate molecule. Notable examples include carbonic anhydrase, which promotes hydrolysis of CO_2; peptidases

Table 1.3
Selected metalloenzyme functions

Function	Enzyme	Reaction
Hydrolytic enzymes	Carboxypeptidase	Removes terminal amino acids from proteins
	Carbonic anhydrase	$H_2O + CO_2 \rightarrow H_2CO_3 \rightleftharpoons H^+ + HCO_3^-$
	Alkaline phosphatase	Hydrolysis of phosphate esters
Protective metalloenzymes	Superoxide dismutase	$2O_2^- + 2H^+ \rightleftharpoons H_2O_2 + O_2$
	Catalase	$2H_2O_2 \rightleftharpoons 2H_2O + O_2$
Dehydrogenases	Liver alcohol dehydrogenase	$CH_3CH_2OH + NAD^+ \rightleftharpoons CH_3CHO + NADH + H^+$
Metal-nucleotide chemistry	Ribonucleotide reductase	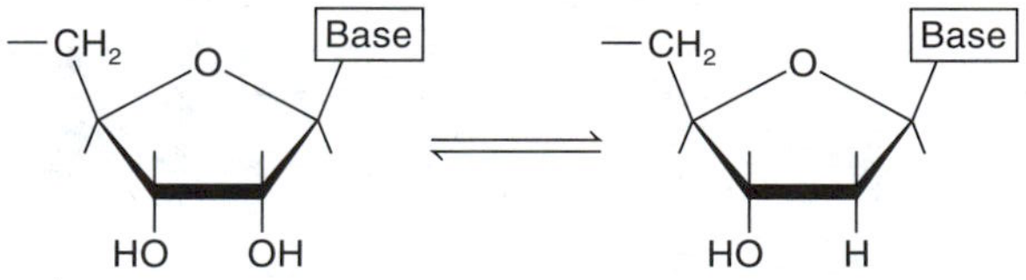 (see structures below)
Isomerization	Glucose isomerase	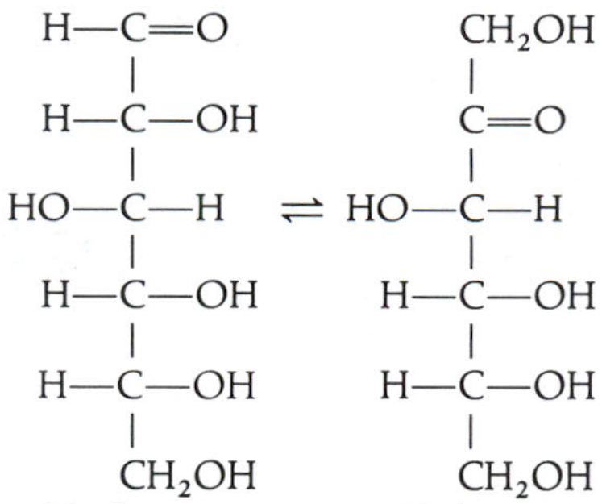 (see structures below)
Nitrogen fixation	Nitrogenase	$N_2 + 8e^- + 8H^+ \rightarrow 2NH_3 + H_2$
Photosynthesis	Photosystem II	$2H_2O \rightarrow O_2 + 4H^+ + 4e^-$

Ribose ⇌ Deoxyribose

D-glucose ⇌ D-fructose

D-glucose:
H—C=O
H—C—OH
HO—C—H
H—C—OH
H—C—OH
CH$_2$OH

D-fructose:
CH$_2$OH
C=O
HO—C—H
H—C—OH
H—C—OH
CH$_2$OH

and esterases, which catalytically hydrolyze carbonyl compounds; and phosphatases, which catalyze the cleavage of phosphate esters. In many hydrolytic enzymes, the active site contains Zn^{2+} ion. This choice optimizes the ability of metals to lower the pK_a of coordinated oxygen-containing ligands, such as the above substrates, while avoiding the potential for undesired electron-transfer chemistry, since divalent zinc does not have any readily accessible redox states. Other metal ions encountered in hydrolytic enzymes, also known as hydrolases, are Mn^{2+}, Ni^{2+}, Ca^{2+}, and Mg^{2+}, further emphasizing the tendency to avoid redox-active species.

1.3.b. Two-Electron Redox Enzymes.   Many metalloenzymes catalyze reactions that involve either oxidation or reduction of substrate. Unlike the electron-transport chemistry discussed in Section 1.2.b., these reactions are generally two-electron redox processes. In addition, they often involve atom or group transfers as well. In some of the most important and common reactions of this type, an oxygen atom is added to substrate. Examples include the oxidation of hydrocarbons to alcohols catalyzed by the iron-porphyrin centers in cytochrome P-450 enzymes, the ortho-hydroxylation of phenolic substrates catalyzed by tyrosinase, which contains a dinuclear-copper active site, and the oxidation of sulfite to sulfate by sulfite oxidase, in which there is a molybdenum atom bound to a pterin ring at the active site. These enzymes differ both in the source of the oxygen added to the substrate and in the identity of the metal ion present. For cytochrome P-450, the oxygen is derived from one of the atoms of dioxygen, and the enzyme requires one mole of the reducing agent NAD(P)H for each mole of substrate oxidized. For tyrosinase, the oxygen again is derived from dioxygen, but for it no additional cofactors are required. For sulfite oxidase, the oxygen atom is derived from water, and two molecules of cytochrome c are reduced for each sulfate ion produced.

Other metalloenzymes remove oxygen atoms from a substrate. Ribo-nucleotide reductase (RR) is one such enzyme, responsible for the first step in DNA biosynthesis. Cells contain very low amounts of deoxynucleotides, the building blocks of DNA. To initiate replication, ribonucleotides are first reduced to deoxyribonucleotides by RR. One subunit of this enzyme contains a stable tyrosyl radical (Figure 1.5) and a dinuclear iron center. The role of the latter is uncertain, but it may serve to generate the radical initially and to regenerate it if it becomes inadvertently reduced. Here we see the close relationship between inorganic and organic functional centers at a protein-active site. Another example is nitrate reductase, which catalyzes the reduction of nitrate, NO_3^-, to nitrite, NO_2^-, a key step in nitrate assimilation by green plants. This enzyme has a molybdenum center related to that at the active site of sulfite oxidase.

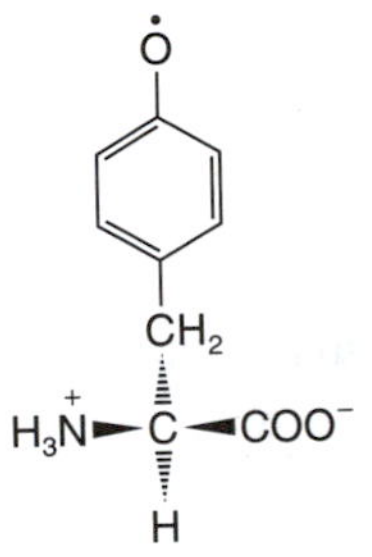

Figure 1.5
Tyrosyl radical.

Dehydrogenations constitute another class of two-electron redox processes. The removal of two electrons and two protons is equivalent to the loss of a molecule of dihydrogen from a substrate. Liver alcohol dehydrogenase, the active site of which contains a zinc(II) ion, catalyzes the formation of acetaldehyde from ethanol. One of the two hydrogens released is added to the organic cofactor NAD^+ (Figure 1.3) to form NADH. Some enzymes actually generate or consume dihydrogen itself. These hydrogenases are complex systems that contain iron-sulfur clusters and, often, nickel.

1.3.c. Multielectron Pair Redox Enzymes.   In the reactions discussed in Section 1.3.b, the oxidation state of the substrate is changed by two electrons (one electron pair). Metalloenzymes also participate in multielectron pair transformations, examples of which are given in Table 1.4. These reac-

Table 1.4
Selected multielectron pair redox reactions

Two-electron pair reactions:

$$O_2 + 4H^+ + 4e^- \rightarrow 2H_2O$$
$$2H_2O \rightarrow O_2 + 4H^+ + 4e^-$$

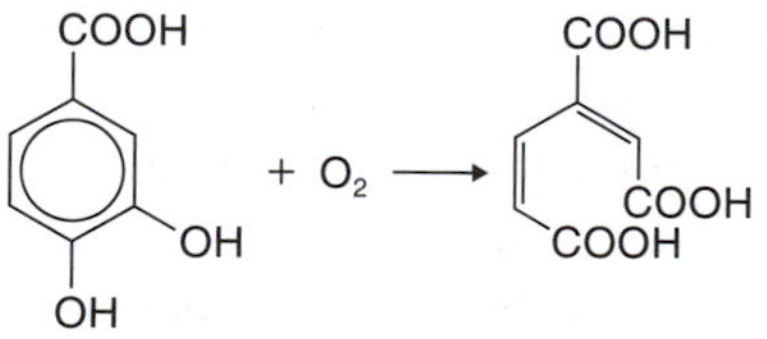

Three-electron pair reactions:

$$N_2 + 6H^+ + 6e^- \rightarrow 2NH_3$$
$$SO_3^{2-} + 8H^+ + 6e^- \rightarrow H_2S + 3H_2O$$
$$NO_2^- + 8H^+ + 6e^- \rightarrow NH_4^+ + 2H_2O$$

tions are of central importance in many fundamental biochemical pathways and, in general, have not been duplicated by non-enzymic chemistry. Two key processes in oxygen metabolism involve the four-electron interconversion of two molecules of water and one molecule of dioxygen. Cytochrome c oxidase, a highly complex enzyme containing two copper and two heme iron centers, catalyzes the reduction of dioxygen to water. The energy released is stored in the form of a proton electrochemical gradient across the cell membrane in which cytochrome c oxidase is embedded, and it can be converted to chemical energy by using the gradient to drive the phosphorylation of ADP to ATP. The reverse reaction, oxidation of water to dioxygen, is also catalyzed by a metalloenzyme, the photosynthetic oxygen-evolving complex (OEC) of photosystem II. The OEC contains four atoms of manganese in its active site in some cluster or clusters of currently unknown structure. Light energy absorbed by the OEC is used to drive the thermodynamically uphill process of coupling two water molecules to form O_2.

Metalloenzymes are involved in other multielectron pair processes involving dioxygen. One example is the iron-containing enzyme catechol dioxygenase, which cleaves the aromatic ring of catechol (1,2-dihydroxybenzene). Multielectron pair processes also play fundamental roles in nitrogen metabolism. Certain green plants are able to obtain nitrogen in useful forms from bacteria that live in their root nodules and have the amazing ability to reduce dinitrogen in the air to ammonia. The enzyme system, called *nitrogenase*, that executes this remarkable six-electron transformation comprises two proteins. The so-called iron-molybdenum protein contains a unique iron-molybdenum cofactor, or FeMoco, made up of a cluster of molybdenum, iron, and sulfur atoms, and the tricarboxylic acid homocitrate. The iron-molybdenum protein is the site of dinitrogen binding and reduction. It also contains several other iron-sulfur clusters involved in transferring electrons. The molybdenum site in FeMoco is distinct from that of the other molybdenum-containing enzymes discussed in Section 1.3.b. The second protein in the nitrogenase system transfers electrons to the iron-molybdenum protein in an ATP-dependent manner and contains a single Fe_4S_4 cluster. Another enzyme involved in the nitrogen cycle, nitrite reductase, catalyzes the six-electron reduction of nitrite, NO_2^-, to ammonia. This enzyme contains a reduced iron-porphyrin complex called siroheme at its active site. A related enzyme catalyzes the six-electron reduction of sulfite to hydrogen sulfide.

1.3.d. Rearrangements. Many biological transformations do not involve any net change in the oxidation state of the substrate. Enzymes that

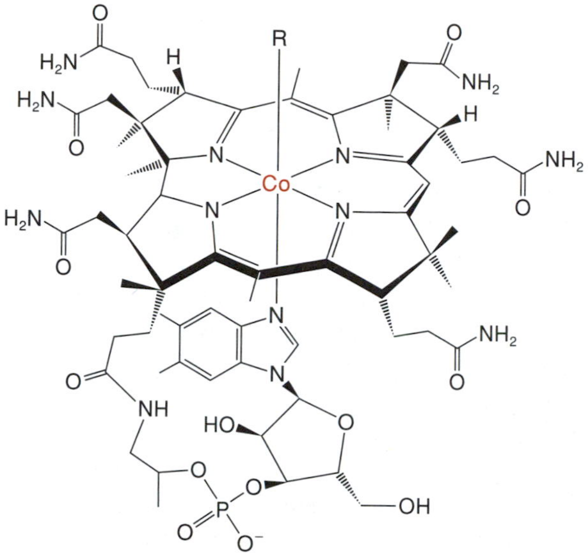

R = CN (Vitamin B$_{12}$)
R = Adenosyl (Coenzyme B$_{12}$)
R = Me

Figure 1.6
Cobalamin structures.

catalyze 1,2 carbon shifts frequently require vitamin B-12 or one of its derivatives (Figure 1.6) as a cofactor. This cofactor is an alkyl cobalt(III) complex of a substituted corrin (Figure 1.3), and is the best-studied organometallic unit in biology, where *organometallic* refers to any species having a direct metal-to-carbon bond. Many of the reactions that use the B-12 coenzyme take place by free-radical pathways initiated or catalyzed by homolytic cleavage of the Co–C bond. The connection between this chemistry and that of transition-metal-generated organic radicals mentioned previously is underscored by the fact that both B-12 and tyrosyl radical-dependent ribonucleotide reductases (see Section 1.3.b) exist and catalyze the same ribo- to deoxyribonucleotide transformation. Another important rearrangement is the conversion of citrate to isocitrate in the Krebs (tricarboxylic acid) cycle, catalyzed by aconitase. This enzyme is activated by the addition of iron and a reducing agent, and recent work has shown that such activation is required for the conversion of an Fe_3S_4 to an Fe_4S_4 cluster in the active site of the protein. Interestingly, although such clusters are characteristic of proteins involved in electron-transfer processes, there is no evidence of any redox process in the mechanism of the citrate to isocitrate interconversion.

1.4. Communication Roles for Metals in Biology

Metal ions are used in biology as magnetic compasses, as triggers for specific cellular functions, and to regulate gene expression. Studies of these cellular communication roles are an exciting frontier area in bioinorganic chemistry. Magnetotactic bacteria use magnetite, Fe_3O_4, as an internal compass by which to navigate. These microorganisms swim north but also downward along the Earth's magnetic field lines in order to avoid the higher oxygen concentrations of surface waters, which to them are toxic. They orient on the Earth's magnetic pole and, when transported to the opposite hemisphere, become disoriented and swim upward! Some bees, homing pigeons, and even humans are also believed to use magnetite in their brains for orientation purposes. Alkali and alkaline earth ions, especially Na^+, K^+, and Ca^{2+}, are used in biology to trigger cellular responses. The firing of neurons by rapid influx of sodium ions across the cell membrane and the regulation of intracellular functions by calcium-binding proteins such as calmodulin are two examples of the phenomenon. In fact, Ca^{2+} has been referred to as a "second messenger," since primary signals, such as the binding of hormones to the cell surface, are converted into changes in the intracellular concentration of this ion. Among the most recently discovered metal-ion classes in biology are the zinc fingers that occur in many proteins that regulate transcription (see Section 1.2.c). We do not now know whether zinc plays more than a structural role in these proteins or whether Zn^{2+} concentrations are used in some manner to regulate gene expression.

1.5. Interactions of Metal Ions and Nucleic Acids

Metal ions also interact directly with DNA and RNA. Some of these interactions are rather nonspecific, for instance, the stabilization of nucleic-acid structures by Na^+ and Mg^{2+} ions through electrostatic interactions that shield the charged phosphate groups from one another. Recently, more specific binding of metal ions to nucleic acids has been discovered. Thus, Mg^{2+} and other divalent metal ions serve as cofactors for activating catalytic RNA molecules; and monovalent cations such as K^+ stabilize the structure of telomeres, units that terminate the DNA double helix at the ends of chromosomes. Telomere structures (Figure 1.7) are characterized by non-Watson-Crick base-pairing interactions and the clustering of eight guanine bases around the central monovalent metal ion. The relative stability of these structures may be dic-

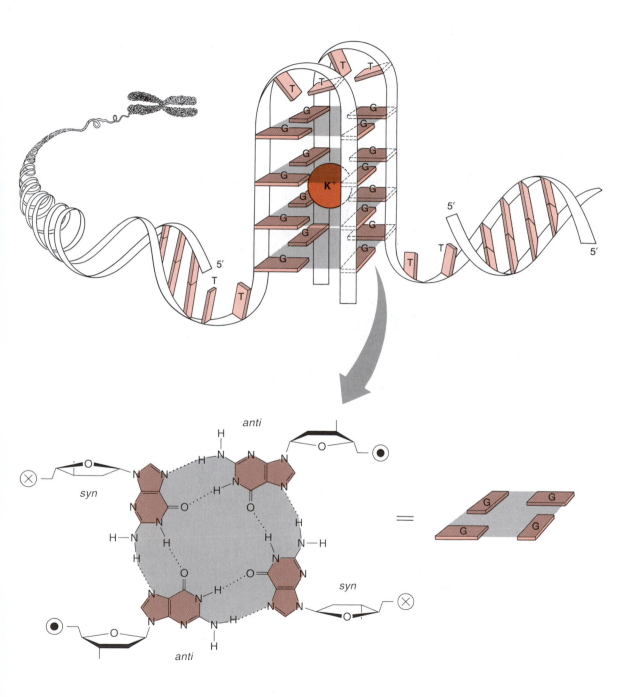

Figure 1.7
Postulated structure for the potassium-stabilized guanine tetrad found in telomeres,
units at the ends of chromosomes.

tated by the concentrations of Na^+ and K^+ in the cell. Finally, some inorganic-based drugs such as cisplatin act by coordinating directly to DNA (see Section 1.7), and metal complexes have been used as cleaving agents to probe the tertiary structures of nucleic acids.

1.6. Metal-Ion Transport and Storage

How do metal ions get into cells and how are they stored? This topic is an active area of investigation in bioinorganic chemistry, although it cannot be classified with the others according to metal function. The most thoroughly studied metal in this respect is iron. Iron enters bacterial cells following chelation by low-molecular-weight compounds called *siderophores* that are excreted by the bacteria. In mammals, iron is bound and transported by the serum protein transferrin, and it is stored by ferritin in most life forms. The nearly spherical, hollow shell of this latter protein has the capacity to bind up to 4,500 Fe^{3+} ions. Details about how iron is passed among these protein systems are incomplete and under active investigation. Copper is transported by the serum protein ceruloplasmin, and another such protein, albumin, is also known to bind and transport metal ions. Metallothionein is a cysteine-rich protein that is expressed in large amounts when excess quantities of certain metal ions, including toxic ones such as Cd^{2+} or Pb^{2+}, are present in cells. Metallothionein thus serves a protective role and may also be involved in the control of metal transport, storage, and concentration under more normal conditions.

1.7. Metals in Medicine

Many first encounter the use of metals in medicine as well as metal toxicity through literature. The use of iron and copper can be traced to the ancient Greeks and Hebrews through their writings. Lewis Carroll's Mad Hatter suffered from mercury poisoning. Among metal ions commonly used over the centuries were Hg^{2+} for the treatment of syphilis, Mg^{2+} for intestinal disorders, and Fe^{2+} for anemia. These early examples represent crude approaches, the refinement of which did not, until recently, begin to match in sophistication or efficacy the contributions of organic chemists, who introduced sulfa drugs, penicillin, and mechanism-based inhibitors such as methotrexate. Today, however, inorganic chemistry is beginning to have a major impact in modern medicine. Three important inorganic pharmaceuticals are depicted in Figure 1.8.

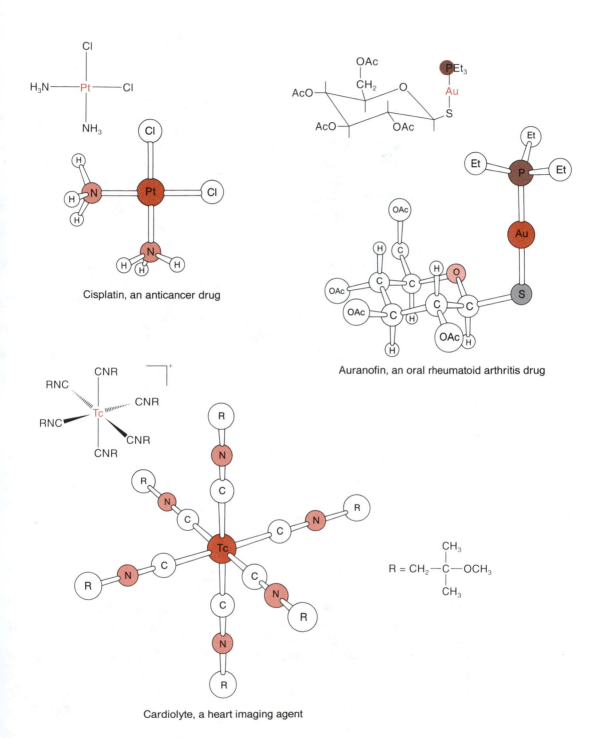

Cisplatin, an anticancer drug

Auranofin, an oral rheumatoid arthritis drug

Cardiolyte, a heart imaging agent

R = CH₂—C(CH₃)₂—OCH₃

Figure 1.8
Structures of three inorganic compounds used in modern medicine.

One of the leading anticancer drugs is *cis*-[Pt(NH$_3$)$_2$Cl$_2$], cisplatin, administered by intravenous injection for the treatment of testicular, ovarian, and head and neck tumors. Cisplatin cures testicular cancer, when diagnosed early enough, in more than 90 percent of the cases. Auranofin, [Au(PEt$_3$)(ttag)], where ttag is tetra-O-acetylthioglucose, is the first orally administered drug for the treatment of rheumatoid arthritis. It is an important member of the class of antiarthritic gold agents, the others of which are injected. The third example is [Tc(CNR)$_6$]$^+$, in which the technetium is supplied as its ^{99m}Tc radioisotope. This class of complexes is selectively taken up by myocardial tissue and has proved to be excellent for imaging the heart. Platinum, gold, and technetium, three nonessential transition elements, have found a place in medicine. Given the range and variety of inorganic compounds, the potential applications of inorganic chemistry to improving human health are boundless. This aspect of bioinorganic chemistry, now in its infancy, seems destined to be an area of rapid and significant growth.

1.8. Organization of This Book

The discussion in this chapter has touched on many essential components of bioinorganic chemistry as practiced today, providing an overview and background for subsequent discussion. Topics have been presented according to the functions of metal ions in biology, an organizational scheme common in treatments of the field. In this book, however, we use a different approach. We begin with three introductory chapters that provide minimal background for those less familiar with inorganic chemistry, biochemistry, and/ or the physical methods commonly employed by bioinorganic chemists. In Chapter 2 are presented features of coordination chemistry that are particularly relevant to the functions of metal ions in biology. Chapter 3 contains similar information about proteins, nucleic acids, and other biopolymers. A brief discussion of the physical methods is contained in Chapter 4. None of these three chapters is intended to provide a rigorous treatment of these topics, for which many excellent texts are already available, and they can easily be omitted by the more experienced reader. The next eight chapters, 5 to 12, which constitute the core of the book, attempt to distill principles out of the most thoroughly documented studies of bioinorganic systems. The organizational scheme adopted in these chapters is vectorial, moving from questions of metal-ion availability to living cells to the most intimate details of the molecular mechanisms of bioinorganic chemistry. In these eight chapters, we intend the principles to transcend the particular descriptive biochemistry from which they were derived. Finally, a chapter is devoted to a discussion of future challenges for research in bioinorganic chemistry, at least as they appear in 1994.

Study Problems

1. An important experiment in the history of biochemistry was the demonstration by Sumner that an enyzme, jack bean urease, could be crystallized. The fact that this crystalline enzyme was a pure protein was crucial to arguments that all enzymes were proteins. Some fifty years later, it was discovered that urease contains two Ni^{2+} ions per protein molecule. Comment on how this observation might have affected biochemistry had it been made fifty years earlier.

2. Your pet lobster (an arthropod) has been sluggish and has been found to be anemic. Do you think that a good treatment would be to supplement its diet with iron? Suggest an alternative.

3. A novel electron-transport protein has been isolated. Elemental analysis suggests that it contains only zinc. Comment on this observation.

4. When you try to buy a sample of the heart-imaging agent cardiolyte at your local chemistry stockroom, the clerk informs you that he does not carry this compound, because it has no shelf life. Explain. How does this property help make it a good radiopharmaceutical?

Bibliography

General References

A. W. Addison, W. R. Cullen, D. Dolphin, and B. R. James, eds. 1977. *Biological Aspects of Inorganic Chemistry*. Wiley-Interscience, New York.

I. Bertini, H. B. Gray, S. J. Lippard, and J. S. Valentine. 1994. *Bioinorganic Chemistry*. University Science Books, Mill Valley, CA.

J. J. R. Frausto da Silva and R. J. Williams. 1991. *The Biological Chemistry of the Elements: The Inorganic Chemistry of Life*. Oxford University Press, Oxford.

R. W. Hay. 1984. *Bio-Inorganic Chemistry*. Ellis Horwood, Chichester, United Kingdom.

M. N. Hughes. 1981. *The Inorganic Chemistry of Biological Processes*, 2nd ed. John Wiley and Sons, Chichester, United Kingdom.

W. Kaim and B. Schwederski. 1991. *Bioanorganische Chemie*. Teubner, Stuttgart.

E. Ochiai. 1977. *Bioinorganic Chemistry, an Introduction*. Allyn and Bacon, Boston.

R. J. P. Williams. 1990. "Bio-Inorganic Chemistry: Its Conceptual Evolution." *Coord. Chem. Rev.* 100, 573–610.

Book Series on Bioinorganic Chemistry

G. L. Eichhorn, ed. 1973. *Inorganic Biochemistry*, Volumes 1 and 2. Elsevier Scientific, Amsterdam.

G. L. Eichhorn and L. G. Marzilli, eds. 1979–. Volumes in *Advances in Inorganic Biochemistry*. Elsevier Biomedical, New York.

R. W. Hay, J. R. Dilworth, and K. B. Nolan, eds. 1991–. *Perspectives on Bioinorganic Chemistry*. JAI Press, London.

S. J. Lippard, ed. 1973 and 1990. Volumes 18 and 38 in *Progress in Inorganic Chemistry*. Wiley-Interscience, New York.

H. Sigel, ed. 1973–. Volumes in *Metal Ions in Biological Systems*. Marcel Dekker, New York.

T. G. Spiro, ed. 1980–. Volumes in *Metals in Biology*. Wiley-Interscience, New York.

CHAPTER 2

Principles of Coordination Chemistry
Related to Bioinorganic Research

This chapter presents aspects of inorganic coordination chemistry that are of particular relevance to the roles of metal ions in biology. The special properties of metal ions distinguish them from surrounding organic functional groups in a biological milieu. In order to understand the functions of metal centers, a minimal knowledge of these special properties is required. This chapter provides information about the chemical reactions and electronic structural features of metal ions; the measurement of physical properties is treated separately, in Chapter 4.

2.1. Thermodynamic Aspects

2.1.a. The Hard-Soft Acid-Base Concept. Metal ions in biology most frequently bind to donor ligands according to preferences dictated by the hard-soft theory of acids and bases. Table 2.1 lists the hard-soft character of essential metal ions listed previously (Table 1.1) and several of those used as biological probes and pharmaceuticals. In this classification scheme, the term "soft" refers to species that are large and fairly polarizable, whereas "hard" species are small and less easily polarized. Metal ions are considered to be Lewis acids. Also listed in the table are the hard-soft preferences of Lewis bases, ligand atoms that coordinate to the metal centers. In a biological medium, these ligands are provided by protein side chains, the bases of the nucleic acids, small cellular cytoplasmic constituents, organic cofactors (see Figure 1.3), and, of course, water. Although exceptions exist, the general rule is that hard acids bind preferentially to hard bases and soft acids to soft bases. For example, when a crystal of a protein is soaked in a solution of K_2PtCl_4 in

Table 2.1

Hard-soft acid-base classification of metal ions and ligands important to bioinorganic chemistry

Metals	Ligands
Hard	
H^+ Mn^{2+} Cr^{3+} Na^+ Al^{3+} Co^{3+} K^+ Ga^{3+} Fe^{3+} Mg^{2+} Ca^{2+} Tl^{3+}	H_2O CO_3^{2-} NH_3 OH^- NO_3^- RNH_2 $CH_3CO_2^-$ ROH N_2H_4 PO_4^{3-} R_2O RO^- $ROPO_3^{2-}$ $(RO)_2PO_2^-$ Cl^-
Borderline	
Fe^{2+} Ni^{2+} Zn^{2+} Co^{2+} Cu^{2+}	NO_2^- N_2 SO_3^{2-} ⬡—NH₂ Br^- N_3^- (imidazole)
Soft	
Cu^+ Pt^{2+} Pt^{4+} Au^+ Tl^+ Hg^{2+} Cd^{2+} Pb^{2+}	R_2S R_3P RS^- CN^- RSH RNC $(RS)_2PO_2^-$ $(RO)_2P(O)S^-$ SCN^- CO H^- R^-

order to prepare a heavy-atom derivative to help provide phasing in an X-ray structure determination, the soft Pt(II) ion will bind most avidly to the exposed soft ligands, most commonly cysteine sulfhydryl groups or methionine thioether linkages. Binding to exposed glutamate or aspartate carboxylate groups is far less likely. Alkali and alkaline earth metals like Ca^{2+} are most commonly coordinated by carboxylate oxygen atoms, Fe^{3+} by carboxylate and phenoxide oxygen donors, and Cu^{2+} by histidine nitrogen atoms (Table 2.1). One of the best illustrations of the hard-soft acid-base principle in bioinorganic chemistry is provided by the metallothionein proteins (Figure 2.1). Nearly 30—35 percent of the amino acids of this class of small proteins are cysteine residues, the sulfhydryl groups of which bind avidly to soft metal ions such as Cd^{2+}, Hg^{2+}, Pb^{2+}, and Tl^+. One of the biological functions of metallothionein is to protect cells against the toxic effects of these metal ions.

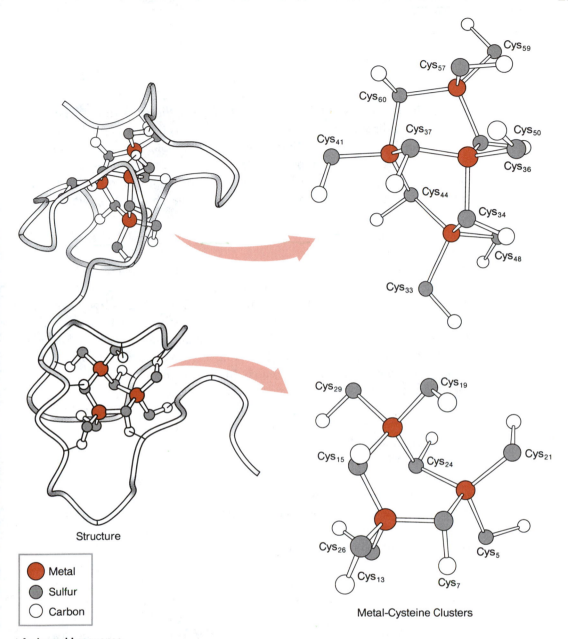

Metal
Sulfur
Carbon

Metal-Cysteine Clusters

Structure

Amino acid sequence

CH₃CONH−Met Asp Pro Asn **Cys** Ser **Cys** Ala Thr Asp Gly Ser **Cys** Ser **Cys** Ala Gly Ser **Cys** Lys **Cys** Lys Gln **Cys** Lys **Cys** Thr Ser Cys Lys-Lys Ser **Cys Cys** Ser **Cys Cys** Pro Val Gly **Cys** Ala Lys **Cys** Ser Gln Gly **Cys** Ile **Cys** Lys Glu Ala Ser Asp Lys **Cys** Ser **Cys Cys** Ala-COO⁻

Figure 2.1
Amino acid sequence and three-dimensional structure of metallothionein and its tetrametallic (top) and trimetallic (bottom) clusters.

2.1.b. The Chelate Effect and the Irving-Williams Series.

Chelation refers to coordination of two or more donor atoms from a single ligand to a central metal atom. The resulting metal-chelate complex has unusual stability derived in part from the favorable entropic factor accompanying the release of nonchelating ligands, usually water, from the coordination sphere. This phenomenon is illustrated in Equation 2.1 for the chelation of nickel(II) in

$$[Ni(OH_2)_6]^{2+} + H_2edta^{2-} \rightarrow [Ni(edta)]^{2-} + 4H_2O + 2H_3O^+ \qquad (2.1)$$

aqueous solution by the hexadentate ligand ethylenediaminetetraacetate (H_2edta^{2-}; shown in its protonated form in Figure 2.2). Ligands such as H_2edta^{2-}, hereafter referred to for simplicity as EDTA in this book, are used in medicine to chelate metal ions that might be present in toxic excess and as food additives to limit the availability of essential metals to harmful bacteria, thereby preventing spoilage. The ligand EDTA itself is commonly added to buffer solutions in biological research to reduce the concentration of free metal ions that could promote undesired reactions.

An important example of the chelate effect in bioinorganic chemistry is afforded by the porphyrin and corrin ligands, illustrated in Figure 1.3. These macrocyclic molecules have four nearly coplanar pyrrole rings with their nitrogen donor atoms directed toward a central metal ion. The resulting metalloporphyrin or -corrin units are thermodynamically very stable, accommodating a variety of metal ions in different oxidation states. As a consequence, these chelating units provide bioinorganic functional groups of widespread occurrence and utility in biology, being found in cytochromes (Fe), chlorophyll (Mg), and vitamin B-12 (Co), to mention but a few examples.

Another useful principle of inorganic chemistry is the binding preference, for a given ligand, of divalent first-row transition-metal ions. These preferences typically follow the stability series $Ca^{2+} < Mg^{2+} < Mn^{2+} < Fe^{2+} < Co^{2+} < Ni^{2+} < Cu^{2+} > Zn^{2+}$, as first delineated by Irving and Williams. This order, known as the Irving-Williams series, is related to the decrease in ionic radii across the series, an effect that leads to stronger metal-ligand bonds.

2.1.c. pK$_a$ Values of Coordinated Ligands.

The positive charge on most metal ions in biology stabilizes the acid anion (conjugate base) of protic ligands bound in the coordination sphere. This effect is best exemplified by coordinated water, but occurs for many other biological ligands, including thiols, imidazole, phenols, alcohols, phosphoric and carboxylic acids, and their derivatives. Table 2.2 lists the pK$_a$ values of selected examples from among these ligand groups in the presence and absence of various metal ions. Tri-

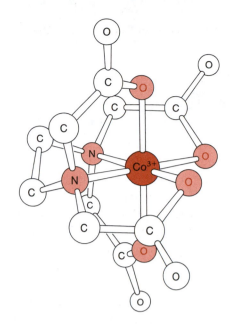

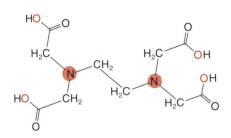

(a) Ethylenediaminetetraacetic acid (H₄EDTA)   (b) [Co(EDTA)]⁻ complex

Figure 2.2
(a) The metal chelating agent ethylenediaminetetraacetic acid (EDTA).
(b) Structure of a Co^{3+} complex of EDTA.

Table 2.2
pK_a values for selected ligands with and without metal ions

Ligand and reaction	Metal ion	pK_a (25°C, 0.1 M)
$H_2O + M^{2+} \rightleftharpoons M-OH^+$	None	14.0
	Ca^{2+}	13.4
	Mn^{2+}	11.1
	Cu^{2+}	10.7
	Zn^{2+}	10.0
$NH_3 + M^{2+} \rightleftharpoons M-NH_2^+$	None	35.0
	Co^{2+}	32.9
	Cu^{2+}	30.7
	Ni^{2+}	32.2
acetic acid $+ M^{2+} \rightleftharpoons M-O-C(=O)CH_3$	None	4.7
	Mg^{2+}	4.2
	Ca^{2+}	4.2
	Ni^{2+}	4.0
	Cu^{2+}	3.0
imidazole $+ M^{2+} \rightleftharpoons M-N$ imidazole	None	7.0
	Co^{2+}	4.6
	Ni^{2+}	4.0
	Cu^{2+}	3.8

valent metal ions are better able to lower the pK_a values of protic ligands than their divalent analogs, as expected on the basis of charge considerations.

Deprotonation of coordinated water to form a hydroxo ligand is a step postulated in several hydrolytic mechanisms to explain metalloenzyme catalysis. Coordination of two or more metal ions to a protic ligand produces an even more dramatic lowering of the pK_a. This effect is illustrated for the hydrolysis of iron(III) in Table 2.3. In $[Fe(OH_2)_6]^{3+}$, water can

Table 2.3
Hydrolysis reactions of Fe(III), 25°[a]

Reaction	pK_a
$Fe^{3+} + H_2O \rightarrow Fe(OH)^{2+} + H^+$	2.2
$2Fe^{3+} + 2H_2O \rightarrow Fe_2(OH)_2^{4+} + 2H^+$	2.9
$Fe(OH)^{2+} + H_2O \rightarrow Fe(OH)_2^{+} + H^+$	3.5
$Fe(OH)_2^{+} + H_2O \rightarrow Fe(OH)_3\downarrow + H^+$	6
$Fe(OH)_3 + H_2O \rightarrow Fe(OH)_4^{-} + H^+$	10

[a] Additional water molecules coordinated to the iron atoms are not shown.

be deprotonated with a pK_a of 2.2, whereas the hydroxide ligand in $[Fe(H_2O)_4(OH)_2]^+$ is deprotonated with $pK_a \sim 6$ owing to the formation of (μ-hydroxo)- and μ-(oxo)diiron(III) units. These results illustrate that aquated iron(III) and many other metal-ion complexes cannot exist at physiological pH values ~ 7 in the absence of supporting ligands. Di- or polymetallation of other protic biological ligands is a fairly common occurrence, examples being the triply bridged (μ_3-sulfido) triiron cluster units found in iron-sulfur proteins (deprotonation of H_2S) and the bridged (μ_2-imidazolato) copper(II)-zinc(II) moiety in bovine erythrocyte superoxide dismutase (deprotonation of imidazole side chain of histidine).

2.1.d. Tuning of Redox Potentials. Alterations in the ligand donor atom and stereochemistry at the metal center can produce great differences in the potential at which electron-transfer reactions will occur. The potentials listed for the aqua ions in Table 1.2 can be altered by more than 1.0 V by these factors. The data in Table 2.4 illustrate the ability of various ligands to tune the Cu(I)/Cu(II) redox potential, one of the best-studied examples in both inorganic and bioinorganic chemistry. Copper(I), a closed shell, d^{10} ion (see Section 2.4), prefers tetrahedral four-coordinate or trigonal three-coordinate

Table 2.4
Effect of ligands on Cu(I)/Cu(II) reduction potential in DMF solution

Compound name	$E_{1/2}$, V^a
Cu(O-sal)$_2$en	-1.21
Cu(Me-sal)$_2$	-0.90
Cu(Et-sal)$_2$	-0.86
Cu(S-sal)$_2$en	-0.83
Cu(i-Pr-sal)$_2$	-0.74
Cu(t-Bu-sal)$_2$	-0.66

Cu(R-sal)$_2$ Cu(X-sal)$_2$en
X = O or S

^{a} Potential at which the complex is half-oxidized and half-reduced.

geometries. Divalent copper(II) complexes, on the other hand, are typically square-planar with perhaps one or two additional, weakly bonded axial ligands. Thus, a ligand environment that produces a tetrahedral geometry will usually stabilize Cu(I) over Cu(II), rendering the latter a more powerful oxidizing agent by raising the redox potential. As can be seen in Table 2.4, addition of bulky R groups in the Cu(R-sal)$_2$ complexes distorts the geometry from planar toward tetrahedral, making it easier to reduce the copper and raising the potential. In addition, Cu(I) is a soft acid, preferring to bind to soft donors such as RS$^-$ or R$_2$S ligands. The placement of soft ligands in the coordination sphere also increases the Cu(I)/Cu(II) reduction potential.

These effects of ligand donor type and stereochemistry on the Cu(I)/Cu(II) potential are manifest not only in the inorganic complexes listed in Table 2.4, but also in several copper-containing proteins. Here, high redox potentials are achieved by the proteins through distortion of the coordination geometry toward trigonal planar and the use of two histidine-imidazole and one cysteine-thiolate side chains as donor ligands. Many other important examples of redox-potential tuning by the local protein environment are encountered in bioinorganic chemistry, including the iron-sulfur clusters and the cytochromes. Sometimes the potential is influenced by the local dielectric constant provided by residues in the vicinity of, but not necessarily coordinated to, the metal atom. This phenomenon is analogous to the influence of solvent on the redox potentials of simple coordination complexes.

2.1.e. Biopolymer Effects.   As the previous discussion has already illustrated, the thermodynamic stability of a metal center in a biological environment is determined not only by the inherent preferences of the metal for a particular oxidation state, ligand donor set, and coordination geometry, but also by the ability of the biopolymer to control, through its three-dimensional structure, the stereochemistry and ligands available for coordination. Noncoordinating residues also contribute such factors as local hydrophilicity or hydrophobicity, steric blockage of coordination sites, and hydrogen-bonding groups that can interact with bonded and nonbonded atoms in the coordination sphere of the metal to increase or reduce stability. These factors, which occur for metals bound to nucleic acids as well as to proteins, must be elucidated in any serious attempt to understand how metals function in biology.

Some of the most dramatic manifestations of the chelate effect in bioinorganic chemistry are the occurrences in many metalloprotein cores of strong and specific metal-binding sites. Nature has possibly the most effective chelating ligands at its disposal in the form of protein chains that can fold (see Chapter 3), orienting amino-acid residue donors to provide virtually any desired stereochemistry at a metal center. A nice example of this phenomenon is the zinc-binding site in bovine erythrocyte superoxide dismutase (Cu_2Zn_2SOD). The zinc-coordinating environment at this site is so favorable that the metal-free (or apo) protein (E_2E_2SOD; E = empty) can remove traces of zinc from phosphate buffer upon dialysis. The binding site is sufficiently specific that when Cu_2Cu_2SOD, in which Cu^{2+} ion occupies the Zn^{2+} site, is treated with excess divalent zinc, the copper in the Zn^{2+} site is displaced. This chemistry is unusual, since it violates the usual preferences of the Irving-Williams series (see Section 2.1.b). Clearly, zinc must be functionally important in this site, which has evolved in a manner that assures the requisite metal-ion specificity.

2.2. Kinetic Aspects

2.2.a. Ligand Exchange Rates.   Table 2.5 lists the water-exchange rates for many essential metal ions. From these values, it is clear that $M-OH_2$ bonds are very labile, breaking and reforming as fast as a billion times per second. The labilities of metal-ligand bonds typically follow the trends for the

Table 2.5
Exchange rates for water molecules from the first coordination sphere of metal ions at 25°C

Ion	k_1, sec^{-1}	Ion	k_1, sec^{-1}	Ion	k_1, sec^{-1}
Li^+	4×10^8	V^{2+}	8×10^1	Sn^{2+}	$> 10^4$
Na^+	7×10^8	Cr^{2+}	1×10^9	Hg^{2+}	4×10^8
K^+	1×10^9	Mn^{2+}	2×10^7	Al^{3+}	1
Be^{2+}	8×10^2	Fe^{2+}	4×10^6	Fe^{3+}	2×10^2
Mg^{2+}	6×10^5	Co^{2+}	3×10^6	Ga^{3+}	4×10^2
Ca^{2+}	3×10^8	Ni^{2+}	4×10^4	Gd^{3+}	2×10^9
Ba^{2+}	2×10^9	Cu^{2+}	1×10^9	Bi^{3+}	$> 10^4$
		Zn^{2+}	2×10^7	Cr^{3+}	2×10^{-6}
				Co^{3+}	$< 10^{-6}$
				Rh^{3+}	6×10^{-9}

aqua complexes in Table 2.5. In general, ligand exchange rates are faster for the less highly charged M^{2+} than for the M^{3+} metal ions. Very inert first-row transition-metal ions such as Cr^{3+} and Co^{3+} are only rarely encountered in bioinorganic chemistry. Second- and third-row transition-metal complexes are much more kinetically inert than their first-row counterparts. For example, once the anticancer drug cis-[Pt(NH$_3$)$_2$Cl$_2$] binds to DNA through loss of chloride ligands, the platinum cannot be exchanged out even upon prolonged dialysis of the platinated biopolymer. Only strong platinum-binding ligands such as cyanide can displace the Pt-DNA adduct. A similar situation occurs for ruthenium bound to amino-acid residues, as encountered in studies of protein electron-transfer reaction kinetics.

The fast metal-ligand exchange rates of first-row transition-metal ions such as Fe^{2+} are markedly diminished when they are bound by multidentate chelating ligands. Metalloporphyrins (Figure 1.3), for example, are kinetically rather inert. The axial ligands, which are not part of the chelate ring, can undergo exchange at the usual fast rates, however. Ligands such as CO, RS$^-$, and CN$^-$ form more inert M–L bonds. Many metalloproteins contain tightly bound metal ions that cannot be exchanged with free metal ions even during prolonged dialysis against good chelating ligands. The kinetic inertness of these protein cores often results from solvent inaccessibility to the metal coordination sphere owing to steric shielding by the protein. If the protein is denatured, for example by heating or adding a solvent such as dimethyl sulfoxide, the metal can usually be released.

2.2.b. Substitution Reactions.   Displacement of one ligand by another in the coordination sphere of a metal ion can occur by either associative (second-order) or dissociative (first-order) pathways, with kinetic and mechanistic features quite analogous to S_N2 and S_N1 substitution reactions in organic chemistry, respectively. Metals with lower coordination numbers (≤ 4) tend to undergo associative ligand displacement reactions, whereas higher-coordinate metals (coordination number ≥ 6) use dissociative pathways. Substitution reactions occurring at metal centers bound to proteins or nucleic acids can be far more complex, owing to interactions of the incoming ligand with other nearby groups and to coupling of these reactions to conformational changes in the macromolecule.

2.2.c. Electron Transfer Reactions.   Two major pathways, designated as inner-sphere and outer-sphere, have been elucidated for transfer of electrons to or from transition metal ions. Inner-sphere electron transfer reactions are characterized by the presence of one or more bridging ligands directly bonded to the coordination spheres of the reactants. A classic example of this reaction pathway is illustrated in Equation 2.2, in which the labile Cr(II) complex transfers an electron to an inert Co(III) receptor to form a labile,

$$[Co(NH_3)_5Cl]^{2+} + [Cr(H_2O)_6]^{2+}$$
$$\xrightarrow{5\,H^+} [Co(H_2O)_6]^{2+} + [Cr(H_2O)_5Cl]^{2+} + 5NH_4^+ \qquad (2.2)$$

reduced Co(II) complex and an inert Cr(III) species. The specific transfer of the chloride ion from the Co(III) to the Cr(II) center in this reaction proves that the transition state for the electron-transfer step consists of the bridged binuclear complex illustrated in Figure 2.3. Since electron transfer in this bridged complex is fast compared to atom transfer, the chloride ion remains in the coordination sphere of the kinetically inert chromium(III) product. In the outer-sphere electron-transfer reaction mechanism, the two redox partners approach one another with their associated solvent molecules to form a so-called "precursor complex." Electron transfer then occurs without any accompanying exchange of ligands between coordination spheres of the oxidant and reductant.

We do not now know of any inner-sphere electron-transfer reactions that involve a pair of metalloproteins. Although such a mechanism cannot be ruled out *a priori*, the steric barrier to formation of the requisite ligand-bridged transition state would appear to be formidable. Electron-transfer reactions between metal centers within and between metalloprotein mole-

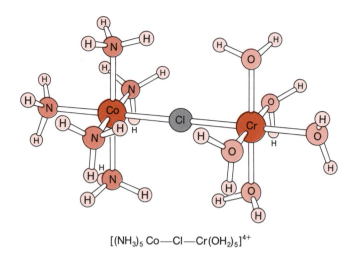

$$[(NH_3)_5\,Co-Cl-Cr(OH_2)_5]^{4+}$$

Figure 2.3
Structure of the chloride-bridged binuclear intermediate formed in the inner-sphere
electron transfer reaction between $[Co(NH_3)_5Cl]^{2+}$ and $[Cr(H_2O)_6]^{2+}$.

cules is a subject of great current interest. Long-range electron transfer occurs
over distances up to ~30 Å at reasonable rates (> 10 sec^{-1}).

Useful theoretical relationships between the equilibrium and rate constants have been derived by Marcus for outer-sphere electron-transfer reactions. The correlation between second-order rate constants predicted by this
theory with those obtained experimentally is remarkably good for small
molecules. Among the predictions of this theory is that electron-transfer
reactions should have an optimal driving force. Making the free energy of a
reaction more favorable than this value decreases rather than increases the
rate. This so-called "inverted region" has been observed experimentally for
small-molecule reactions and, more recently, for electron-transfer reactions
involving metalloproteins as well.

2.3. Electronic and Geometric Structures of Metal Ions in Biology

Table 2.6 lists the common oxidation states of metal ions in biology and, for
the transition metal ions, the corresponding d-electron configuration. The
latter values are obtained by subtracting the formal oxidation state of the
metal from its atomic number (Z) and calculating how many electrons must
be added to the preceding noble-gas element (usually Ar, Z = 18) to achieve

Table 2.6
Common oxidation states of important elements
in bioinorganic chemistry

Metal	Redox state	Number of d electrons
Na	(I)	
K	(I)	
Mg	(II)	
Ca	(II)	
V	(III)	2
	(V)	0
Cr	(III)	3
Mo	(II)	4
	(III)	3
	(IV)	2
	(V)	1
	(VI)	0
Mn	(II)	5
	(III)	4
	(IV)	3
Tc	(I)	6
Fe	(II)	6
	(III)	5
Co	(I)	8
	(II)	7
	(III)	6
Ni	(II)	8
Cu	(II)	9
	(I)	10
Zn	(II)	10

the resulting number. For example, Fe(III) has a d^5 electron count $(26 - 3 - 18 = 5)$, Mo(IV) is a d^2 ion $(42 - 4 - 36 = 2)$, Cu(I) is d^{10} $(29 - 1 - 18 = 10)$, and so forth. In Figure 2.4 are illustrated the most common coordination geometries, for coordination numbers 3 to 6, for metals encountered in bioinorganic chemistry. As previously mentioned, substantial distortions from these idealized structures can and do occur. When a metal ion in a given formal oxidation state is placed at the center of a coordination polyhedron defined by a set of ligands, the energy levels of the d-orbitals housing the metal electrons are altered from those found in the free

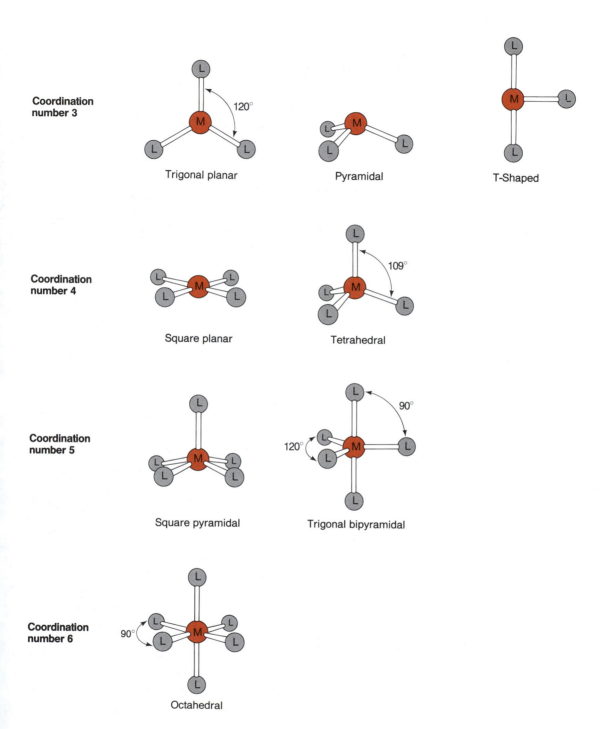

Figure 2.4
Common geometries for coordination numbers 3–6.

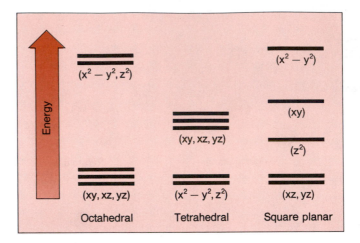

Figure 2.5
Ligand-field splitting diagrams for d orbitals in octahedral, tetrahedral, and square-planar transition metal complexes. The energy separation between the lower and upper sets of orbitals is designated Δ_o for octahedral complexes and Δ_t for tetrahedral complexes.

metal ion. This phenomenon, referred to as ligand-field splitting, is best described in terms of an energy-level diagram that reveals the one-electron orbital energies as a function of the strength of the ligand field.

Ligand-field splitting diagrams for the more important structures adopted by bioinorganic metal centers are provided in Figure 2.5. These diagrams are extremely useful when attempting to correlate the physical properties of metal centers in proteins, such as their optical spectra, magnetism, and electron-spin resonance spectra, with their structures or reactivity. If, for example, one were to encounter a diamagnetic (no unpaired electrons) Ni(II) center in a protein, it most probably would have a square-planar geometry, since both tetrahedral and octahedral d^8 complexes would be expected to have two unpaired electrons and be paramagnetic, as shown in Figure 2.6. The strength of the ligand field at a metal center is largely determined by the set of ligand donor atoms to which it is bonded. The ability of ligands to split the d-orbitals varies according to the "spectrochemical series" in the following order:

$$I^- < Br^- < S^{2-} < Cl^- < NO_3^- < OH^-$$

$$\approx RCOO^- < H_2O \approx RS^- < NH_3$$

$$\approx Im \text{ (imidazole)} < en \text{ (1,2-diaminoethane)}$$

$$< bpy \text{ (2,2'-bipyridine)} < CN^- < CO.$$

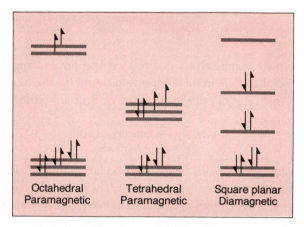

Figure 2.6
Ligand-field splitting diagrams, orbital occupancies, and magnetic properties for
d^8 Ni(II) complexes having octahedral, tetrahedral, and square planar geometries.

This list is valuable for deciding whether a high-spin or low-spin electronic
configuration will occur in the several situations in which such a dichotomy
can exist.

Apart from their characteristic electronic features, many of the metal ions
in bioinorganic chemistry possess nuclear properties that greatly facilitate
their spectroscopic investigation. The experimental methods employed in
such studies will be discussed in Chapter 4. Thus it is possible to highlight a
very localized region of a metalloprotein by measuring the magnetic and
spectroscopic properties of the metal center, which is a unique functional
group embedded in a sea of organic residues that are more difficult to
distinguish.

2.4. Reactions of Coordinated Ligands

The ability of metal ions to alter the reactivity of ligands toward external
substrates is at the heart of their role as catalytic centers in biology. One such
reaction is the enhancement of the acidity of coordinated ligands. As shown
in Table 2.2 and discussed in Section 2.1.c, the pK_a values of coordinated
water and other molecules are lower than those of the free ligands, because
the positively charged metal center can stabilize the conjugate base, in this
case, hydroxide (Equation 2.3). Increasing the susceptibility of substrate mole-

cules toward nucleophilic attack is a related and no less important biological function of metal centers in metalloproteins. Among the biological processes that fall into this category are hydrolyses of acid anhydrides, esters, amides,

$$M(H_2O)_n^{Z+} \longrightarrow M(H_2O)_{n-1}(OH)^{(Z-1)+} + H^+ \qquad (2.3)$$

phosphate esters, and Schiff's bases, carboxylation and decarboxylation reactions, and transaminations. An example of how metal centers can serve as Lewis acids is the hydrolysis of amino-acid esters at neutral pH. As indicated in Equation 2.4, nucleophilic attack on the carbonyl group is facilitated by coordination to the positively charged metal ion. The rates for this reaction

$$(2.4)$$

vary as $M = Cu^{2+} > Co^{2+} > Mn^{2+} > Ca^{2+} \sim Mg^{2+}$, a trend that follows the Irving-Williams series. The uncatalyzed reaction at neutral pH is essentially unobservable.

Additional reactions facilitated by metal centers include the template effect, enhancement of leaving group reactivity, the activation of small molecules such as N_2 and O_2, and the masking of chemical reactivity by coordination. The template effect is a phenomenon by which the metal serves to organize reactive units in condensation reactions, as illustrated in Equation 2.5 for the laboratory synthesis of porphyrin rings. The use of zinc, which prefers tetrahedral stereochemistry, in this reaction facilitates removal of the metal and generation of the porphyrin free base. Equation 2.6 depicts a reaction in which phosphate ester hydrolysis is promoted by coordination of cupric ion to the leaving group. In this biomimetic chemistry, hydrolysis of 2-(imidazol-4-yl)phenyl phosphate is accelerated 10^3 to 10^4 times in the pH 4−7 range by the metal center.

$$(2.5)$$

$$+ [PO_3^-] \xrightarrow{\text{H}_2\text{O, rapid}} \text{H}_2\text{PO}_4^- \quad (2.6)$$

2.5. Model Complexes and the Concept of Spontaneous Self-Assembly

Because of the large size of metallobiopolymers, it is usually difficult to obtain high-resolution structural information about the metal coordination sphere. Other physical properties are sometimes also difficult to measure because of the complexity of the molecule, as is true for the redox potential of a particular metal center in a metalloprotein that has several different metal sites. Investigating the chemical reactivity of a metal ion in a biopolymer can present a similar challenge, since it is difficult to modify the coordinated ligands in a systematic manner to test features of a postulated reaction mechanism, although this objective can now be approached by using site-directed mutagenesis. For these reasons, bioinorganic chemists frequently synthesize and study model complexes designed to replicate as faithfully as possible the physical and chemical properties of the metal center in a biopolymer. If the structure of the latter is known, for example, through an X-ray crystal structure determination, it is possible to design an exact replica of the coordination environment in a model complex. These complexes have been termed *replicative models*. If the environment is unknown, the model approach affords bioinorganic chemists a chance to test postulated structures in what may be called *speculative models*. In both approaches, good judgment must prevail to prevent overinterpretation of information from a model compound in attempting to explain the physical or chemical properties of the biological molecule. It is essential that the structure of the model complex be known, preferably by single-crystal X-ray diffraction work. The model approach has provided many valuable insights into structural and mechanistic metallobiochemistry, including assignment or verification of the charges, and hence the metal-oxidation states, of metal clusters in proteins, the effects of distance and medium on electron-transfer rates, the roles of steric and electronic factors in promoting reversible dioxygen binding to heme iron centers, and the identification of likely intermediates in a variety of enzyme-catalyzed reactions. It should be pointed out, however, that the discovery and characterization of metal centers in biology have had a similar impact on the development of the field of coordination chemistry. In other words, the effects have been symbiotic, and both have been essential parts of the growth of the field of bioinorganic chemistry.

In the design of model complexes, a variety of strategies is available. One of these, which has come to be known as "spontaneous self-assembly," involves reaction of the metal and the simplest ligands containing the known or suspected biological donor atoms to form the desired replica molecule. The rationale behind this approach is that nature may have adopted a similar strategy in assembling such a metal center, borrowing available chemistry from the geosphere during evolution of the biosphere. The spontaneous self-assembly approach has been successful in replicating known metalloprotein core structures, such as the $\{Fe_nS_n\}^{m-}$ clusters in iron-sulfur proteins and the $\{Fe_2O(O_2CR)_2\}^{2+}$ cores in diiron-carboxylate proteins. These successes have inspired some to rephrase "spontaneous self-assembly" as "dumb luck," but such critics have missed an important point. As already discussed, metal-ligand bonds in most bioinorganic centers are labile. Thus, the synthetic model chemistry is nearly always under thermodynamic, rather than kinetic, control. The molecules that are obtained, usually as crystalline solids, often become major components of the mixture of compounds formed between metal and ligands only because of the tactical skills of the synthetic inorganic chemist at the laboratory bench. Undesired side products must be avoided by judicious choice of solvents, counter-ions, ligand steric properties, and temperature. The use of such tactics to assemble the desired model complex from simple ligands as the thermodynamically most stable entity among a potentially complex mixture of products contrasts with the synthetic approach to most organic and organometallic compounds, where reactions are usually under kinetic control owing to the more inert character of carbon-carbon and metal-carbon bonds.

Although spontaneous self-assembly has produced good models for the structural and spectroscopic properties of metalloprotein cores, duplicating the functional chemistry usually requires more complex ligands. The development of compartmentalized ligands such as crown ethers, cryptands, sepulchrates (Fig. 2.7), and other macropolycyclic molecules has afforded a variety of new strategies for mimicking metalloprotein core chemistry. The construction of synthetic peptides is another rapidly developing field that is likely to have important effects on bioinorganic model chemistry. At some point,

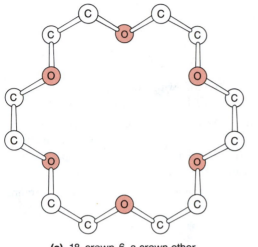

(a) 18-crown-6, a crown ether

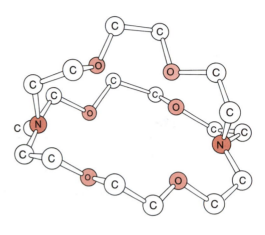

(b) 2,2,2-cryptand, a cryptand

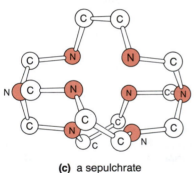

(c) a sepulchrate

Figure 2.7
Macrocyclic ligands.

however, the limitations of the model approach will be reached, because the subtleties of metalloprotein core chemistry cannot fully be duplicated, no matter how good the designer ligand. Nature did, after all, develop complicated molecules to achieve such specific functions as those listed in Table 1.1. Ultimately, there is no substitute for direct studies of the biomolecules.

Study Problems

1. Calculate the number of d electrons for the following complexes:
(a) $[Fe(CN)_6]^{3-}$ (b) $[CoCl_4]^{2-}$ (c) $[Ca(OH_2)_6]^{2+}$ (d) $[Ru(NH_3)_6]^{3+}$

2. Uranyl ion (UO_2^{2+}) is commonly used to form heavy-atom derivatives for X-ray diffraction studies. Given that this is a relatively hard ion, what sorts of functional groups would you expect to be derived? How does this behavior compare to that expected for another heavy-metal derivatizing ion, Hg^{2+}?

3. Match the three d-orbital energy-level diagrams below with the corresponding complexes $[Fe(OH_2)_6]^{2+}$, $[Fe(CN)_6]^{4-}$, and $[FeCl_4]^{2-}$.

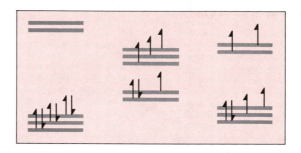

4. Chromous ion, $[Cr(OH_2)_6]^{2+}$, is a good reductant. Treatment of the oxidized form of an electron-transfer protein with this ion results in conversion of the protein to its reduced form. The Cr^{3+} ion produced is found to be tightly associated to the protein. Explain this observation.

5. Which of the two copper complexes shown below (I or II) will have the lower reduction potential (be harder to reduce), and why?

Me   Bu^t

Cu   Cu

Me   Bu^t

I   II

Bibliography

General Inorganic Chemistry
F. A. Cotton and G. Wilkinson. 1988. *Advanced Inorganic Chemistry*, 5th ed. Wiley-Interscience, New York.

B. Douglas, D. H. McDaniel, and J. J. Alexander. 1983. *Concepts and Models of Inorganic Chemistry*, 2nd ed. Wiley, New York.

J. E. Huheey, E. A. Keiter, and R. L. Keiter. 1993. *Inorganic Chemistry: Principles of Structure and Reactivity*, 4th ed. Harper Collins, New York.

S. M. Owen and A. T. Brooker. 1991. *A Guide to Modern Inorganic Chemistry*. Wiley, New York.

D. F. Shriver, P. Atkins, and C. H. Langford. 1994. *Inorganic Chemistry*, 2nd ed. W. H. Freeman, New York.

Electronic Structure
R. L. Dekock and H. B. Gray. 1989. *Chemical Structure and Bonding*, 2nd ed. University Science Books, Mill Valley, CA.

B. N. Figgis. 1966. *Introduction to Ligand Fields*. Robert E. Krieger, Melbourne, FL.

H. B. Gray. 1973. *Chemical Bonds: An Introduction to Atomic and Molecular Structure*. Benjamin/Cummings, Menlo Park, CA.

L. E. Orgel. 1966. *An Introduction to Transition Metal Chemistry: Ligand Field Theory*, 2nd ed. Wiley, New York.

Stability and Reactivity
F. Basolo and R. G. Pearson. 1967. *Mechanisms of Inorganic Reactions*, 2nd ed. Wiley, New York.

K. Burger, ed. 1990. *Bio Coordination Chemistry: Coordination Equilibria in Biologically Active Systems*. Ellis Horwood, New York.

H. Irving and R. J. P. Williams. 1953. "The Stability of Transition-Metal Complexes." *J. Chem. Soc.*, part III, 3192–3210.

D. Katakis and G. Gordon. 1987. *Mechanisms of Inorganic Reactions*. Wiley-Interscience, New York.

R. G. Wilkins. 1991. *Kinetics and Mechanisms of Reactions of Transition-Metal Complexes*, 2nd ed. VCH, Weinheim.

Coordination Sites in Proteins and Models
J. P. Glusker. 1991. "Structural Aspects of Metal Liganding to Functional Groups in Proteins," in C. B. Anfinsen, J. T. Edsall, D. Eisenberg, and F. M. Richards, eds., *Advances in Protein Chemistry*, Volume 42. Academic Press, San Diego, 1–76.

J. A. Ibers and R. H. Holm. 1980. "Modeling Coordination Sites in Metallobiomolecules." *Science* 209, 223–235.

K. D. Karlin. 1993. "Metalloenzymes, Structural Motifs, and Inorganic Models." *Science* 261, 701–708.

CHAPTER 3

Properties of
Biological Molecules

The ligands in bioinorganic chemistry are most commonly amino-acid side chains or constituents of nucleic acids. The coordination environment presented by these residues depends critically upon the three-dimensional folding of proteins and the tertiary structures of nucleic acids. More complex structural elements, such as the positioning of proteins in the matrix of a cell membrane or the wrapping of DNA around the nucleosome core proteins in chromatin, further modulate the metal-binding capabilities of biological macromolecules. The purpose of this chapter is to provide the basic information about these topics that is necessary to understand the principles developed in later chapters. In addition, fundamental concepts from molecular biology related to gene cloning and expression as well as the basic principles of enzyme kinetics are presented.

3.1. Proteins and Their Constituents

3.1.a. The Naturally Occurring Amino Acids. Proteins are composed of 20 essential amino acids, many of which present donor atoms suitable for coordination to metal ions. The structures of the amino-acid side chains at pH 7 are given in Figure 3.1. The pK_a values for both basic and acidic side chains are summarized in Table 3.1. As described in Chapter 2, metal coordination at these sites can substantially diminish the effective pK_a values. The amino-acid side chains extend from the C_α carbon atoms. The chirality at the α-carbon of naturally occurring amino acids is almost always S (or L). Proteins consist of unbranched chains of amino acids joined by peptide bonds. These peptide

44

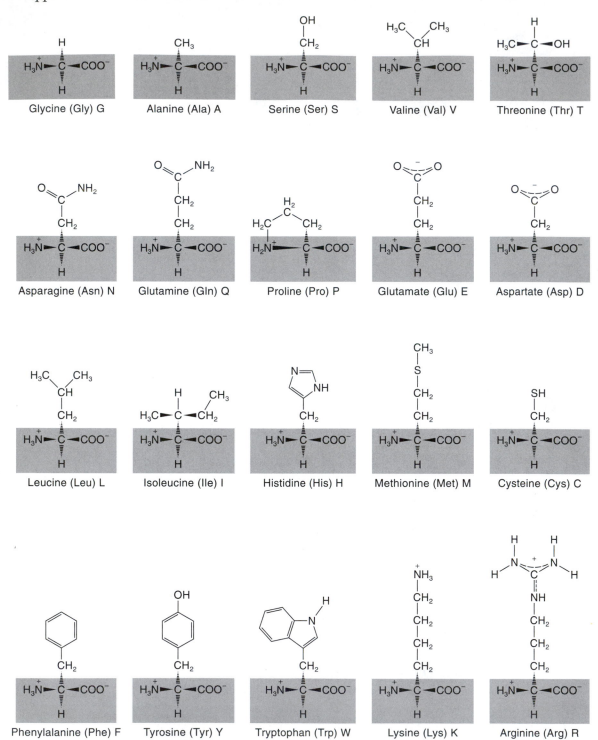

Figure 3.1
Structures of the twenty common amino acids and their three- and one-letter code representations.

Table 3.1
pK values of amino acid side chains

Amino acid	pK_a
Arginine	12.48
Aspartic acid	3.65
Cysteine	8.35
Glutamic acid	4.25
Histidine	6.00
Lysine	10.79
Tyrosine	10.13
α-COOH	~ 2.2
α-NH_3^+	~ 9.5

bonds are essentially planar units that usually exist in the *trans* configuration, as shown here.

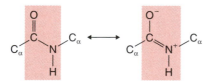

Amino-acid residues in proteins are numbered sequentially, starting from the first residue at the amino terminus. Each amino acid is given a single-letter code in addition to the usual three-letter one (e.g., histidine = His or H; aspartic acid = Asp or D); these are shown in Figure 3.1. The one-letter codes have become increasingly more common as a result of advances in DNA sequencing technology, which, through application of the genetic code, have greatly expanded the number of known protein sequences. With this information explosion, it was necessary to shorten the earlier, perhaps more familiar, three-letter abbreviations.

3.1.b. Proteins as Ligands.   The amino acids that commonly function as ligands and their modes of interaction are illustrated in Figure 3.2. The most prevalent side-chain ligands are the thiolate of cysteine, the imidazole of histidine, the carboxylates of glutamic and aspartic acids, and the phenolate group of tyrosine. With the exception of tyrosine, each of these amino acids has been observed to act as a bridging ligand between two metal ions and to serve as a terminal ligand to a single ion. Less frequently encountered metal donors are the hydroxyl groups of serine and threonine, the thioether group

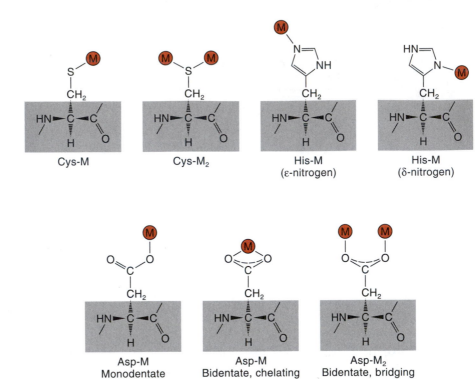

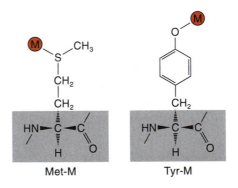

Figure 3.2
Amino acid side chain−metal ion binding modes.

of methionine, the carboxamide groups of glutamine and asparagine, the amino group of lysine, and, perhaps, the guanidine group of arginine. In addition to the donor atoms provided by side chains, metal ions can also bind to peptide carbonyl groups, deprotonated peptide-bond nitrogen atoms, and the N-terminal amino and C-terminal carboxyl groups.

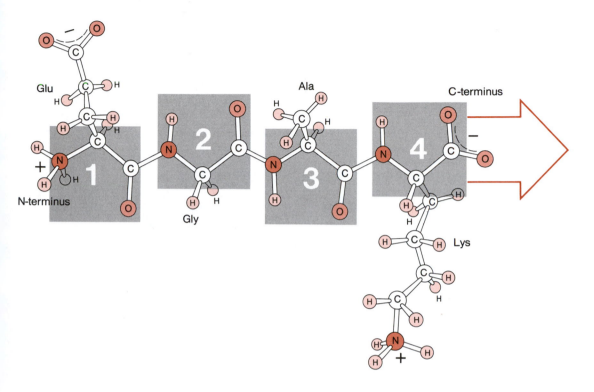

Figure 3.3
Protein primary structure. Amino acids along the polypeptide chain are numbered from the amino (N) to the carboxyl (C) terminus.

3.1.c. Protein Structure. The orientation of a set of ligand donor atoms in a metalloprotein is critically dependent upon its three-dimensional structure. This structure is determined entirely by the amino-acid sequence of the protein. The folding of a polypeptide into a unique three-dimensional structure depends on a number of forces. Chief among these are the packing of hydrophobic surfaces against one another, usually within the interior of the protein, to avoid interaction with the more polar solvent environment; the formation of specific hydrogen bonds and ionic interactions; and sometimes the existence of covalent cross-links effected by disulfide bonds and coordination of amino-acid side chains to metal ions. Protein structure can be described at four levels. The *primary structure* refers to the covalent bonds that are present, determined by the amino-acid sequence, as illustrated in Figure 3.3. *Secondary structure* refers to local conformation extending over groups

48

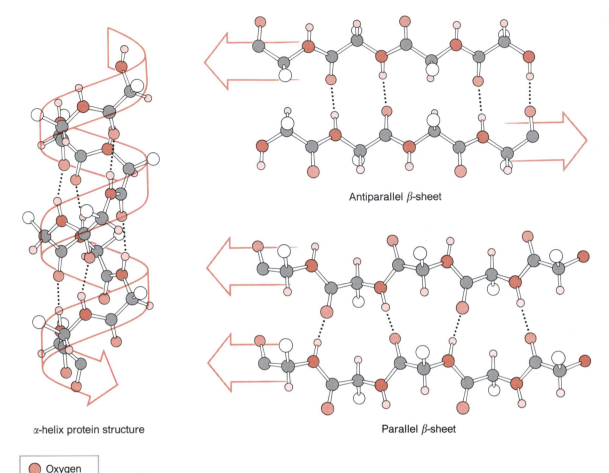

Antiparallel β-sheet

Parallel β-sheet

α-helix protein structure

● Oxygen
● Nitrogen
● Carbon
○ R-group
○ Hydrogen

Figure 3.4
Protein secondary structure. The most common motifs, the α-helix and antiparallel and parallel β-sheets, are depicted.

of a few adjacent amino acids, often determined by specific hydrogen-bonding patterns. The most commonly found elements of secondary structure are the right-handed α-helix, the antiparallel and parallel β-pleated sheets, and various types of turns. These elements are illustrated in Figure 3.4. *Tertiary structure* is the term used to describe the overall course of a biopolymer chain that generates "folds" that can be quite elaborate. The tertiary structures of several metalloproteins are illustrated in Figure 3.5. Finally, the quaternary structure refers to the assembly of different polypeptide chains to form larger units. These higher-order structures are discussed in Section 3.1.d.

The metal-dependence of metalloprotein folding varies tremendously. For the copper protein plastocyanin, structural studies of the apo protein, the form that lacks the metal ion, have revealed that very little structural modification occurs upon metal removal. The metal-binding site is highly preorganized by the protein tertiary structure, which is quite stable even in the

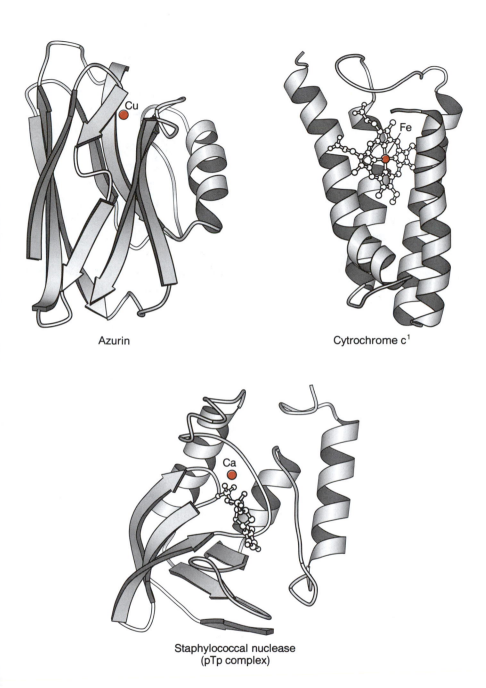

Azurin

Cytrochrome c[1]

Staphylococcal nuclease
(pTp complex)

Figure 3.5
Protein tertiary structure. The folded structures of three metalloproteins are shown.

absence of bound metal. In contrast, for peptides corresponding to "zinc finger" domains, spectroscopic and chemical data indicate that the apo peptides exist as random coils with no unique secondary or tertiary structure. In the presence of Zn^{2+} or some other metal ions, a highly stable three-dimensional structure is adopted. Finally, for Ca^{2+}-binding proteins such as calmodulin and troponin C, an intermediate situation occurs, in which the binding of metal ions promotes a transition from a partially folded form with a relatively well defined tertiary structure to a somewhat different, more completely folded form.

While it is tempting and sometimes possible to predict the metal-binding site from the amino-acid sequence in a metalloprotein, experience has shown that errors can result from such an approach. For example, in Figure 3.6, we see that the cysteine sulfur atoms that bind and hold the $\{Fe_4S_4\}$ cores in 8-iron ferredoxin proteins do not correspond to the two distinct clusters of four Cys residues in the primary structure.

3.1.d. Higher-Order Structures.

Proteins can be made up of several individually folded polypeptide chains, termed *subunits*, that interact with one another to form larger assemblies. The quaternary structure of a protein may be essential for its function and can sometimes profoundly influence the structural properties at metal-binding sites and *vice versa*. Many metalloproteins and enzymes contain more than one subunit. For example, Cu—Zn superoxide dismutase is a dimer of two identical subunits, whereas the Mo—Fe protein of nitrogenase is a tetramer consisting of two α subunits and two β subunits. We shall see in Chapter 7 how zinc binding to one subunit of the enzyme aspartate transcarbamoylase is required for assembly of the intact enzyme. And, in Chapter 11, we examine how subunit interactions endow hemoglobin with its special ability to bind four molecules of dioxygen in a cooperative manner.

3.1.e. The Manipulation of Proteins by Gene Cloning and Expression.

The composition and sequence of amino acids in proteins ultimately determine their structure and properties. Amino-acid sequences are controlled by the sequence of genes, which are made of DNA. The pathways by which this exquisite genetic apparatus works are shown in Figure 3.7. The DNA sequence is first transcribed into RNA by the enyzme RNA polymerase. A variety of signals determines which sequences of DNA are transcribed. In prokaryotes, cells lacking a nucleus, the RNA produced acts directly as a message that is translated into protein. This translation takes place on ribosomes, large assemblies of specific RNA and protein molecules that function as the protein synthetic machinery. In eukaryotes, cells containing a

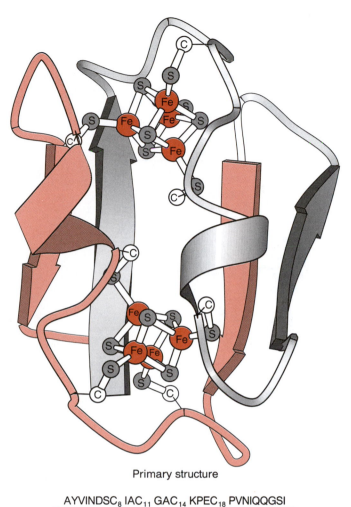

Primary structure

AYVINDSC$_8$ IAC$_{11}$ GAC$_{14}$ KPEC$_{18}$ PVNIQQGSI
YAIDADSC$_{35}$ IDC$_{38}$GSC$_{41}$ ASVC$_{45}$ PVGAPNPED

Figure 3.6
Primary and tertiary structures of an eight-iron ferredoxin. Although the primary structure includes two separate CXXCXXCXXXC sequences, each of the Fe$_4$S$_4$ cubes is bonded to cysteine residues from both cys-rich sequences.

nucleus, the initial RNA transcript is processed into a mature messenger RNA. This procedure is called *splicing* and involves removal of intervening sequences within the RNA, followed by connecting of the coding regions. Splicing is necessary because the genomic DNA includes regions, termed *exons*, that encode the final protein and that are interspersed with regions, called *introns*, that do not. The mature messenger RNA is then translated as in prokaryotes.

Fragments of DNA can be cloned, a process that involves their incorporation into a *vector*, a piece of DNA that is capable of replication within an

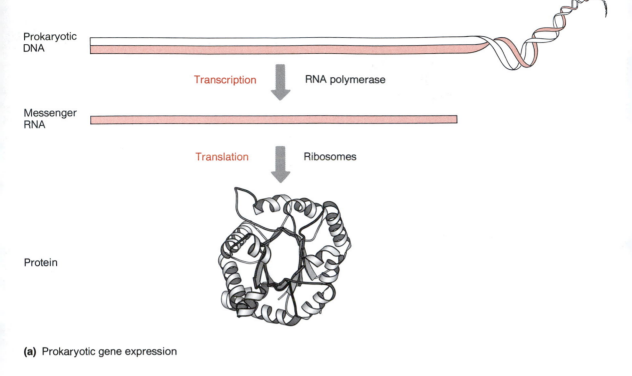

Prokaryotic
DNA

Transcription   RNA polymerase

Messenger
RNA

Translation   Ribosomes

Protein

(a) Prokaryotic gene expression

Figure 3.7
Processes involved in gene expression in prokaryotes and eukaryotes.

appropriate type of cell. These vectors are generally either plasmids, relatively small (4,000 to 8,000 base pairs) circular pieces of DNA, or viruses such as bacteriophage lambda. Cloning allows isolation of regions of DNA for sequencing and other characterization. Cloning of a specific gene is usually accomplished with the aid of a *library*. A library is a collection of vectors into which have been cloned different DNA fragments. The library may be introduced into cells in such a way that each cell contains a single member of the library. The member or members of the library that include the gene of interest can then be identified, a process called *screening*, by one of several methods that recognize differences in nucleotide sequence, protein structure, or protein function. For example, members of a library can be introduced into a strain of cells lacking the activity of the sought-after gene in such a manner that the genes contained in the library are *expressed*; that is, the corresponding protein is produced. Any cell that has obtained a member of the library that encodes the desired gene will now show the missing activity. This approach, cloning by complementation, is a powerful method for cloning genes whose function is known but for which no other information is available.

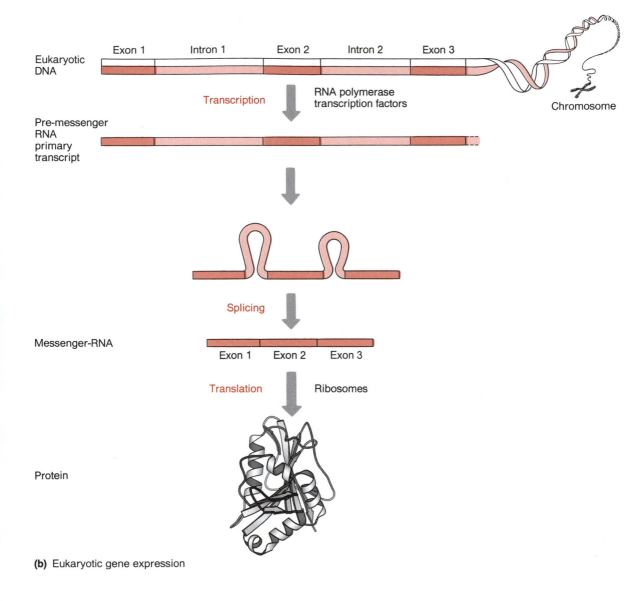

(b) Eukaryotic gene expression

Cloned genes can often be expressed in cells in a manner that facilitates the production of large amounts of protein for study, as summarized in Figure 3.8. The process of expression involves cloning the gene into a vector that has all of the signals needed to cause high-level transcription of the gene and subsequent translation of the messenger RNA into protein. Sometimes genes from eukaryotes can be expressed in functional form in bacteria, making them much more accessible for detailed biochemical, structural, and spectroscopic studies.

The expression of a gene also allows structure-function investigations to be carried out by the method of *site-directed mutagenesis*. Methods have

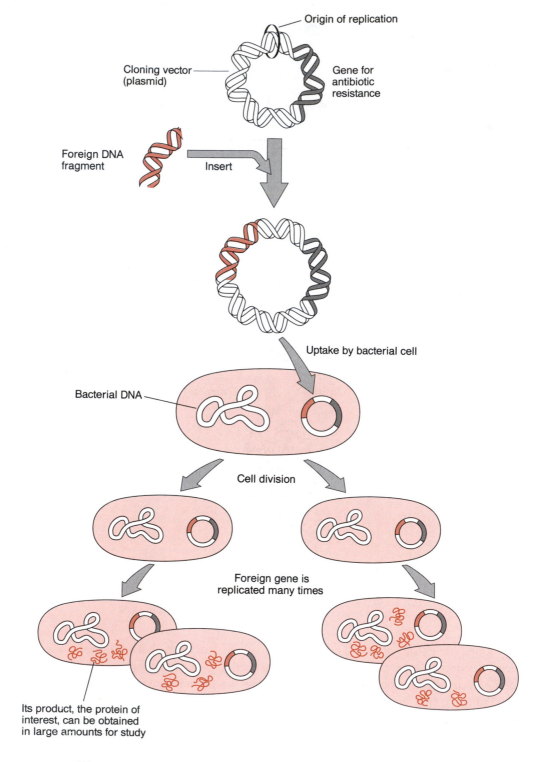

Figure 3.8
Processes involved in protein overexpression via a plasmid containing a cloned DNA fragment.

been developed for introducing specific changes into the nucleotide sequence of a gene in such a way that virtually any desired modification into the nucleic-acid sequence, and hence in the corresponding amino-acid sequence, can be made. For example, phenylalanine in position 137 of a polypeptide chain could be changed to tyrosine. The mutant would be denoted Tyr137Phe or Y137F. This process allows the bioinorganic chemist the opportunity to alter, virtually at will, any residue in the protein, including amino acids at or near the metal-binding site. Although this approach is not without its limitations, the methodology has provided a powerful tool to probe hypotheses about the structures and reactivity of metal sites in proteins.

3.1.f. The Kinetic Analysis of Enzyme Reactions.

Many proteins are enzymes; that is, they are catalysts that promote biochemical transformations. Enyzme-catalyzed reactions share some unifying properties. The most important of these is the phenomenon of substrate-saturation of the rate of the reaction, which is most simply described by the Michaelis-Menten model shown in Equation 3.1. The substrate, S, first binds to the enzyme, E, to form an

$$E + S \rightleftharpoons ES \rightleftharpoons EP \longrightarrow E + P \qquad (3.1)$$

enzyme-substrate complex, ES. The chemical transformation then takes place to form an enzyme-product complex (EP), which then dissociates to the product (P) and the free enzyme. This scheme explains the observed substrate-saturation behavior. The kinetics of such a reaction can be described by two parameters obtained from a plot of reaction velocity (V) as a function of substrate concentration, as illustrated in Figure 3.9. These parameters are

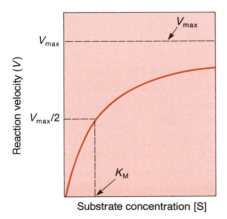

Figure 3.9
Dependence of reaction velocity on substrate concentration for enzyme kinetics corresponding to the Michaelis-Menten model. The parameters V_{max} and K_M are defined.

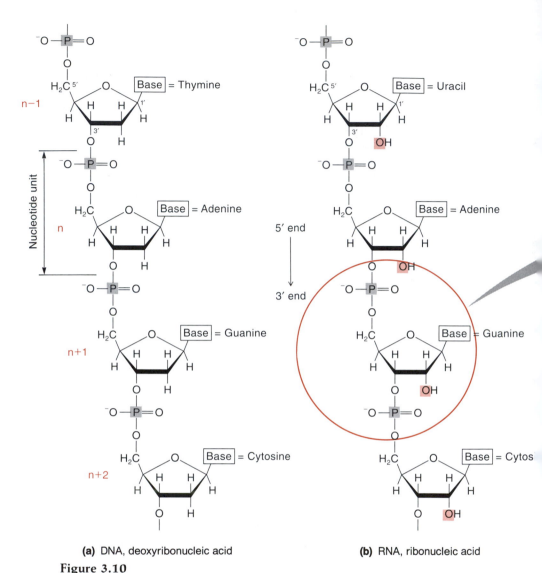

(a) DNA, deoxyribonucleic acid **(b)** RNA, ribonucleic acid

Figure 3.10
Structures of the phosphodiester backbones of DNA and RNA, showing one nucleotide unit with the torsional angles that are used to describe its conformation.

V_{max}, the maximal rate of the reaction in the presence of saturating levels of substrate, and K_M, the concentration of substrate at which the reaction proceeds at half the maximal rate. Although it is tempting to think of K_M in terms of the equilibrium constant for the enzyme-substrate complex, it is important to realize that K_M can also be affected by the rates of subsequent steps in the reaction. The ratio V_{max}/K_M is often used as a measure of the effective second-order rate constant for an enzyme-catalyzed reaction. This approximation is often useful for comparison of enzyme-catalyzed reactions with non-enzymatic reactions that do not show substrate-saturation kinetics.

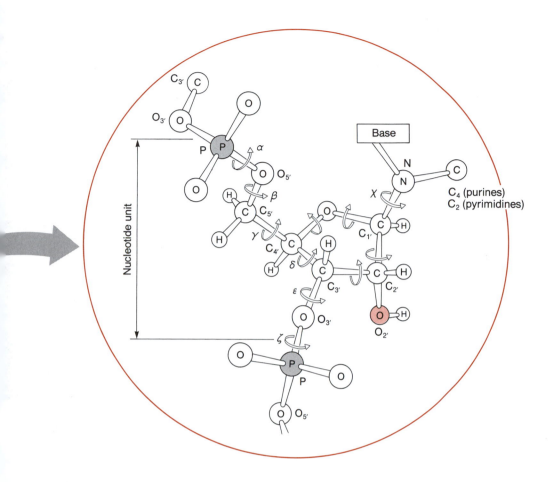

3.2. Nucleic Acids and Their Constituents

3.2.a. RNA, DNA, and Their Building Blocks. Nucleic acids are polymers of ribonucleotides or deoxyribonucleotides in RNA and DNA, respectively. As shown in Figure 3.10, phosphodiester linkages join adjacent residues to form a sugar-phosphate polymer backbone. Linked to each sugar through glycosidic bonds is one of five nucleotide bases, the purines, adenine and guanine, and the pyrimidines, cytosine and thymine (for DNA) or uracil (for RNA). Unlike the planar peptide linkages in proteins, the phosphodiester bonds in nucleic acids can experience a variety of conformations, defined by the torsional angles depicted in Figure 3.10. Additional conformational complexity arises from the torsion angles within the sugar rings. Two basic structures prevail, one found in RNA, where the C3' carbon atom is on the same side of the plane defined by C4'—O—C1' as the C5' carbon (C3'-endo), and the other, often found in DNA, where the C2' is endo to C5'. In addition, the bases can adopt *syn* and *anti* conformations with respect to the glycosidic

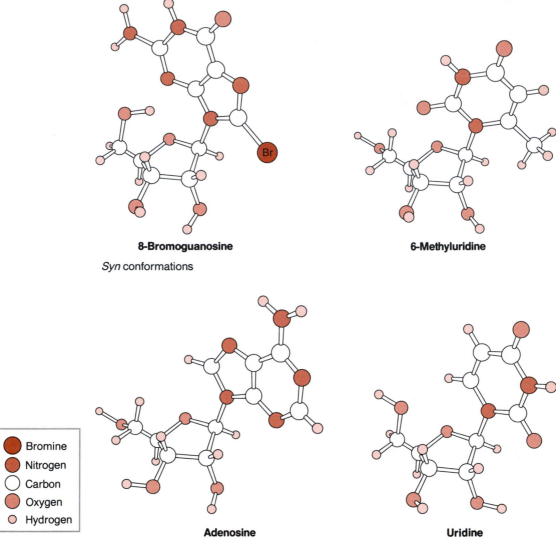

8-Bromoguanosine

Syn conformations

6-Methyluridine

Bromine
Nitrogen
Carbon
Oxygen
Hydrogen

Adenosine

Anti conformations

Uridine

Figure 3.11
Structures of representative nucleosides and nucleoside derivatives illustrating *syn* and *anti* conformations with respect to the glycosidic bond.

linkage (Figure 3.11). These seemingly subtle changes in sugar pucker can have dramatic consequences on the tertiary structure, and hence the metal-binding properties, of the biopolymers, as will be discussed.

The unit comprising only a sugar and a base is referred to as a *nucleoside*, whereas *nucleotides* contain additional 5'- or 3'-phosphate groups. Examples of nucleotides include ATP, adenosine triphosphate, dGDP, deoxyguanosine diphosphate, and 3'-UMP, uridine monophosphate. In the last case, the phosphate ester link is made to the 3'-hydroxyl group. In Figures 3.12 and 3.13 are drawn the structures of the five nucleoside bases, their numbering schemes, and Watson-Crick base-pairing interactions. The last of these fea-

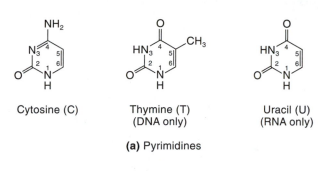

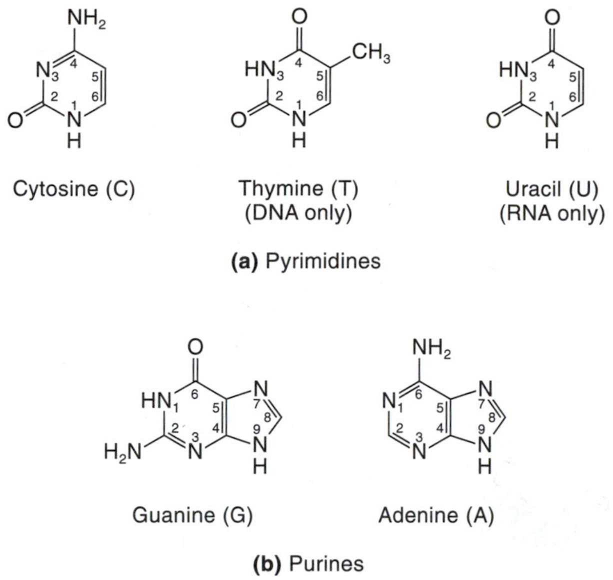

Cytosine (C)

Thymine (T)
(DNA only)

Uracil (U)
(RNA only)

(a) Pyrimidines

Guanine (G)

Adenine (A)

(b) Purines

Figure 3.12
Chemical structures and numbering schemes of nucleoside bases
found in DNA and RNA.

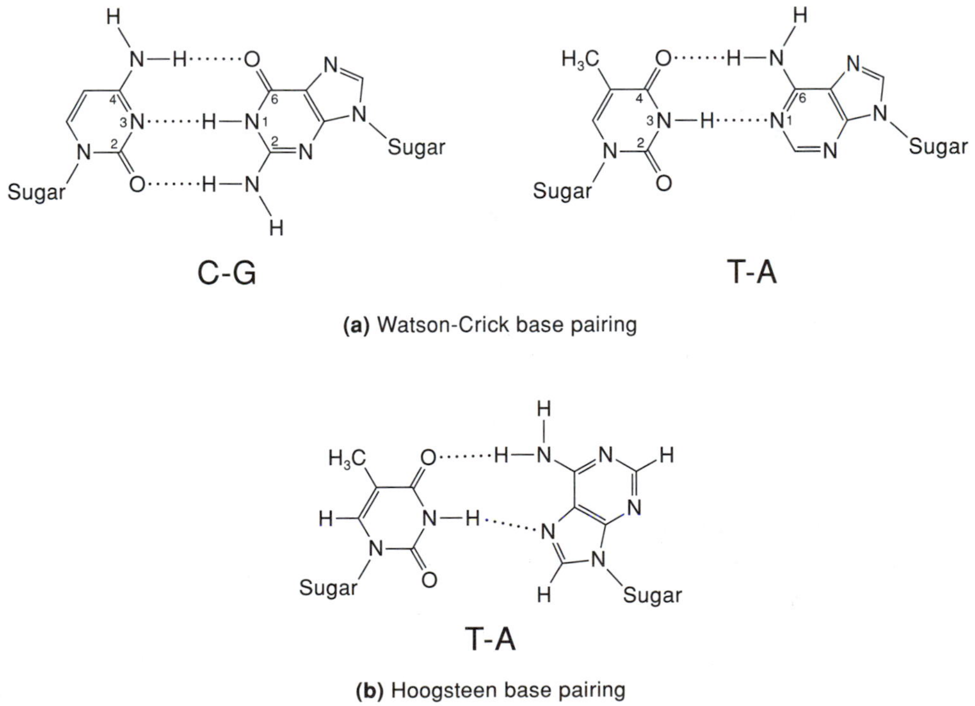

C-G

T-A

(a) Watson-Crick base pairing

T-A

(b) Hoogsteen base pairing

Figure 3.13
Base-pairing schemes that occur in nucleic acids.

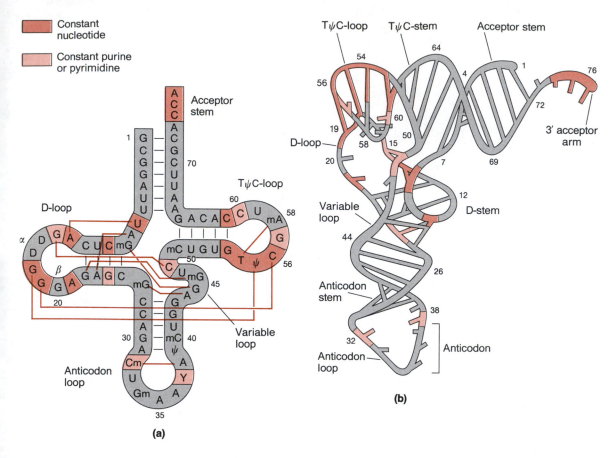

Figure 3.14
Structure of a transfer RNA molecule, yeast phenylalanine tRNA. (a) Primary and secondary structures. (b) Tertiary structure.

tures is responsible for linking the two DNA strands into a double helix and for much of RNA secondary structure. An alternative scheme sometimes encountered is Hoogsteen base pairing, which is also shown in Figure 3.13.

3.2.b. RNA Structure. Ribonucleic acids are classified according to their functions in the cell. Messenger RNA, mRNA, carries genetic information from DNA in the nucleus to ribosomes in the cytoplasm, where protein synthesis occurs. A constituent of ribosomes is ribosomal RNA, rRNA, whereas transfer RNA (tRNA) molecules bind amino acids to their 3' termini and carry them to the ribosome, where they are transferred to the growing protein chain. The best-known RNA structure is that of tRNA; several tRNA structures have been characterized by X-ray crystallography. The structure of yeast phenylalanyl tRNA, tRNA[Phe], is depicted in Figure 3.14. The molecule is drawn in cloverleaf form to illustrate the A−U and G−C Watson-Crick hydrogen-bonded base pairs that determine its secondary structure. The secondary structures of several other RNA molecules have been determined. One of the most powerful tools for the determination of RNA secondary structure is phylogenetic comparison. Homologous RNA sequences from several different organisms are compared. Often regions that appear to be capable of forming secondary structures, such as hairpin loops, can be identified because the Watson-Crick base pairing is conserved even if the actual sequence is not. The secondary structure of the 16S ribosomal RNA from *E. coli* deduced from such comparisons is shown in Figure 3.15. This secondary structure folds up to form a well-defined tertiary structure. It appears that the tertiary structures of RNA molecules may be distinctive and functionally very important, much like those for proteins, and RNA molecules are known to have enzymatic type functions. These RNA enzymes, termed *ribozymes*, require metal ions for their activity, as discussed further in Section 7.2.b. The elucidation of the three-dimensional structures of RNA molecules, the development of rules for RNA folding, and understanding how ribozymes function are among the more important problems in molecular biology today.

3.2.c. DNA Structure.   Compared with RNA, the structures adopted by DNA are less complex, but nonetheless interesting. Both single- and double-stranded forms exist, but we discuss only the latter, which are more common and better understood. The molecular weights, lengths, and base composition of DNAs from a variety of sources are given in Table 3.2.

Table 3.2
Size and composition of various DNAs

Organism	Size (kilobases, kb)	MW (10^6 Daltons)	GC/AT content (mole %)
Viruses			
Polyoma	5.1	3.3	41/59
Lambda	48.6	33	49/51
Vaccinia	190	125	37/63
Bacteria			
Mycoplasma	760	502	33/67
E. coli	4,000	2,640	50/50
Eukaryotes			
Yeast	13,500	8,910	37/63
Drosophilia	165,000	109,000	45/55
Human	2.9×10^6	1.9×10^6	40/60

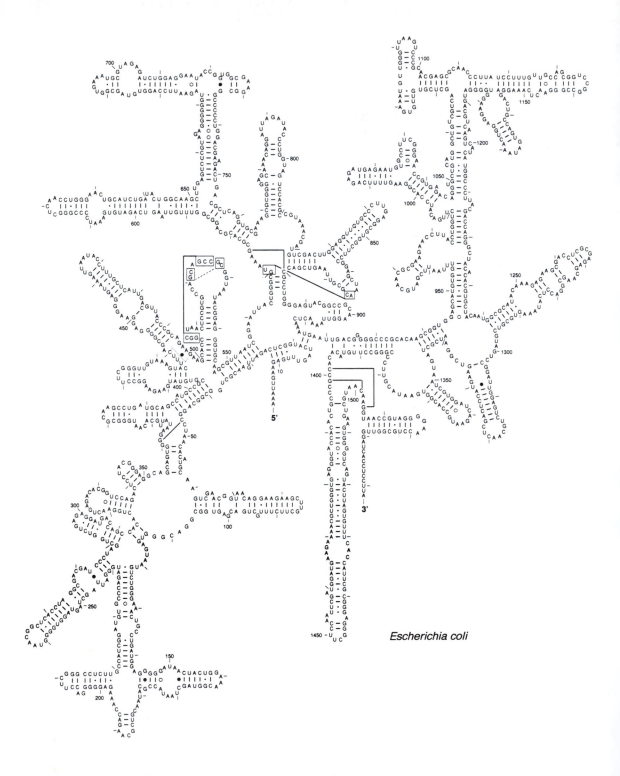

Figure 3.15
Secondary structure of bacterial 16S ribosomal RNA deduced by comparison of sequences from a wide variety of species. (Courtesy of R. Gutell.)

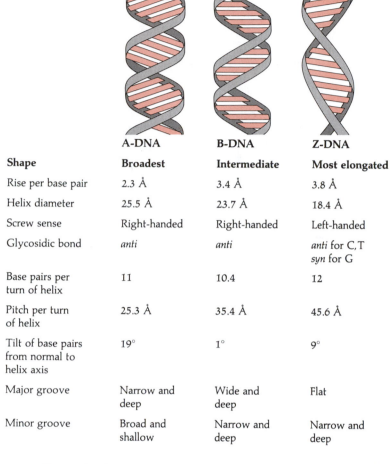

	A-DNA	B-DNA	Z-DNA
Shape	**Broadest**	**Intermediate**	**Most elongated**
Rise per base pair	2.3 Å	3.4 Å	3.8 Å
Helix diameter	25.5 Å	23.7 Å	18.4 Å
Screw sense	Right-handed	Right-handed	Left-handed
Glycosidic bond	*anti*	*anti*	*anti* for C,T *syn* for G
Base pairs per turn of helix	11	10.4	12
Pitch per turn of helix	25.3 Å	35.4 Å	45.6 Å
Tilt of base pairs from normal to helix axis	19°	1°	9°
Major groove	Narrow and deep	Wide and deep	Flat
Minor groove	Broad and shallow	Narrow and deep	Narrow and deep

Figure 3.16
Structural features and parameters of A-, B-, and Z-DNA.

The structure of the most common, or B-form, of DNA is a right-handed double helix characterized by major and minor grooves, parallel Watson-Crick base pairs stacked on one another at a 3.4 Å spacing, a pitch of 35 Å corresponding to 10.4 bp per turn, C2'-endo sugar puckers, and *anti* conformations at the individual nucleotides (Figure 3.11), as illustrated in Figure 3.16. A second type of DNA is the A-form, having a C3'-endo sugar pucker, a deep major groove, and parallel base pairs that are no longer perpendicular to the helix axis. Z-DNA is a fascinating left-handed variant seen for sequences having alternating stretches of purine-pyrimidine nucleosides, for example, $\{d(CG)\}_3$, *syn* conformations for the guanosine nucleosides, and no deep grooves. A- and Z-DNA are shown in Figure 3.16, which also compares the characteristics of all three types of structures. It is likely that sequence-

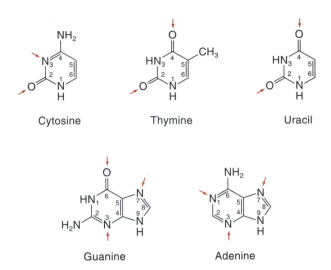

Figure 3.17
Most common metal-binding sites on nucleoside bases,
depicted by arrows.

dependent local variations in both structure and deformability occur in all natural DNAs. These variations can dramatically modulate both covalent and noncovalent interactions of metal complexes with DNA and are also important for interactions between DNA and proteins.

3.2.d. Metal Binding and Nucleic Acid Structure. The metal-binding sites on the nucleoside bases, shown in Figure 3.17, consist primarily of the endocyclic nitrogen atoms. Because their lone pairs are largely delocalized into the ring by resonance, the exocyclic amino groups of guanosine, adenosine, and cytidine have relatively poor metal-coordinating abilities unless deprotonated to form the very nucleophilic RNH^- ion. Metal complexes of this species rarely occur at neutral pH in DNA or RNA, except under special circumstances. Experimental and theoretical studies show the purine N_7 atoms to be the best nucleophiles among the nucleic-acid base heteroatoms and, consequently, the most likely to be metal-binding sites.

Apart from the heteroatoms of the nucleoside bases, which form strong complexes with transition-metal ions such as Cu(II), Cr(III), and Pt(II), the negatively charged oxygen atoms of the phosphodiester groups also bind metal ions, especially hard metals, such as those of the alkali and alkaline earth groups (Na^+, K^+, Mg^{2+}). In ATP, for example, various metal chelates of the α, β, and γ phosphate groups have been identified by X-ray crystallography and by ^{31}P NMR spectroscopy. Two of the known structures are depicted in Figure 3.18. Enzymes requiring ATP frequently need Mg^{2+} as well, the Mg-ATP complex serving as a substrate or cofactor. It is estimated that more than

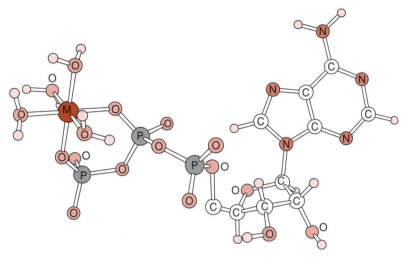

M(ATP) coordinated to β and γ phosphates,
based on the crystal structure of [Co(III)(NH$_3$)$_4$ (triphosphate)]

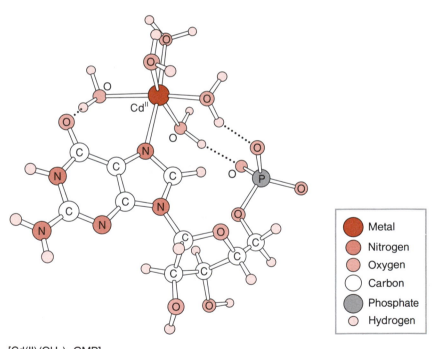

[Cd(II)(OH$_2$)$_5$ GMP]

Figure 3.18
Structures of two metal-nucleoside complexes.

90 percent of the ATP in cells is bound to Mg^{2+} ion, the association constant being 3.8×10^2 M^{-1}. Magnesium also promotes the hydrolysis of the high-energy phosphate bond in ATP, and binds to the ADP thus formed. Many mechanistic details of how Mg-ATP, and M-NTP in general, function as substrates or cofactors in biochemistry remain to be established.

Magnesium ions are tightly bound to the sugar-phosphate backbone of tRNA, stabilizing its tertiary structure. Both Pb(II) and *trans*-[Pt(NH$_3$)$_2$Cl$_2$] adducts of tRNA have been crystallographically characterized. The lead(II) derivative can promote the cleavage of the sugar-phosphate backbone in yeast tRNA^{Phe}, as indicated in Figure 3.19; this figure also illustrates the binding sites of the platinum derivative. From studies reported to date, it is apparent that RNA has excellent metal-binding capacities. How metal-RNA complexes function in biology is a largely unexplored area, but it is likely that nature has evolved a bioinorganic chemistry using these constituents.

3.2.e. Chromosome Structures.

In eukaryotic cells, DNA is packaged into chromosomes. When partially unfolded, the structure of the resulting chromatin, as viewed in the electron microscope, resembles a string of beads (Figure 3.20). The beads are termed *nucleosomes*, a nucleoprotein complex composed of eight histone proteins at the core, around which are wrapped about 146 base pairs of DNA in ~ 1.75 turns of a shallow left-handed helix. The string is called linker DNA and contains 40 to 80 bp, depending on the species from which the chromatin is isolated. The role of metal ions in stabilizing or destabilizing these structures is largely unexplored. When DNA functions in replication (DNA synthesis) or transcription, these structures further unfold, and the double helix separates, exposing its genetic information in the form of chemically distinguishable nucleobases.

3.2.f. Nucleic-Acid Synthesis in Vitro.

A recent advance in biotechnology has been the ability to synthesize fairly long deoxyribonucleotides, up to ~ 100 bases long, for use in site-specific mutagenesis studies or even in total gene synthesis. In addition, similar chemistry for producing oligoribonucleotides is developing rapidly. This methodology also affects bioinorganic chemistry, affording the opportunity to prepare oligonucleotides with unique metal-binding properties. Some site-specifically modified DNAs have been inserted into full-length genomes and studied *in vivo*. The specificity of metal binding in the synthetic DNA fragment can be achieved by various strate-

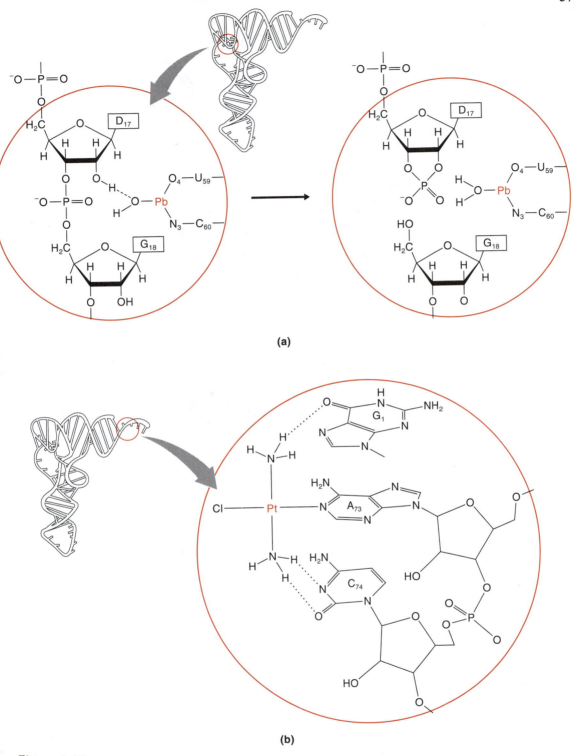

Figure 3.19
Structures of two metal derivatives of tRNA^Phe. (a) A Pb(II) complex that promotes cleavage of the tRNA backbone. D represents dihydrouracil. (b) A complex derived from *trans*-[Pt(NH₃)₂Cl₂].

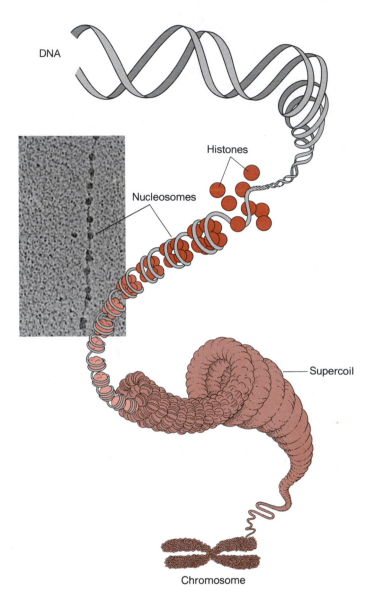

DNA

Histones

Nucleosomes

Supercoil

Chromosome

Figure 3.20
Compaction of DNA in chromosomes. (Electron micrograph courtesy of Victoria Foe).

gies, for example, by introduction of a sulfur atom using the thionucleoside or phosphorothioate chemistry described below, or by the use of special unique sequences, such as d(GpG), which forms cis-$\{Pt(NH_3)_2\}^{2+}$ intrastrand cross-links. Thus the application of site-specific mutagenesis to bioinorganic problems in metalloprotein chemistry has a parallel in metal/nucleic-acid chemistry, both being driven by technical advances in molecular biology.

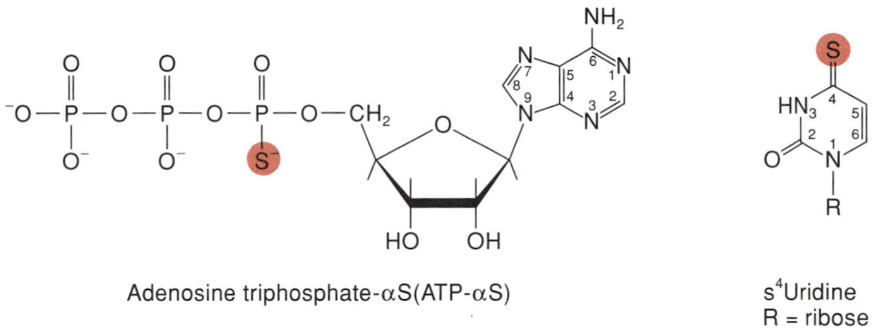

Figure 3.21
Two sulfur-containing nucleic acid components.
R = ribose

3.2.g. Sulfur Atoms in Nucleic Acids.   Apart from the five standard bases shown in Figure 3.12, several modified nucleosides have been found, especially in RNA molecules. One of the more interesting of these for bioinorganic chemistry is 4-thiouridine, s^4U, the structure of which is given in Figure 3.21. Found as a naturally occurring constituent of certain transfer RNA molecules, the sulfur atom in s^4U has been shown to be a good metal-binding site, especially for soft metal ions such as Hg(II) or Pt(II). The specificity of such binding is nearly absolute, since no other DNA or RNA donor atom can coordinate with such high affinity to a soft-metal center. Similar specificity can be achieved by incorporation of phosphorothioate nucleotides into DNA or RNA. These nucleotides, such as ATP-αS, are accepted in place of their normal counterparts by many enzymes. It is therefore possible to synthesize DNA or RNA molecules containing sulfur atoms at the 5′ position of any nucleotide, again providing a very selective binding site for a soft-metal atom. Chemistry of this kind has been used to probe nucleic-acid structure using both electron-dense heavy metals, for study by X-ray diffraction or electron microscopic techniques, and paramagnetic metal ions, to assist in magnetic-resonance studies. The potential of this approach in solving problems in bioinorganic chemistry is far from being exhausted, however.

3.3. Other Metal-Binding Biomolecules

3.3.a. Prosthetic Groups.   Many metalloproteins contain organic components in addition to the protein itself. These are often referred to as *prosthetic groups* and may contain bound metal ions. Prosthetic groups include protoporphyrin IX (shown in Figure 1.2), which as the iron complex is commonly referred to as heme. Another important class of prosthetic groups

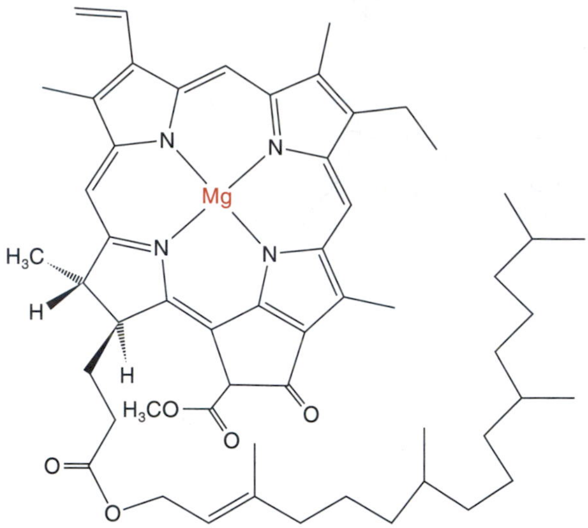

Figure 3.22
Chemical structure of a chlorophyll molecule.

consists of the chlorophylls (Figure 3.22), which are magnesium complexes essential for photosynthesis. Finally, the pterin ring system (Figure 1.3), a constituent of folic acid, is a subject of current interest, because a derivatized pterin has been implicated as the molybdenum-binding unit in the so-called molybdenum cofactor (Mo-co). Units of this type appear to be common to all molybdenum-containing enzymes except nitrogenase. Further discussion of these units may be found in Chapter 5.

3.3.b. Coenzyme B-12.   Coenzymes are low-molecular-weight, non-protein molecules that assist in enzymatic catalysis as true substrates. The best studied metallo-coenzyme is cobalamin, or vitamin B-12, which, as its name implies, contains cobalt. The vitamin is obtained as the cyanide adduct, cyanocobalamin (Figure 1.6), which is reduced in a two-electron reaction by FAD to form the Co(I) complex. This powerful nucleophile reacts with ATP to give the adenylated B-12 coenzyme. The Co-C bond in coenzyme B-12 undergoes homolytic cleavage to form radical species believed to participate in several enzymatically catalyzed hydrogen-transfer reactions. Methylcobalamin can also transfer a CH_3^- ion to Hg, Pb, and Sn salts in aqueous solution, a biomethylation reaction that probably contributes to the toxicity of these elements.

3.3.c. Bleomycin and Siderophores.   Bleomycin is an antitumor antibiotic isolated from the fungus *Streptomyces*. Its structure is shown in Figure 3.23. Bleomycin binds to DNA and, using oxygen redox chemistry asso-

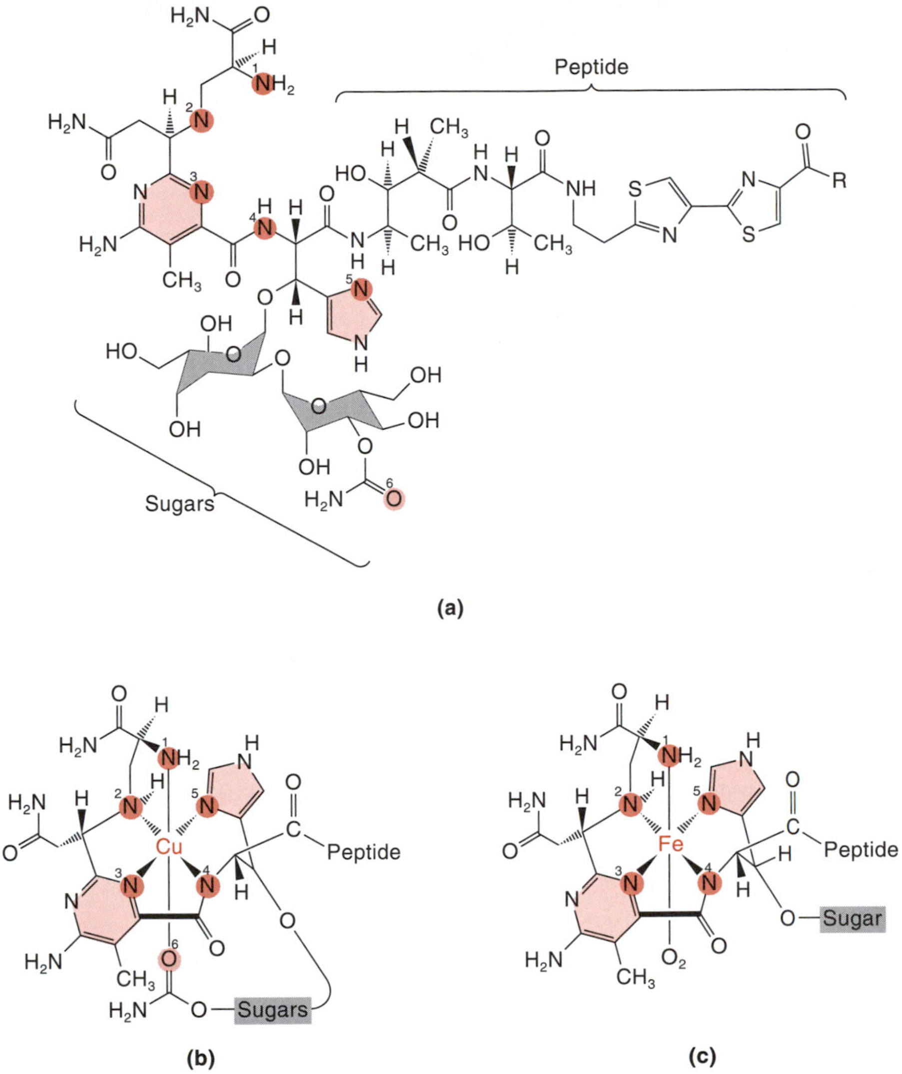

Figure 3.23
Structures of (a) bleomycin and (b), (c) two postulated metal ion–bleomycin complexes.

ciated with its metal-binding polypeptide moiety, catalyzes the cleavage of the sugar-phosphate backbone. Owing to its glycoprotein character, metal-coordination properties, and DNA-binding and -cleaving reactivities, bleomycin is a treasure chest of bioinorganic chemistry.

Siderophores are small molecules secreted by microorganisms into their environment that facilitate the specific uptake of iron. They function by binding ferric ion with high affinity to produce soluble complexes that can be taken up by the organism. Examples include the cyclic triester enterobactin

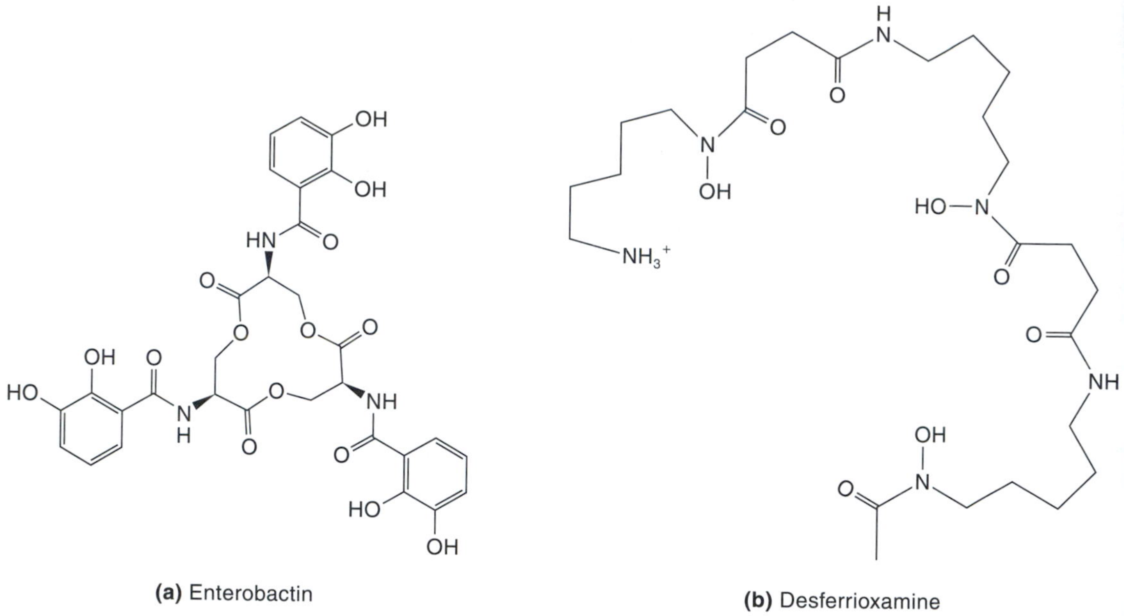

(a) Enterobactin

(b) Desferrioxamine

Figure 3.24
Chemical structures of siderophores enterobactin and desferrioxamine.

and linear molecules such as desferrioxamine, shown in Figure 3.24. Siderophores are discussed in more detail in Section 5.1.

3.3.d. Complex Assemblies. Many interesting metal-binding assemblies arise from complexes containing several biological macromolecules.
These macromolecular assemblies are necessarily more complex than the
structures considered previously in this chapter, and, consequently, we have
only the most primitive knowledge about their coordinating properties. Cell
membranes, viruses, and intracellular compartments such as the ribosome, the
mitochondrion, and the endoplasmic reticulum are all examples of such assemblies. Important bioinorganic components are often embedded in membrane
structures, such as the chain of cytochromes involved in the electron-transfer
steps and oxidative phosphorylation reactions required for respiration. Another key example is the photosynthetic reaction center found in bacterial
and plant membranes. Metal-ion transport through membranes, a crucial phenomenon for neurological and intracellular triggering reactions, is a further
example of where knowledge of complex biological structures affects bioinorganic chemistry.

Study Problems

1. Amines such as ammonia and ethylenediamine are common ligands in inorganic chemisty; yet the ε-amino group of the side chain of lysine only rarely coordinates to metal ions in proteins. Propose an explanation for why the lysine ε-amino group is not a common ligand for aqueous solutions of metalloproteins.

2. It has been proposed that transcription factor IIIA, a protein that stimulates gene transcription, might bind to similar sequences in both double-stranded DNA and double-stranded RNA. This proposal has been controversial. Explain the potential difficulties of having one protein that binds both types of double-stranded nucleic acids.

3. Assuming that the groups around each freely rotating single bond in the backbone of a protein can exist with the usual three rotational conformations, calculate the number of such conformations for a protein made up of 100 amino acids. Repeat this calculation for a 100 base polynucleotide. Comment on the magnitude of the values that you have computed.

4. Draw the structure of an antiparallel beta sheet formed by the pair of hexapeptides NH_2-A-L-D-V-S-F-COOH and NH_2-K-T-N-C-R-Y-COOH.

5. Give the single-letter equivalents for all amino acids in the following sequence (note the message!), and circle the three residues most likely to bind a soft transition-metal ion.

His-Glu-Leu-Pro-Met-Glu-Ile-Met-Thr-Arg-Ala-Pro-Pro-Glu-Asp-Ile-Asn-Ala-Gly-Glu-Asn-Glu.

Bibliography

Protein Structure

C. Branden and J. Tooze. 1991. *Introduction to Protein Structure*. Garland, New York.

T. E. Creighton. 1993. *Proteins: Structure and Molecular Properties*, 2nd ed. W. H. Freeman, New York.

G. E. Shultz and R. H. Schirmer. 1979. *Principles of Protein Structure*. Springer-Verlag, New York.

Nucleic Acid Structure

R. L. P. Adams, J. T. Knowler, and D. P. Leader. 1986. *The Biochemistry of Nucleic Acids*, 10th ed. Chapman and Hall, London.

D. Freifelder, ed. 1978. *The DNA Molecule: Structure and Properties*. W. H. Freeman, San Francisco.

W. Saenger. 1988. *Principles of Nucleic Acid Structure*, 2nd corrected printing. Springer-Verlag, New York.

Molecular and Cell Biology

B. Alberts, D. Bray, J. Lewis, M. Ruff, K. Roberts, and J. D. Watson. 1989. *Molecular Biology of the Cell*. Garland, New York.

B. Lewin. 1990. *Genes IV*. Oxford University Press, Oxford.

L. Stryer. 1988. *Biochemistry*, 3rd ed. W. H. Freeman, New York.

Enzymes

A. Fersht. 1985. *Enzyme Structure and Mechanism*, 2nd ed. W. H. Freeman, New York.

W. P. Jencks. 1969. *Catalysis and Chemistry in Enzymology*. McGraw-Hill, New York.

C. Walsh. 1979. *Enzymatic Reaction Mechanisms*. W. H. Freeman, San Francisco.

Prosthetic Groups

S. P. Cramer and E. I. Stiefel. 1985. "Chemistry and Biology of the Molybdenum Cofactor," in T. G. Spiro, ed., *Molybdenum Enzymes*. Wiley, New York, 411–442.

J. Dabrowiak. 1982. "Bleomycin." *Adv. Inorg. Biochem.* 4, 69–113.

D. Dolphin. 1982. B_{12}, Volumes 1 and 2. Wiley, New York.

A. B. P. Lever and H. B. Gray. 1983. *Iron Porphyrins*, Volumes 1–3. Addison Wesley, New York.

K. N. Raymond. 1984. "Complexation of Iron by Siderophores: A Review of their Solution and Structural Chemistry and Biological Function." *Top. Curr. Chem.* 123, 49–102.

CHAPTER 4

Physical Methods in Bioinorganic Chemistry

4.1. Purpose of this Chapter

In order for progress to be made in experimental science, one must use the appropriate methods to address a specific question of interest. The purpose of this chapter is to provide a critical survey of the landscape of physical methods currently available to the bioinorganic chemist. No attempt will be made to teach the principles behind each of the techniques, since more detailed texts are readily available. Rather, we discuss how the various physical methods have affected the development of bioinorganic chemistry and, interestingly, how bioinorganic chemistry has occasionally led to improved methodologies. The advantages and limitations of the physical methods are discussed for the specific applications of molecular structure determination and characterization of electronic structures. The electronic properties and high electron densities of metal ions as a class of functional groups in biology render them especially amenable to study by physical techniques such as EXAFS, Mössbauer, resonance Raman, and electron-paramagnetic resonance spectroscopy, methods rarely employed by the bioorganic chemist.

4.2. A Comment on Time Scales

In order to understand the nature of a transition-metal center in biology, it is often useful to obtain knowledge about its electronic states and magnetic properties. These properties are frequently correlated with the coordination capabilities and molecular geometry of the metal center, both of which affect the reactions that can occur at such a site. In addition, electronic and magnetic phenomena form the bases for most spectroscopic techniques that are used to

Table 4.1
Time scales for commonly used spectroscopic analysis techniques

Technique	Time
Electron diffraction	$\sim 10^{-20}$ sec
Neutron diffraction	$\sim 10^{-18}$ sec
X-ray diffraction	$\sim 10^{-18}$ sec
UV spectroscopy	$\sim 10^{-15}$ sec
Visible spectroscopy	$\sim 10^{-14}$ sec
Infrared spectroscopy	$\sim 10^{-13}$ sec
EPR spectroscopy	$\sim 10^{-4}-10^{-8}$ sec
NMR spectroscopy	$\sim 10^{-1}-10^{-9}$ sec
NQR spectroscopy	$\sim 10^{-1}-10^{-8}$ sec
Mössbauer spectroscopy	$\sim 10^{-7}$ sec
Molecular-beam spectroscopy	$\sim 10^{-6}$ sec
Experimental separation of isomers	$> 10^2$ sec

monitor the chemical reactivity at metal sites in biology. The various forms of spectroscopy all have in common the irradiation of a sample with light of a given frequency or frequency range. This light can be either scattered or absorbed, giving rise to a change in intensity that forms the basis for the physical method. The frequency of light, and its associated wavelength, may be used to evaluate the time scale at which a chemical phenomenon can be probed by a given spectroscopic method. Table 4.1 summarizes this information for the methods commonly employed in bioinorganic chemistry.

The relationship between the frequency of light used to probe a chemical reaction and the rate of that reaction may be appreciated by a simple analogy. Consider making a motion picture with a camera in which the shutter opens and closes 40 times each second. A wagon wheel with spokes is being photographed with this camera. If the wheel is rotating slowly, the motion picture will provide an accurate account of the circular motion of the individual spokes. As the wheel rotates more and more rapidly, the film will show a slowing of the spokes, no motion at all, a backward motion, and finally a blur of spokes. Some readers will have seen this phenomenon in commercial films. The speed of the rotating wheel is analogous to a chemical reaction rate and the shutter speed to the time scales tabulated in Table 4.1. For accurate kinetic studies, the appropriate technique must be selected. For example, it is generally true that electron-transfer reactions will require faster methods than atom-transfer chemistry. Electronic spectroscopy is a fast method. Light of wavelength $\lambda = 500$ nm has a frequency ($v = c/\lambda$) of $\sim 6 \times 10^{14}$ sec^{-1} and so can resolve chemical changes occurring on the femtosecond time scale. On the other hand, ^{1}H NMR spectroscopy, where the ability to resolve dynamic phenomena depends on the difference in resonant frequencies of protons in disparate chemical environments, is a much slower method. Typically, the time scale of this technique is on the order of milliseconds.

4.3. X-ray Methods

X-rays have wavelengths that are on the order of 1 Å, that is, the same order of magnitude as interatomic distances. This fact makes X-ray sources extremely useful for the determination of molecular structure. Two fundamentally different sorts of methods using X-rays are currently widely used: diffraction and absorption.

4.3.a. X-ray Diffraction.   The diffraction process depends on the availability of a translationally ordered sample, that is, a single crystal. When suitable crystals are available, X-ray crystallography offers the most powerful structural probe of macromolecular structure, including metal coordination geometries. Small molecule structures can often be determined with uncertainties in the bond distances on the order of 0.01–0.001 Å. The precision of macromolecular structures is usually an order of magnitude less than that for small inorganic complexes. Protein crystals do not diffract X-rays as well as small molecule crystals do, largely because of their high water content, often ~ 50 percent, which is usually highly disordered. Resolution, defined by Bragg's law (Equation 4.1), where θ is half the scattering angle between

$$d = \lambda/(2\sin\theta) = 1.5418\ \text{Å}/(2\sin\theta)\ \text{(for Cu radiation)} \qquad (4.1)$$

the incident and diffracted X-ray beams, is limited in the best cases to slightly less than 2 Å, compared to less than 1 Å for most coordination complexes. The limited resolution, readily estimated from the spots farthest from the center on X-ray photographs of a single crystal (largest θ values), makes it difficult to discern individual atoms. Thus, for metalloproteins, it is usually necessary to know the amino-acid sequence in order to assign the electron density to specific amino-acid residues. This requirement, in addition to the need for heavy atom derivatives to determine the phase angles for the thousands of measured reflections from which the maps are generated (isomor-

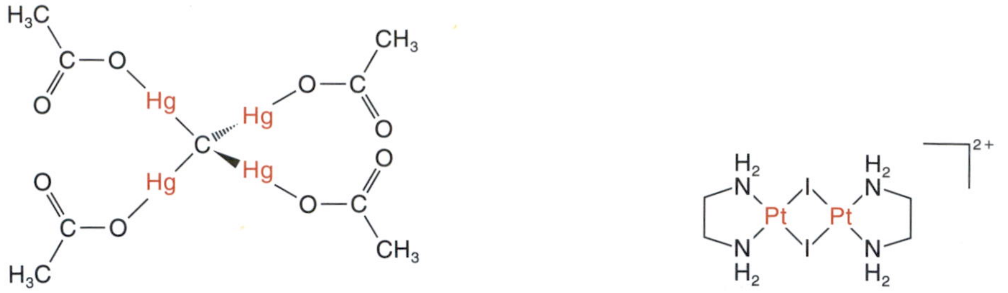

(a) Tetrakis(acetoxymercuri)methane, TAMM

(b) Di-μ-iodobis(ethylenediamineplatinum(II)), PIP

Figure 4.1
Structures of two highly electron-dense complexes used for preparing heavy atom derivatives of macromolecules for X-ray crystallographic studies.

phous replacement technique), makes the method very labor intensive. Often months or years will pass between the first demonstration of X-ray diffraction by a metalloprotein single crystal and the final determination of a refined molecular structure, although such times are being shortened by advances in data collection and structure refinement methods. These advances include the development of area detectors that allow rapid measurement of diffraction patterns, as well as the use of synchrotron radiation, which allows the use of smaller crystals. X-ray diffraction methods can also be employed for determining the structures of nucleic acids, although suitably diffracting nucleic-acid single crystals are somewhat more difficult to obtain. These problems notwithstanding, the final results are usually well worth the effort, having afforded the best structural information currently available about metallo-proteins and, to a limited extent, metals bound to nucleic acids.

The methods generally used for macromolecular structure determination by diffraction methods include a bioinorganic component. Derivatives with electron-dense atoms or groups must be prepared, usually by soaking crystals in solutions of heavy-metal complexes. Standard reagents in common use contain single heavy atoms such as platinum, mercury, or gold. As the size and complexity of crystallizable biomolecules have increased, so has the demand for heavy-atom derivatives with sufficiently high electron density to provide adequate phasing by the isomorphous replacement techniques. For example, the compounds tetrakis(acetoxymercuri)methane, TAMM, and di-μ-iodobis(ethylenediamineplatinum(II)), PIP (see Figure 4.1), were used in solving the structure of the nucleosome core particle, a complex of eight polypeptide chains and 146 base pairs of DNA (see Section 3.2.e). A portion of the electron-density map and the deduced model are shown in Figure 4.2. These water-soluble, stable, and commercially available reagents are among the growing list of multi-heavy-atom inorganic complexes available to the crystallographer studying macromolecular assemblies.

The X-ray diffraction technique has also been used to study polycrystal-line fibers of DNA. Such data, of course, were of great significance in the

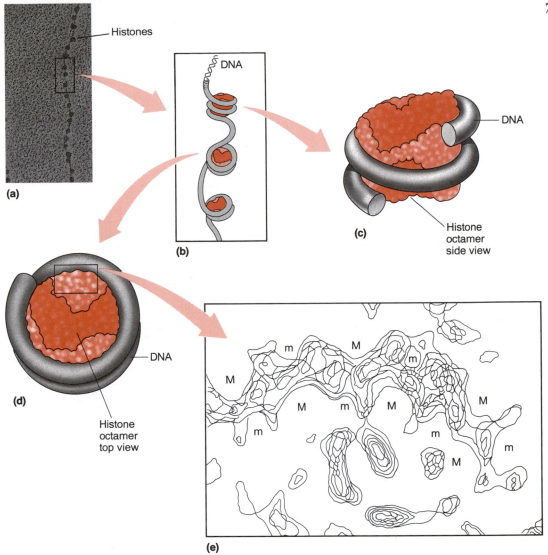

Figure 4.2
Structure of the nucleosome core particle. (a) Electron micrograph showing
nucleosomes. (b) Schematic view of the nucleosomes and linker DNA. (c) Side view
of nucleosome core particle showing DNA wrapped around the histone octamer.
(d) Top view of the nucleosome core particle. (e) Portion of electron density map
showing the major (M) and minor (m) grooves of DNA. Phases for generating the
map were provided through the use of the complexes shown in Figure 4.1. (Part (e)
adapted from T. J. Richmond et al., *Nature* **311**, 532–537 (1984)).

discovery of the double-helical structure for DNA. Electron-dense metal reagents have proved to be valuable for certain studies of this type. Fused-ring
organic heterocycles such as ethidium bromide bind to duplex DNA by
intercalation, occupying every other interbase pair site at saturation (Figure
4.3). Direct proof of this "neighbor exclusion" binding model was provided

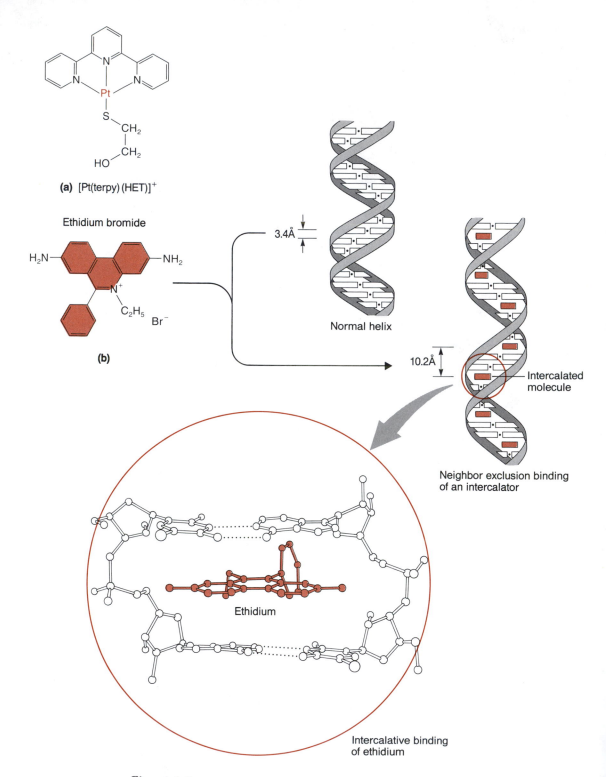

(a) [Pt(terpy)(HET)]⁺

Ethidium bromide

(b)

3.4Å

Normal helix

10.2Å

Intercalated molecule

Neighbor exclusion binding of an intercalator

Ethidium

Intercalative binding of ethidium

Figure 4.3
Binding to DNA by intercalation. (a) Structure of the metallointercalator, [Pt(terpy)(HET)]⁺. (b) Structure of the organic intercalator, ethidium. The binding of intercalators to DNA is shown, illustrating the neighbor exclusion principle. At saturation, only every other potential site is occupied, the resulting intercalators being spaced at 10.2 Å intervals. Also depicted is a close-up view of the ethidium molecule intercalated into DNA.

Table 4.2
Some metalloproteins for which crystallographic data are available[a]

Protein	Metal(s)
Ascorbate oxidase	Copper
Azurin	Copper
Cytochrome c	Iron
Cytochrome P-450	Iron
Ferredoxin	Iron
Ferritin	Iron
Galactose oxidase	Copper
Hemerythrin	Iron
Hemocyanin	Copper
Hemoglobin	Iron
Insulin	Zinc
Lactoferrin	Iron
Liver alcohol dehydrogenase	Zinc
Metallothionein	Cadmium/zinc
Myoglobin	Iron
Nitrite reductase	Copper
Nitrogenase	Iron, molybdenum
Photosystem I reaction center	Magnesium, iron
Plastocyanin	Copper
Protocatechuate 3,4-deoxygenase	Iron
Ribonucleotide reductase	Iron
Rubredoxin	Iron
Superoxide dismutase	Manganese, iron, or copper/zinc

[a] For a more complete listing, see T. Creighton. 1993. *Proteins: Structures and Molecular Properties*, 2nd ed. W. H. Freeman, New York.

by the X-ray diffraction patterns of polycrystalline DNA fibers containing intercalated [Pt(terpy)(HET)]. The platinated sample exhibited clear 10.2 Å reflections along the meridian in the X-ray diffraction pattern corresponding to the nearest Pt-Pt repeat distance along the helix axis in the intercalated structure, whereas X-ray studies of fibers containing intercalated ethidium revealed only 3.4 Å meridional reflections, since the scattering power of the pure organic intercalator is indistinguishable from that of the nucleotide base pairs.

Once a set of atomic coordinates is available from X-ray diffraction work, it is usually deposited in a readily accessible data bank, such as the Brookhaven Protein Data Bank. From these coordinates may be generated computer graphics and other models as well as lists of bond distances and angles. Summarized in Table 4.2 are some of the metalloproteins for which X-ray structures are available.

4.3.b. X-ray Absorption Spectroscopy (XAS).   The absorption of X-rays can excite the 1s (K edge) or 2s,2p (L edge) electrons of an element to empty localized orbitals or, for higher-energy X-ray photons, the continuum. This phenomenon, which has been known for decades, became of great utility following the availability of high-intensity, tunable X-ray beams from synchrotron radiation sources. Prior to this development, the intensity of X-rays from conventional sources was too weak to produce interpretable X-ray absorption signals for all but the most concentrated samples. By examining the energy of the X-ray absorption edge, where the onset of absorption occurs, one can often determine the oxidation state of the metal ion of interest. Transitions from an inner shell of electrons to the valence level allow one also to get electronic structural information in favorable cases. Modulation of the X-ray absorption-energy spectrum by back-scattering from neighboring atoms produces an extended X-ray absorption fine structure (EXAFS) from which details about metal-coordination geometries can be extracted. The method has been widely applied in metalloprotein structural studies and has the great advantage that noncrystalline solid and even solution samples can be studied. Geometric information derived from fitting EXAFS data to a model structure can be reliable to ± 0.01 Å, as shown by comparison with single-crystal X-ray diffraction of small molecule complexes investigated by both methods.

An early triumph of EXAFS spectroscopy, one that afforded it great credibility among the bioinorganic chemical community, involved the iron-sulfur protein rubredoxin. The experimental EXAFS spectrum for this protein and the fit obtained by using a single Fe–S distance are shown in Figure 4.4. An analysis of the data demonstrated that all four Fe–S(cysteine) distances in rubredoxin are equal to within ± 0.02 Å. This result contradicted earlier single-crystal X-ray diffraction work that had indicated that one Fe–S bond was significantly shorter, ~ 0.25 Å, than the average of three long ones. Subsequent refinement of the single-crystal X-ray diffraction data led to a more symmetric structure, consistent with the EXAFS study.

EXAFS results for metallobiopolymers tend to be most accurate for symmetric sites in which the metal ion has relatively heavy elements $(Z > 14)$ within the first coordination sphere. In this respect, the $[Fe(SR)_4]^-$ rubredoxin case was nearly ideal. For less symmetric sites with several different atom types and distances in the first coordination sphere, the data can be more difficult to fit unambiguously to a model structure. Comparisons with well-defined model compounds can be very helpful in this process. Low temperatures (to 4.2 K), commonly employed to minimize atomic thermal motion, may require frozen solutions or lyophilized powders. These conditions some-

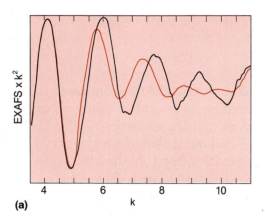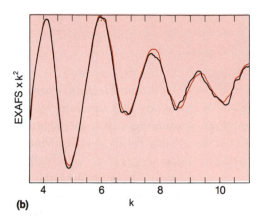

Figure 4.4
Extended X-ray absorption fine structure (EXAFS) spectra of the iron protein
rubredoxin. (a) Experimental spectrum (color) and the spectrum calculated from the
original crystallographic coordinates (black). (b) Comparison of the experimental
spectrum (color) and that calculated by assuming identical iron-sulfur distances
(black). (Adapted from R. G. Shulman et al., *Proceedings of the National Academy of
Sciences USA* **72**, 4003–4007 (1975)).

times alter protein-active-site structures, a possibility that must be checked in
control experiments. It is also desirable to perform control experiments to
assure that subjecting a biological sample to high-intensity X-irradiation in
the EXAFS experiment has not altered its structure, biological activity, or
chemical nature. Measurement of the activity of an enzymatic sample in
solution after an EXAFS experiment is one way to assure its viability during
the run.

4.4. Magnetic Resonance Methods

When a particle with a magnetic moment is placed in a magnetic field, the
states that have the moment oriented in different directions with respect to
the applied field have different energies. Transitions between these states
may be induced by absorption of light of the appropriate frequency. This
phenomenon is called *magnetic resonance*. If the particles are electrons, the
appropriate light frequencies are in the microwave region, using readily
achievable magnetic field strengths, and the method is called *electron paramagnetic resonance* spectroscopy. If the particles are nuclei, the appropriate
frequencies are in the radiofrequency region, and the method is called *nuclear
magnetic resonance* spectroscopy.

4.4.a. Electron Paramagnetic Resonance (EPR) Spectroscopy. This method requires samples with unpaired electrons and is ideally suited for studying many metalloproteins, including those containing Cu(II), Co(II), Fe(II), Fe(III), Mn(II), Mn(III), Mo(V), and metal clusters such as $\{Fe_2O\}^{3+}$ and $\{Fe_4S_4\}^{+,3+}$. The method is very sensitive, being able to detect high-spin ferric ions at concentrations as low as the μM range. Interestingly, EPR spectroscopy was used in the isolation and purification of iron-containing proteins, such as the ferredoxins. Since these electron-transfer proteins have no measurable catalytic activity, their purity could not be established by the classic biochemical approach of enzymatic activity measurements. Instead, the appearance and integrated signal intensity of their EPR spectra were monitored following the usual column-chromatographic purification steps.

The value of EPR spectroscopy in establishing electronic structures and their dependence on metal coordination-sphere composition and geometry may be illustrated by its application to several aspects of copper bioinorganic chemistry. Copper(II) is especially amenable to EPR investigation, because it has only one unpaired electron. In addition, the ^{63}Cu and ^{65}Cu (nuclear spin $I = 3/2$) nuclei give rise to a characteristic four-line pattern because of electron-nuclear-spin hyperfine interactions, denoted by the parameter A. Moreover, the electronic spin-relaxation time is sufficiently long at room temperature that spectra can be observed in fluid solution. This last property permits biological samples to be monitored at physiologically relevant temperatures. Frequently, however, EPR spectra of frozen solutions are investigated to stop molecular tumbling, which averages out the anisotropy of the g-tensor. For copper(II) centers, which commonly have tetragonal geometry, the g-tensor is usually axially symmetric, with $g_{zz} > g_{xx} = g_{yy} > 2.0023$, the free-electron g-value, where the z-axis is taken to be perpendicular to the plane of copper and its four ligands. In addition, anisotropy of the hyperfine interaction is manifest by components parallel and perpendicular to the molecular symmetry axis, designated $A_{\parallel}$ and $A_{\perp}$, respectively. The EPR spectrum of a typical Cu(II) complex, $[Cu(H_2O)_6]^{2+}$, is shown in Figure 4.5, together with spectra from the copper proteins plastocyanin and superoxide dismutase.

At least three spectroscopically distinct classes of copper occur in biology, designated as type 1, 2, or 3 (Figure 4.5). Values for the g-tensor and the hyperfine splitting parameters extracted from EPR spectra for these centers are reported in Table 4.3, together with related data for a variety of other important paramagnetic transition-metal ions in bioinorganic chemistry. The EPR signal of type-1 copper is exemplified by that of plastocyanin (Figure 4.5), which has an uncommonly small $A_{\parallel}$ value for a cupric center, indicating

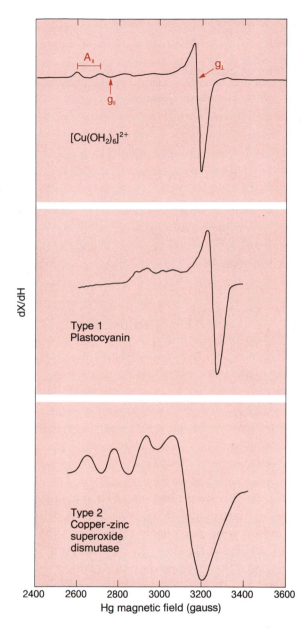

Figure 4.5
Electron paramagnetic resonance spectra of (top) $[Cu(OH_2)_6]^{2+}$, (middle) a type 1 copper site in plastocyanin, and (bottom) a type 2 copper site in copper-zinc superoxide dismutase.

significant delocalization of the metal unpaired-electron density. This result, together with a very intense ($\varepsilon \approx 3,000$ M^{-1} cm^{-1}) blue ($\lambda \approx 625$ nm) optical absorption band, was used to define the type-1 copper center in plastocyanin and several related copper redox and oxidase proteins. X-ray structural work on plastocyanin revealed the presence of a highly distorted copper-coordination environment comprising two histidine nitrogen and one cysteine thiolate sulfur donors lying nearly in a plane containing the metal, as well as a long, out-of-plane axial Cu–S bond to a thioether unit of a methionine

Table 4.3
EPR signatures for biologically interesting transition-metal centers

Metal	Details of spectrum
Mn(II), high-spin	6 lines spaced at 90 gauss, centered on $g = 2$
Cu(II), type 1	$g_x = 2.036, g_y = 2.058, g_z = 2.227, A_z = 45, A_x = A_y = 4.5$ gauss
Cu(II), type 2	$g_\parallel = 2.242, g_\perp = 2.053, A_\parallel = 170$ gauss
Cu(II), type 3	EPR silent
Mo(V)	$g = 1.96 +$ (hyperfine splitting ($I = 5/2$))
Mo(III)	$g_\parallel \sim 2, g_\perp \sim 4 +$ (hyperfine splitting)
Ni(III), low-spin	$g_\perp \sim 2.2, g_\parallel \sim 2.02$
Co(II), low-spin	$g \sim 2$, 8 line hyperfine, axial
Fe(III), high-spin, rhombic	$g \sim 4.3$
Fe(III), high-spin, axial	$g_\perp = 6, g_\parallel = 2$ (e.g., hemes)
Fe(III), low-spin, rhombic	$g_x \sim 2.4, g_y \sim 2.2, g_z \sim 1.8$
$Fe_4S_4(SR)_4^{3-}$	$g \sim 1.88, 1.92, 2.06$
$Fe_4S_4(SR)_4^{-}$	$g \sim 2.04, 2.12$
$Fe_2S_2(SR)_4^{3-}$	$g \sim 1.89, 1.95, 2.04$

residue. The second type of copper(II) EPR signal found in bioinorganic chemistry, illustrated for Cu–Zn superoxide dismutase in Figure 4.5, has g-values characteristic of most cupric complexes. The third type of copper is EPR silent, because of antiferromagnetic exchange coupling (see below) and/or rapid electron-spin relaxation between two neighboring Cu(II) centers. Proteins such as laccase and ascorbate oxidase contain all three types of copper, and the elucidation of such a complex situation by EPR and optical spectroscopic studies was a major triumph in the formative stages of bioinorganic chemistry. The recent X-ray crystallographic determination of ascorbate oxidase reveals that the structural conclusions drawn from the spectroscopic studies are essentially correct. The four copper ions are quite close together, as shown in Figure 4.6.

As already indicated, EPR techniques have been important for studying iron as well as copper in bioinorganic chemistry. For iron, both Fe(II) and Fe(III) redox states can exist in low- and high-spin forms. EPR spectroscopy readily distinguishes these configurations for ferric ion, the g parameters providing additional information about the likely coordinated ligands (Table 4.3). Extensive EPR spectral studies of heme iron and iron-sulfur clus-

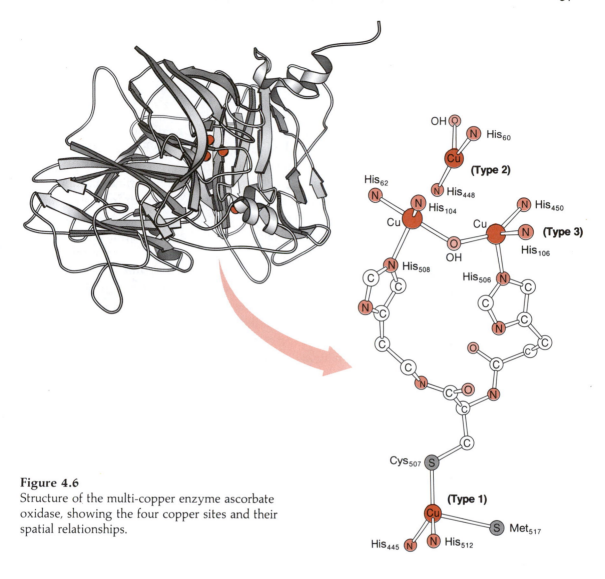

Figure 4.6
Structure of the multi-copper enzyme ascorbate oxidase, showing the four copper sites and their spatial relationships.

ters have been carried out. EPR spectra have also proved to be invaluable in studying Mn(II) ($I = 5/2$, giving a six-line hyperfine pattern), polymanganese clusters involved in photosynthesis, Mo in nitrogenase and oxotransferases, Co in vitamin B-12 and related complexes, and, most recently, Ni in the hydrogenase component of methane-generating bacteria as well as other enzymes. In addition, EPR signals are not infrequently encountered in organic residues adjacent to metal centers in proteins, for example, the stable tyrosyl radical in ribonucleotide reductase, and in substrates such as the super-oxide ion. Radical intermediates formed during enzymatic reactions can also be observed on occasion by their EPR spectra.

Two important variants of the EPR method have recently found significant applications in bioinorganic chemistry. One of these is *electron-nuclear double resonance* spectroscopy (ENDOR), in which the EPR transition is saturated while a second irradiation signal is scanned across the NMR frequency range of an isotope in the complex that might cause hyperfine splitting of the EPR lines. Usually the latter cannot be directly resolved. Irradiation of the sample at specific NMR frequencies causes the EPR signal to becomes unsaturated, and an ENDOR signal results. A typical application is study of ^{17}O-labeled substrates, such as H_2O, O_2, or H_2O_2, with EPR active metal centers, such as iron porphyrins. The second method, called *electron-spin-echo envelope modulation* (ESEEM) spectroscopy, also facilitates identification of ligand hyperfine interactions, which are sometimes called superhyperfine to distinguish them from hyperfine coupling with the metal nucleus that bears the unpaired spin. Here a series of two or sometimes three microwave pulses is applied to the sample in the EPR cavity, and the effect of varying the time delay between pulses on the "spin-echo" signal so generated is monitored. The principles are similar to those involved in NMR spin-echo experiments. This method can be used to detect superhyperfine interactions, for example, to the imidazole N—H proton of a histidine ligand coordinated to Cu(II) or Fe(III), where observing the effects due to deuteration provides a clear confirmation of the spectral assignment. Another application is to detect ^{14}N and ^{15}N nuclear superhyperfine interactions. From the magnitude of superhyperfine coupling constants, the extent of electron delocalization from the metal center out onto coordinated ligands can be estimated. This information may be valuable in deducing where sites of reactivity are on bound substrate molecules.

4.4.b. Nuclear Magnetic Resonance (NMR) Spectroscopy.

Nuclear magnetic resonance spectroscopic methods have been highly useful in bioinorganic chemistry, as they have in other branches of chemistry, for many years. With the development of Fourier-transform spectrometers that employ two or more time variables (two- or higher dimensional NMR), the applicability of NMR spectroscopy, especially for macromolecular structural studies, has increased still further. From ^1H and, to a lesser extent, ^{13}C, ^{15}N, and ^{31}P NMR spectra can be obtained distance (from nuclear Overhauser studies) and torsion angular (from coupling constants) information about the three-dimensional structure of biopolymers, including bound metal ions. Indeed, the complete three-dimensional structures of small to medium-sized proteins can now be determined by NMR methods. Such studies require the full assignment of NMR spectra containing hundreds of different resonances, and the measurement of about as many interproton distances and torsional angles. One emerging area of bioinorganic chemistry that has been strongly influenced by these techniques involves zinc-binding domains from proteins that interact with nucleic acids. The structures of six distinct classes of such domains have now been determined by NMR spectroscopy. In all cases, the structures were determined before or contemporaneously with structural analyses of similar units by X-ray diffraction.

Paramagnetic metal ions, having electron-spin relaxation times that are fast enough not to broaden nearby proton resonances beyond recognition, can produce shifted NMR signals that facilitate identification of amino-acid side-chain ligands. This method is especially useful for naturally occurring Fe, Co, and Cu centers. Line broadening without chemical shift changes occurs for ligands surrounding Mn(II) and Gd(III) and can be used to extract distance information when these metals are substituted for diamagnetic metals such as Mg^{2+}, Ca^{2+}, or Zn^{2+} in metalloproteins. Line broadening by paramagnetic transition metals is also useful for determining the accessibility of water to the active site of a metalloenzyme.

Nuclear magnetic resonance spectroscopic studies of metal nuclei provide yet another powerful dimension of this method. Table 4.4 lists the nuclear properties of metals that occur naturally in biology as well as metals used as probes or drugs. The most useful are metal nuclei having $I = 1/2$, for which no quadrupolar line broadening occurs. The very low resonance frequencies of most naturally occurring metal isotopes lead to poor sensitivity and inability to observe signals at achievable protein concentrations. Moreover, apart from ^{57}Fe, which suffers from very low sensitivity, all the nuclei are quadrupolar ($I > 1/2$) and thus present unacceptably broad lines. For these reasons, the most useful results are expected for ^{111}Cd, ^{113}Cd, ^{203}Tl, ^{205}Tl, ^{195}Pt, and ^{207}Pb nuclei, all of which have $I = 1/2$. Examples include ^{195}Pt NMR structural studies of platinum anticancer drugs and their adducts with DNA and ^{113}Cd NMR spectral investigations of metallothionein.

Finally, NMR methods are often quite useful for kinetic studies. The extremely high resolution and small line widths of NMR methods, in which differences in resonance positions of nuclei are often measured in fractions of a part per million (ppm), allow signals to be distinguished or line broadening to be observed for species that are interconverting on a wide variety of time scales. For example, the resonances of two species that differ from one another by 1 ppm at 300 MHz can be resolved as long as the interconversion rate is below 2×10^3 sec^{-1}. Such NMR-based kinetic methods have been used to measure the electron self-exchange rate for $\{Fe_4S_4(SR)_4\}^{2-,3-}$ cores, revealing electron-transfer rate constants of 3×10^6 $M^{-1}sec^{-1}$ (See Section 9.1.a.).

Table 4.4
Nuclear properties of selected elements

Element	Atomic number	Atomic mass	Spin	Natural abundance (%)	NMR frequency MHz (10 kgauss)
H	1	1	1/2	99.985	
		2	1	0.015	
C	6	13	1/2	1.108	
N	7	14	1	99.63	
		15	1/2	0.37	
Na	11	23	3/2	100	11.262
Mg	12	25	5/2	10.13	2.6054
P	15	31	1/2	100	
S	16	33	3/2	0.76	
K	19	39	3/2	93.10	1.9869
		40	4	0.01	2.470
		41	3/2	6.88	1.0915
Ca	20	43	7/2	0.145	2.8690
V	23	50	6	0.24	4.245
		51	7/2	99.76	11.19
Cr	24	53	3/2	9.55	2.4065
Mn	25	55	5/2	100	10.501
Fe	26	57	1/2	2.19	3.8047
Co	27	59	7/2	100	11.285
Ni	28	61	3/2	1.19	12.089
Cu	29	63	3/2	69.09	2.663
		65	3/2	30.91	
Zn	30	67	5/2	4.11	
Y	39	90	2	Radioactive	
Mo	42	95	5/2	15.72	
		97	5/2	9.46	
Tc	43	99	9/2	Radioactive	
Cd	48	111	1/2	12.75	
		113	1/2	12.26	
In	49	111	9/2	Radioactive	
Gd	64	154−160		100	
Pt	78	195	1/2	33.8	
Tl	81	203	1/2	29.5	
		205	1/2	70.5	
Pb	82	207	1/2	22.6	

4.5. Mössbauer Spectroscopy

This method, in which γ-rays emitted from the nuclei of a source element in an excited nuclear state are absorbed by the same element in a sample, is especially valuable for studies of ^{57}Fe in bioinorganic chemistry. From the isomer shift (δ, in mm sec^{-1}), analogous to the chemical shift in an NMR experiment, can be obtained some information about the metal oxidation and spin states (Table 4.5) as well as the types of ligands coordinated to iron.

Table 4.5
**Mössbauer isomer shifts for classes
for iron compounds**

Iron oxidation state	Isomer shift δ, mm/sec
Fe^{2+} (high-spin)	~1.3
Fe^{3+} (high-spin)	~0.5–0.7
Fe^{2+} (low-spin)	~0.1
Fe^{3+} (low-spin)	~0

More powerful from a structural point of view, however, is the quadrupole splitting of the signal, ΔE_Q, which reveals the asymmetry of the electric field surrounding the metal center. These parameters are defined in Figure 4.7, which presents data for a model of the $\{Fe_2O(O_2CR)_2\}^{2+}$ core in iron-oxo proteins, which contain short, μ-oxo-iron(III) bonds. This feature leads to an axial distortion of the electric field, producing large quadrupole splittings (~ 1.5–2.0 mm sec^{-1}) characteristic of this chemical unit. Protonation of the μ-oxo bond to form the μ-hydroxo species lengthens the Fe−O distance to ~ 1.96 Å, with a concomitant diminution in ΔE_Q to ~ 0.4 mm sec^{-1} (Figure 4.7).

One drawback in the use of ^{57}Fe Mössbauer spectroscopy in bioinorganic chemistry is the low natural abundance of this isotope (Table 4.4), often necessitating enrichment either by reconstitution of the metalloprotein from the apo form or, if this route is not feasible, by isolation from a growth medium containing iron salts enriched in ^{57}Fe. For clusters of iron in biology, Mössbauer spectroscopy can be used to study magnetic coupling phenomena and, for mixed valence species, to provide an estimate of the rate of internal electron-transfer reactions. The time scale of the Mössbauer method (Table 4.1), 10^{-7} sec, is well suited for this task. ^{57}Fe Mössbauer spectroscopy has been especially valuable in sorting out the various kinds of iron-sulfur and Fe−Mo−S clusters in nitrogenase and the iron formal oxidation states. These

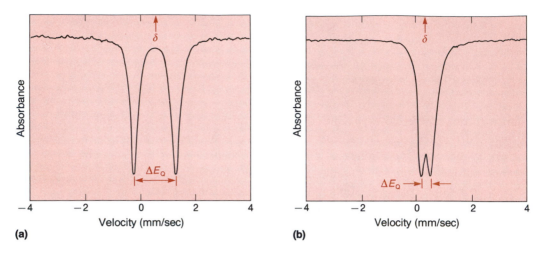

Figure 4.7

Mössbauer spectra of oxo- and hydroxo-bridged diiron(III) complexes illustrating the isomer shift (δ) and quadrupole splitting (ΔE_Q) parameters. (a) Spectrum of oxo-bridged species, $[Fe_2O(O_2CCH_3)_2(HBpz_3)_2]$, where $HBpz_3$ is hydrotris(pyrazolyl)borate. (b) Spectrum of the hydroxo-bridged species, $[Fe_2(OH)(O_2CCH_3)_2(HBpz_3)_2]$. Note the larger ΔE_Q in (a) due to its less symmetrical iron(III) coordination geometry.

studies are facilitated by temperature-dependent spectral measurements, down to 4.2 K, in the presence of applied external magnetic fields, and by computer simulation of the spectrum to provide isomer-shift, quadrupole-splitting, and nuclear-electron magnetic hyperfine coupling parameters. Although ^{57}Fe Mössbauer spectroscopy has dominated bioinorganic applications of the technique, other possible nuclei are ^{40}K, ^{61}Ni, and ^{67}Zn.

4.6. Electronic and Vibrational Spectroscopy

Transitions between various electronic states lead to the absorption of energy in the ultraviolet, the visible, and, for many transition-metal complexes, the near infrared regions of the electromagnetic spectrum. Transitions between vibrational states occur at lower energies, entirely within the infrared region. Both electronic and vibrational spectra can be of tremendous utility in evaluating the structural and electronic properties of metallobiomolecules.

4.6.a. Electronic Spectra.   There are three major sources of electronic spectra in metal complexes, namely, internal ligand bands such as are found

Table 4.6
General characteristics of metalloprotein spectra

Chromophore	Principal bands	Assignment
Hemes [Fe(II)/Fe(III)]	400–500 nm (Soret)	$\pi \to \pi^*$
	500–600 nm (α, β)	$\pi \to \pi^*$
Fe(III)–O_xN_y	Geometry dependent	$d \to d$
		$O \to Fe(III)$
		(LMCT)
Iron-sulfur ($Fe^{II/III}$–S_4)	350–600 nm	$S \to Fe(II)$
	near IR	(LMCT)
		$d \to d$
Probes [Co(II)/Ni(II)]	Geometry dependent	$d \to d$
		(LMCT)
Copper blue [$Cu(II)N_2S_2$]	~ 600 nm	$S \to Cu(II)$
	near IR	(LMCT)
		$d \to d$

in porphyrins, transitions associated purely with metal orbitals such as d-d transitions, and charge transfer bands between metal and ligand. Some general features of selected metalloprotein chromophores are given in Table 4.6. These spectral properties have been used to assign metal oxidation states, follow reactions, and identify chemical species in newly discovered systems. They remain among the most valuable signatures of the coordination environment in metalloenzymes. When several metals are present, for example, the two heme and two copper centers in cytochrome c oxidase, the task becomes to identify individual spectroscopic bands with specific metal centers. Depleting the protein selectively of one or more such centers, or selectively changing the oxidation state and hence the electronic spectrum of a specific chromophore, are two approaches to making the assignments. Although in general it is not simple to assign the various optical bands to specific centers in a polymetallic protein, this information, if available, can be used to map the path of electron transfer through the metal chromophores to substrate. In cytochrome c oxidase, for example, optical and other spectroscopic studies revealed that electrons from cytochrome c enter first via the copper A center, and that the four-electron reduction of dioxygen to water occurs at a dinuclear heme a_3-copper B site.

In addition to the more conventional UV-visible absorption spectroscopy, there exist several related methods for measuring electronic transitions that provide insight about the electronic structures of metals in biology. One that we have already encountered is near absorption edge XAS, in which X-ray photons excite a 1s electron to empty, bound states near the continuum. This technique has been used successfully to assign oxidation states in dinuclear iron-oxo proteins such as methane monooxygenase, where Fe(II)-Fe(II), Fe(II)Fe(III), and Fe(III)(Fe(III) redox levels have been delineated. The XAS method measures the absorption of variable-wavelength photons by a metal-containing sample. A related experiment is to use a fixed-wavelength source of either X-rays or high-energy ultraviolet light to ionize core electrons and to analyze the energy of the emitted photoelectrons. This approach forms the basis for XPS, X-ray photoelectron spectroscopy, or ESCA, electron spectroscopy for chemical analysis, and for UPS, ultraviolet photoelectron spectroscopy, in which X-rays and a vacuum ultraviolet light, respectively, are used as a source. For some elements, especially Fe, Co, Ni, and Cu, the core electron-binding energies can experience shifts from 8–11 eV as a result of changes in the chemical environment. The ESCA and UPS methods have not yet been used to their full potential in bioinorganic chemistry.

4.6.b. Vibrational Spectroscopy.

As might be imagined, the vibrational spectra of proteins are extremely complex. Metalloproteins in which the metal center has associated electronic transitions offer a special advantage, however, producing *resonance Raman spectra* in which the ligand vibrations are markedly enhanced. With modern instrumentation, including the availability of tunable lasers that can excite over a wide range of visible and near-ultraviolet wavelengths, resonance Raman spectral studies have provided important structural insights about the coordinated ligands in metalloproteins. Especially valuable are molecules in which intense ligand-to-metal charge-transfer electronic transitions occur, such as in metal-thiolate, metal-phenoxide, and metal-azide complexes. Resonance Raman spectroscopic investigations of those chromophores can be used to identify cysteine, tyrosine, or exogenous azide (N_3^-) coordination, respectively. If the vibrational frequency of a coordinating ligand falls in a spectral region having little or no overlap with protein bands, or where the band intensity is very strong, both infrared and resonance Raman methods can be employed. Examples include studies of carbon monoxide and dioxygen binding to O_2 transport proteins such as hemoglobin and hemocyanin. Another technical advance that has facilitated the application of vibrational spectroscopy to bioinorganic structure determination has been the development of on-line computers and Fourier-transform methodologies. Solvent and protein matrix background signals can often be averaged out by difference spectroscopy, where spectra of the metalloprotein in two forms are subtracted from one another. The two forms might be with and without added substrate or inhibitor molecules, or isotopically substituted ligand derivatives such as $^{16}O-^{16}O$ and $^{18}O-^{18}O$, or H_2O and D_2O. As with EXAFS spectroscopy, care must be taken in laser Raman studies that the chemical integrity of a biological sample be maintained during and after irradiation, preferably by activity measurements if feasible.

4.6.c. Circular Dichroism and Magnetic Circular Dichroism Spectroscopy.

Circular dichroism (CD) is related to the difference in absorption of left- and right-circularly polarized light by an optically active sample. CD spectroscopy has several important applications in bioinorganic chemistry. First, CD can be useful for detecting or resolving electronic transitions that are more difficult to isolate by simple absorption spectroscopy. This situation occurs because the selection rules for CD and absorption spectroscopy are different, and because CD bands can be either positive or negative, so that overlapping bands in an absorption spectrum may be clearly delineated in the CD spectrum if the signs are different. Second, CD is a very powerful tool for examining the secondary structure of biopolymers. Different types of polypeptide secondary structure have distinct types of CD spectra in the ultraviolet region corresponding to absorption by the peptide carbonyl groups. Thus, it is possible to estimate the amount of alpha helix, beta sheet, and other structures from the CD spectrum of a protein. In addition, CD can be used to follow folding-unfolding transitions in polypeptides. CD can also be used to examine the secondary structure of nucleic acids, clearly distinguishing, for example, B-form and Z-form DNA.

Whereas only molecules that are chiral have CD spectra, all molecules exhibit CD spectra in the presence of an applied magnetic field. The technique that measures this phenomenon is called *magnetic circular dichroism* (MCD) spectroscopy. Whereas detailed theoretical analyses of the spectra obtained in most real applications are complex, the MCD method provides a powerful "fingerprint" approach to identifying bound ligands in certain well-defined cases. A useful application has been to use MCD to assign the presence of a thiolate ligand coordinated to the low-spin ferric heme centers in cytochrome P-450 and in chloroperoxidase. To illustrate the value of this approach, the MCD spectra of one of these heme centers and a useful model complex are shown in Figure 4.8. When the metal is spectroscopically silent, such as Zn^{2+}, substitution of Co^{2+} has provided useful MCD spectra. The structures of the Zn^{2+} centers in carboxypeptidase, carbonic anhydrase, and alkaline phosphatase have all been probed as their Co^{2+}-substituted analogs. Because MCD can be used as a fingerprint method, comparison with well-designed model complexes is essential. In addition to its value as a fingerprint method, MCD has also been used to characterize paramagnetic metal centers in metallo-

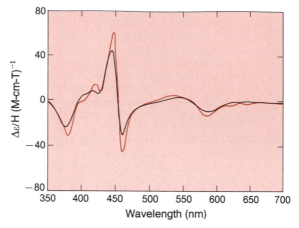

Figure 4.8
Magnetic circular dichroism spectra of a heme protein, the iron(II)-carbon
monoxide derivative of a cytochrome P-450 (color), and an appropriate model
complex, the benzene thiolate complex of ferrous protoporphyrin IX bound to
carbon monoxide. (Adapted from J. H. Dawson and M. Sono, *Chemical Reviews* **87**,
1255−1276 (1987))

proteins. In this application, information about the magnetic spin states is
obtained by monitoring the temperature- and field-dependence of an MCD
spectrum at low temperatures. Such an approach has been extensively applied
to iron-sulfur and heme proteins.

4.7. Magnetic Measurements

Closely related to the characterization of the electronic states of metal ions in
biology by optical, EPR, and Mössbauer spectroscopy is the direct measurement of the magnetic moment. Until recently, this method could be applied
only to small-molecule models of metalloprotein cores because of the large
diamagnetic contribution to the magnetism by the nonmetallic protein and
high water content of the sample. Higher sensitivities provided by the
SQUID (superconducting quantum interference device) susceptometers, however, have made direct studies of metalloprotein magnetic properties more
feasible. Since the magnetic susceptibility (χ) of most substances having
unpaired electrons varies with temperature (T), the magnetic moment, and
hence the number of unpaired electrons, is usually deduced from temperature-
dependent studies. In addition to direct magnetic-moment measurements on
the SQUID or other magnetometers, it is also possible to derive the χ versus
T dependence of bioinorganic samples by measuring the EPR signal inten-

sity or the shift in the NMR spectrum of a nucleus experiencing the magnetic field of a paramagnetic center.

When two or more magnetic centers are connected by bridging ligands, their electron spins can annul or reinforce one another, phenomena referred to as antiferromagnetic and ferromagnetic coupling, respectively. Such behavior is reasonably common in bioinorganic chemistry, especially the former phenomenon. Depending on the magnitude of such coupling, which is quantum-mechanically analogous to spin-spin coupling in NMR spectroscopy, it should be possible to determine the magnetic exchange coupling parameter J. Measured in cm^{-1} units, $J < 0$ for antiferromagnetic spin exchange and $J > 0$ for ferromagnetic interactions, where the spin Hamiltonian is $H = -2JS_1 \cdot S_2$. The spin vectors S_1 and S_2 refer to the two metal centers being coupled. For antiferromagnetic coupling, the ground-state total spin (S_T) in an even-electron system is zero; in an odd-electron system, $S_T = 1/2$. At liquid He temperature, the former system is diamagnetic, exhibits no EPR or magnetic splitting in the Mössbauer spectrum, and has a magnetic susceptibility indistinguishable from that of the apo protein. The latter, $S_T = 1/2$, system has magnetic properties characteristic of an isolated metal center with one unpaired electron. As $|J|$ becomes smaller and the temperature rises, the magnetic properties reflect occupancy of higher-lying states in the electron-spin manifold. This behavior can be directly investigated by the temperature dependence of χ and fit to theoretical models of the electronic structure of the polymetallic center. Such studies have been valuable in characterizing dimetallic centers such as $\{Cu(II)\}_2$ in hemocyanin and the Cu_2Cu_2 derivative of superoxide dismutase, $\{Fe(III)\}_2$ and $\{Fe(III)Fe(II)\}$ in dinuclear iron-oxo proteins, hemerythrin, ribonucleotide reductase, methane monooxygenase, and purple acid phosphatase, $Fe_nS_n^{q-}$ clusters in iron-sulfur proteins, and the heme iron-copper center in cytochrome c oxidase.

4.8. Reduction Potential Measurements

Many of the physical and chemical properties of metallobiomolecules reflect the accessibility of several redox states that are required for biological functions. Examples include both mononuclear systems, such as $Cu(I)/Cu(II)$, $Fe(II)/Fe(III)$, $Co(I)/Co(II)/Co(III)$, and $Mo(IV)/Mo(V)/Mo(VI)$, and polymetallic centers. In order to understand the bioinorganic chemistry of these redox-active species, we usually want to know their redox potentials. The classic approach to measuring the redox potentials of metalloprotein cores is to carry out a potentiometric titration with a redox agent of similar potential, observing changes in the optical absorption or EPR spectrum of the protein as a function of added reagent. In order to facilitate electron transfer to the metal center, it is sometimes necessary to employ a mediator, a small redox-active organic molecule that can carry electrons to the protein center, although care must be taken that the mediator not alter the redox properties of the protein. More recently, direct electron transfer of redox-active proteins has been achieved at various electrodes including pyrolytic graphite, ruthenium oxide, and modified gold. Earlier attempts to measure redox potentials electrochemically were thwarted by strong adsorption of proteins to electrode surfaces as happens, for example, with cytochrome c at mercury electrodes. A fast, quasi-reversible electrode reaction with cytochrome c occurs at a gold electrode containing adsorbed 4,4'-bipyridyl. The presence of such adsorbed species at the electrode surface presumably facilitates rapid protein association and dissociation, necessary for the electrochemical measurements. Since 4,4'-bipyridyl is not redox active, it is viewed as a promoter of electron transfer rather than as a mediator. Not all metalloproteins may be studied directly by bioelectrochemical methods, since the redox-active centers may not be close enough to the surface of the protein to allow rapid electron transfer at an electrode surface. The bioelectrochemical methods used to measure protein redox potentials also have potential applications in the field of biosensors. For example, electrodes to which ferrocene has been covalently attached have been used to monitor cholesterol levels. Cholesterol oxidase, a flavoprotein, is reduced by cholesterol and subsequently reoxidized by a ferricinium ion attached to the electrode. The current required to reoxidize the bound ferrocene is related to the cholesterol level. Several related applications have been described, and more can be expected.

4.9. Electron Microprobe Analysis

This method detects the emission of X-rays following sample bombardment by a highly focused beam of electrons. Spatial resolution of 1 micrometer can be achieved, as little as 10^{-15} g of elements of biological interest can be detected, and the X-ray emission spectrum is characteristic of the element being probed. Applications of this method include the determination of the calcium content of mitochondria and endoplasmic reticulum in rat liver cells and the localization of copper in certain compartments of the cornea of patients suffering from Wilson's disease, a metabolic inability to clear copper from cells. Proton-induced X-ray emission (PIXE) is a related method in which a focused beam of protons is used to excite atoms instead of electrons. These and related techniques are certain to receive more attention in bioinorganic chemistry as problems involving compartmentalization of specific metal ions in tissues are attacked.

Study Problems

1. Calculate the approximate time scale in seconds for an ^{57}Fe Mössbauer experiment using 14.4 keV γ-rays, recalling that $E = hc/\lambda$ and given the conversion factor 1 keV $= 8.07 \times 10^6$ cm^{-1}. Reconcile the difference in the value you obtain from this calculation with that given in Table 4.1.

2. In studying the bioinorganic chemistry of nitrogenase, suppose that you trapped an intermediate having an EPR spectrum due to Mo(V). Both ^{95}Mo and ^{97}Mo have $I = 5/2$ and together are ~ 25 percent abundant. Sketch what you would expect the EPR spectrum to look like. Be sure to label your axes and indicate the expected g-value(s). If you could make a 1 : 1 adduct with ^{13}C-enriched cyanide ion ($I = 1/2$), how might it affect the spectrum?

3. The substitution of Co^{2+} for Zn^{2+} in metalloproteins is an example of a substitution probe, greatly facilitating the characterization of the metal center by EPR, optical, and other spectroscopic methods. Discuss the kind of information you might expect to obtain through such an approach.

4. A coworker has just isolated a copper enzyme that catalyzes the conversion of oil sludge into soluble alcohols in the presence of O_2. There are two Cu atoms per protein, which consists of a single polypeptide chain. As the bioinorganic chemist on the project, you are given unlimited quantities of the protein for the purpose of determining the active site structure. You have at your disposal a number of physical techniques, including NMR and EPR spectrometers, a magnetic susceptometer, a Mössbauer instrument, an X-ray absorption beam line, a UV-VIS spectrophotometer, a Raman spectrometer, a magnetic circular dichroism instrument, but, alas, no X-ray diffractometer. You have time to complete measurements by only three techniques before you have to give a report to your colleague. Describe what measurements you would make, in what order you would make them to get the most out of your time, what results you might expect, and how you would use this information to characterize structurally the dicopper center. Include in your discussion the kinds of ligands that you would be looking to identify and how the methods might allow you to make such a determination.

Bibliography

General References

C. R. Cantor and P. R. Schimmel. 1980. *Biophysical Chemistry*, Volumes 1–3. W. H. Freeman, San Francisco.

R. S. Drago. 1992. *Physical Methods for Chemists*, 2nd ed. Saunders College Publishing, Fort Worth, TX.

J. R. Wright, W. A. Hendrickson, S. Osaki, and G. T. James. 1986. *Physical Methods for Inorganic Biochemistry*. Plenum Press, New York.

Time Scales

R. G. Bryant. 1983. "The NMR Time Scale." *J. Chem. Ed.* 60, 933–935.

E. L. Muetterties. 1965. "Stereochemically Nonrigid Structures." *Inorg. Chem.* 4, 769–771.

X-ray Methods

R. E. Dickerson. 1992. "A Little Ancient History." *Pro. Sci.* 1, 182–186.

J. P. Glusker and K. N. Trueblood. 1985. *Crystal Structure Analysis: A Primer*, 2nd ed. Oxford University Press, New York.

R. G. Shulman, P. Eisenberger, W. E. Blumberg, and N. A. Stombaugh. 1975. "Determination of the Iron-Sulfur Distances in Rubredoxin by X-Ray Absorption Spectroscopy." *Proc. Natl. Acad. Sci. USA* 72, 4003–4007.

G. H. Stout and L. H. Jensen. 1989. *X-Ray Structure Determination*, 2nd ed. Wiley-Interscience, New York.

B. K. Teo. 1986. *EXAFS: Basic Principles and Data Analysis*. Springer-Verlag, New York.

Magnetic Resonance Methods

I. Bertini and C. Luchinat. 1986. *NMR of Paramagnetic Molecules in Biological Systems*. Benjamin/Cummings, Menlo Park, CA.

A. J. Hoff, ed. 1989. *Advanced EPR-Application in Biology and Biochemistry*. Elsevier, Amsterdam.

B. M. Hoffman. 1991. "Electron Nuclear Double Resonance (ENDOR) of Metalloenzymes." *Acc. Chem. Res.* 24, 164–170.

D. J. Lowe. 1992. "ENDOR and EPR of Metalloproteins." *Prog. Biophys. Molec. Biol.* 57, 1–22.

C. A. Reed and R. D. Orosz. 1993. "Spin Coupling Concepts in Bioinorganic Chemistry," in C. J. O'Connor, ed., *Research Frontiers in Magnetochemistry*. World Scientific Publishing, London, 351–393.

H. Sigel, ed. 1986. *Applications of NMR to Paramagnetic Species. Metal Ions in Biological Systems*, Volume 21. Marcel Dekker, New York.

H. Sigel, ed. 1987. *ENDOR, EPR, and Electron Spin Echo for Probing Coordination Sphere. Metal Ions in Biological Systems*, Volume 22. Marcel Dekker, New York.

J. E. Wertz and J. R. Bolton. 1986. *Electron Spin Resonance: Elementary Theory and Practical Applications*. Chapman and Hall, New York.

K. Wüthrich. 1986. *NMR of Proteins and Nucleic Acids*. Wiley-Interscience, New York.

Mössbauer Spectroscopy

G. M. Bancroft. 1973. *Mössbauer Spectroscopy: An Introduction for Inorganic Chemists and Geochemists*. McGraw-Hill, Maidenhead, U.K.

D. P. E. Dickson and F. J. Berry, eds. 1986. *Mössbauer Spectroscopy*. Cambridge University Press, New York.

P. Gütlich, R. Link, and A. Trautwein. 1978. *Mössbauer Spectroscopy and Transition Metal Chemistry*. Springer Berlin.

G. J. Long, ed. 1984–1989. *Mössbauer Spectroscopy Applied to Inorganic Chemistry*, Volumes 1–3. Plenum Press, New York.

Electronic and Vibrational Spectroscopy

W. C. Johnson, Jr. 1990. "Protein Secondary Structure and Circular Dichroism: A Practical Guide." *Proteins: Struct., Func., and Gen.* 7, 205–214.

A. B. P. Lever. 1984. *Inorganic Electronic Spectroscopy*, 2nd ed. Elsevier, New York.

K. Nakamoto. 1986. *Infrared and Raman Spectra of Inorganic Coordination Compounds*, 4th ed. Wiley-Interscience, New York.

T. G. Spiro, ed. 1987. *Biological Applications of Raman Spectroscopy*, Volumes 1 and 2. Wiley-Interscience, New York.

Circular Dichroism and Magnetic Circular Dichroism Spectroscopy

M. K. Johnson, A. E. Robinson, and A. J. Thomson. 1982. "Low-Temperature Magnetic Circular Dichroism Studies of Iron-Sulfur Proteins," in T. G. Spiro, ed., *Iron-Sulfur Proteins*. Wiley-Interscience, New York, 367–406.

Magnetic Measurements

C. J. O'Connor. 1982. "Magnetochemistry: Advances in Theory and Experimentation." *Prog. Inorg. Chem.* 29, 203–283.

Redox Potential Measurements

F. Armstrong, S. J. George, A. J. Thomson, and M. G. Yates. 1988. "Direct Electrochemistry in the Characterization of Redox Proteins: Novel Properties of *Azotobacter* 7Fe Ferredoxin." *FEBS Lett.* 234, 107–110.

J. E. Frew and H. A. O. Hill. 1988. "Direct and Indirect Electron Transfer Between Electrodes and Redox Proteins." *Eur. J. Biochem.* 172, 261–269.

H. A. O. Hill. 1987. "Bio-Electrochemistry." *Pure and Appl. Chem.* 59, 743–748.

Electron Microprobe Analysis

A. J. Morgen. 1985. *X-Ray Microanalysis in Electron Microscopy for Biologists*. Oxford University Press, New York.

J. Sommerville and U. Scheer, eds. 1987. *Electron Microscopy in Molecular Biology: A Practical Approach*. IRL Press, Oxford.

CHAPTER 5

Choice, Uptake, and Assembly of Metal-Containing Units in Biology

---

Principles: *Nature uses relatively abundant, kinetically labile, and thermo-dynamically stable units in metalloprotein-active centers. The lability facilitates rapid assembly and disassembly of the metal cores as well as rapid association and dissociation of substrates. Selection of a low-abundance metal ion for a specific function is possible via energy-driven processes. Entry into the cell can occur either by passive diffusion or through ion-specific channels. Metal-binding cofactors have evolved to enhance the bioavailability of metal ions that are insoluble under physiological conditions and to sequester functionally useful labile metal centers for insertion as "bioinorganic chips" into metalloprotein-active sites. Self-assembling metal clusters found in the geosphere have similarly been adopted for special functions in bioinorganic systems.*

It is interesting that, of all the inorganic elements in the periodic table, only a small number are utilized in biological systems. Most frequently encountered are the alkali- and alkaline-earth ions Na^+, K^+, Mg^{2+}, and Ca^{2+}, required to neutralize the charge of simple inorganic ions such as phosphate and sulfate as well as more complex species such as nucleoside triphosphates, DNA, and RNA. Many first-row transition-metal ions are also of key importance in biology, especially iron, copper, and zinc. The bioinorganic chemistry of vanadium, manganese, cobalt, nickel, and molybdenum, the only naturally occurring second-row transition element, is also being rapidly developed as we become increasingly aware of the important roles of these less-common elements. Tungsten has also been recently added to the list of metals found in biology. How is it that nature chooses to utilize the properties of these specific elements rather than others that, from their known inorganic chemistry, would seem to be of potential value? Why, for example, is there not a respiratory protein with cobalt as the dioxygen carrier, since complexes of this element are well known to bind O_2 reversibly?

As will be seen from the examples discussed in this chapter, one factor that seems to dictate the use of metal ions is their relative abundance. Iron is the most abundant transition metal in the earth's crust, and it is perhaps not surprising that iron is an important constituent of many bioinorganic systems. A second factor can be deduced from the fact that the active centers of metalloproteins consist of kinetically labile and thermodynamically stable units. The lability facilitates rapid assembly and disassembly of the metal cores as well as rapid association and dissociation of substrates. Metal ions such as Cr^{3+} and Co^{3+}, well known in inorganic chemistry for their kinetic inertness, are rarely utilized. For similar reasons, the more inert second- and third-row transition elements play almost no role in bioinorganic chemistry, despite the fact that they are extremely valuable as homogeneous and heterogeneous catalysts in the chemical industry. In fact, if present in cells, these heavier transition elements can be toxic. As a consequence, some attention has been paid to their mechanisms of uptake and transport. Since, as stated in Chapter 1, complexes of technetium, platinum, and gold have all found a place in medicine, their entry and metabolism in cells are also matters of considerable interest.

When a specific metal ion is required for a given purpose, it can be concentrated by energy-driven processes in a cell. A striking example of this phenomenon is the uptake of vanadium into certain types of sea squirts. With the use of an ATP-requiring reaction, the element is concentrated by more than 5 orders of magnitude from sea water into the blood cells, termed vanadocytes, of ascidians and tunicates. These cells have ~ 27 g of V per kg dry weight, more than 100 times the amount of Fe. Some ascidians accumulate other rarely encountered transition metals in biology, including Ti, Cr, and even Nb. The functions of V and these other elements in tunicates are unknown but currently under investigation.

The properties of metal ions in biology are attenuated by their local environment to optimize specific functions. Sometimes, however, the available amino-acid residues are unable to perform this task. One way in which this deficiency has been remedied is by the evolution of cofactors containing specific organic functionalities that modulate the properties of their inorganic guests. As briefly mentioned in Chapter 3, these cofactors, such as porphyrins and corrins, are encountered in many different proteins where additional ligands from the peptide chain can further attenuate their properties. Such units thus serve as valuable "bioinorganic chips," much like the solid-state components in modern electronics equipment, being inserted when their specific functional capabilities are required. Another type of bioinorganic chip is the metal-atom cluster, a unit quite familiar to the inorganic chemist. Iron-sulfur clusters such as $\{Fe_4S_4\}^{2+}$ are commonly found in proteins, where they serve important electron-transfer functions. In addition, species such as Fe_3O_4, or magnetite, have been utilized in nature as homing and orientation devices for bacteria, pigeons, bees, and probably humans as well. The ability

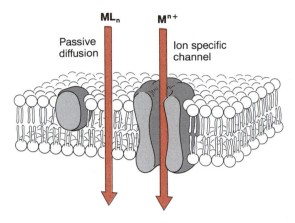

Figure 5.1

Scheme depicting two pathways for absorption of metal ions by cells, passive diffusion and transport through ion-specific channels.

of nature to borrow chemistry from the geosphere for use in the biosphere is fascinating, and constitutes a central principle by which we can begin to understand the choice of metal-containing units in biology.

5.1. Bioavailability of Metal Ions

Table 5.1 lists the relative abundance of inorganic elements found in the Earth's crust and in sea water. Comparison of this information with that in Table 1.1 indicates that the essential metals in biology are relatively more abundant than the nonessential ones. Certain highly abundant metal ions, such as titanium, have no known function in biology. This underutilization may be related to the difficulty in solubilizing and mobilizing the common ion of this element, Ti^{4+}, in biological fluids at pH 6–8.

How does nature manage to extract metal ions from their mineral forms and insert them into their bioenvironments? For many ions, such as Na^+, Mg^{2+}, or Zn^{2+}, solubility presents no particular problems at the millimolar concentrations required. Some elements such as iron, however, are too insoluble at neutral pH (see Table 2.3) to be readily bioavailable. Apart from solubilization, metal ions must be absorbed by cells, a phenomenon that is incompletely understood. Two possible pathways are illustrated in Figure 5.1. Some ions such as sodium pass through special channels and pumps in the cell membrane. We shall return to this interesting phenomenon in Chapter 6. For other ions, there are special chelating ligands that facilitate transport across the cell membrane.

Table 5.1
Relative abundance of elements in the Earth's crust and in the oceans[a]

Element	Crustal average (ppm)	Seawater (mg/l = ppm)	Element	Crustal average (ppm)	Seawater (mg/l = ppm)
H	1.40×10^3	1.1×10^{-5}	Rh	5×10^{-3}	
Li	20	0.17	Pd	1×10^{-2}	
Be	2.8	6×10^{-7}	Ag	7×10^{-2}	3×10^{-4}
B	10	4.5	Cd	0.2	1×10^{-4}
C	200	28	In	0.1	$<2 \times 10^{-2}$
N	20	0.7	Sn	2	8×10^{-4}
O	4.66×10^5		Sb	0.2	3×10^{-4}
F	625	1.3	Te	1×10^{-2}	
Na	2.83×10^4	1.1×10^4	I	0.5	6×10^{-2}
Mg	2.09×10^4	1.35×10^3	Cs	3	3×10^{-4}
Al	8.13×10^4	1×10^{-3}	Ba	425	2×10^{-2}
Si	2.77×10^5	3	La	30	3×10^{-6}
P	1.05×10^3	9×10^{-2}	Ce	60	1×10^{-6}
S	260	9×10^{-2}	Pr	8.2	6×10^{-7}
Cl	130	1.9×10^4	Nd	28	3×10^{-6}
K	2.59×10^4	3.9×10^{-2}	Sm	6.0	5×10^{-7}
Ca	3.63×10^4	4.1×10^{-2}	Eu	1.2	1×10^{-7}
Sc	22	$<4 \times 10^{-6}$	Gd	5.4	7×10^{-7}
Ti	4.40×10^3	1×10^{-3}	Tb	0.9	1.4×10^{-6}
V	135	2×10^{-3}	Dy	3.0	9×10^{-6}
Cr	100	5×10^{-4}	Ho	1.2	2×10^{-6}
Mn	950	2×10^{-3}	Er	2.8	9×10^{-6}
Fe	5×10^4	3×10^{-3}	Tm	0.5	2×10^{-6}
Co	25	4×10^{-4}	Yb	3.4	8×10^{-6}
Ni	75	7×10^{-3}	Lu	0.5	1×10^{-6}
Cu	55	3×10^{-3}	Hf	3	$<8 \times 10^{-6}$
Zn	70	1×10^{-2}	Ta	2	$<3 \times 10^{-6}$
Ga	15	3×10^{-5}	W	1.5	1×10^{-4}
Ge	1.5	7×10^{-5}	Re	1×10^{-3}	8×10^{-6}
As	1.8	2.6×10^{-3}	Os	5×10^{-3}	
Se	5×10^{-2}	9×10^{-9}	Ir	1×10^{-3}	
Br	2.5	67	Pt	1×10^{-2}	
Rb	90	0.12	Au	4×10^{-3}	1×10^{-5}
Sr	375	8	Hg	8×10^{-2}	2×10^{-4}
Y	33	1×10^{-6}	Tl	0.5	$<1 \times 10^{-4}$
Zr	165	3×10^{-5}	Pb	13	3×10^{-5}
Nb	20	1×10^{-5}	Bi	0.2	2×10^{-5}
Mo	1.5	1×10^{-2}	Th	7.2	$<5 \times 10^{-7}$
Ru	1×10^{-2}		U	1.8	3×10^{-3}

[a] Source: Compiled from B. Mason and C. B. Moore. 1982. *Principles of Geochemistry*, 4th ed. Wiley, New York (1982), Tables 3.5 and 9.3.

Best understood are the solubilization and mobilization of iron by bacteria, for which has evolved an elaborate system that utilizes many concepts of classic coordination chemistry. We shall therefore discuss this example in some detail, since the principles involved are likely to apply to the bioavailability of other metal ions. As already indicated, only very low levels of the free aquated iron are available for uptake by cells. Thus, in order to solubilize iron, microorganisms have evolved chelating agents called *siderophores* that bind iron(III) tightly. The siderophores are released by the bacterium into its environment. They sequester iron, solubilizing it in complexed forms that are specifically taken up into the cell, where it is subsequently released. One of the best studied siderophores is the catechol-based compound enterobactin (ent), shown in Figure 3.24. Enterobactin binds Fe(III) very tightly with an overall formation constant K_f, given in Equation 5.1,

$$K_f = [Fe(ent)^{3-}]/([Fe^{3+}][ent^{6-}]) \tag{5.1}$$

of 10^{49}. This value corresponds to an apparent dissociation constant of 10^{-25} at pH 7.0, taking into account protonation of the free enterobactin ligand. Such tight binding requires the development of mechanisms for release of iron within cells. The trianionic complex of iron produced is soluble and, as we shall discuss below, is actively transported into cells.

Information about the structure of the iron-enterobactin complex has been deduced by a series of experiments. The ferric complex of enterobactin has an optical absorption spectrum quite similar to that of tris(catecholato)-iron(III), indicating that the ligands coordinate as catecholate dianions rather than via one of the amide oxygens and one of the ring oxygens from each aromatic ring. Furthermore, this complex is optically active, which indicates that one of two possible absolute configurations at the iron center predominates. This situation arises because of the formation of Λ and Δ isomers, depicted below. Since the trilactone ring of enterobactin has chiral centers, all

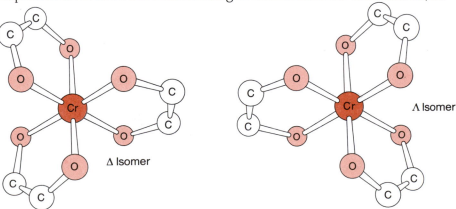

Stereoisomers of octahedral tris-chelate complexes.

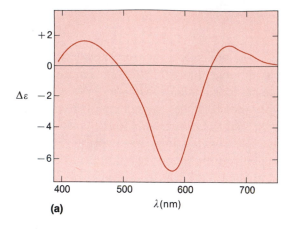

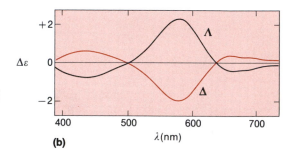

Figure 5.2
Circular dichroism spectra of (a) chromium(III)
enterobactin and (b) the two optical isomers
of tris(catecholato)chromium(III). Note that
the spectrum of the Δ isomer matches that of
the enterobactin derivative. (Adapted from
S. S. Isied et al., *Journal of the American Chemical
Society* **98**, 1763–1769 (1976)).

in the S conformation, the Λ and Δ isomers at iron are diastereomers rather
than enantiomers. Although it has not yet been possible to obtain suitable
crystals of ferric enterobactin for X-ray diffraction studies, the conformation
at iron was deduced by comparison to the enterobactin complex of chro-
mium(III). This task was accomplished by studying tris(catecholato)chrom-
ium(III), a complex that is substitution inert and can be resolved into optical
isomers by chromatography. Comparison of the circular dichroism spectra
of chromium(III) enterobactin with those of the Λ and Δ isomers of tris-
(catecholato)chromium(III) revealed that the spectrum of the latter isomer
resembles that of the enterobactin complex (Figure 5.2). Given the very
similar structures of the iron(III) and chromium(III) tris(catecholato) complexes, as well as the identical chromatographic properties of Cr(III) and
Fe(III) enterobactin, it was concluded that the latter has a predominantly Δ
conformation. This conclusion has recently been supported by the crystal
structure analysis of the vanadium(IV) analogue, [V(enterobactin)]$^{2-}$. The
structure is depicted in Figure 5.3.

Ferric enterobactin interacts with specific receptors on the *E. coli* outer
and inner membranes and is transported into cells via an energy-dependent
mechanism as shown in Figure 5.4. Its uptake has been followed by using

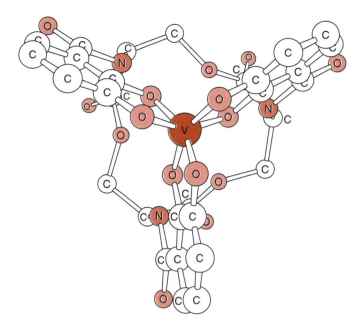

Figure 5.3
Structure of vanadium(IV) enterobactin determined by X-ray crystallography.

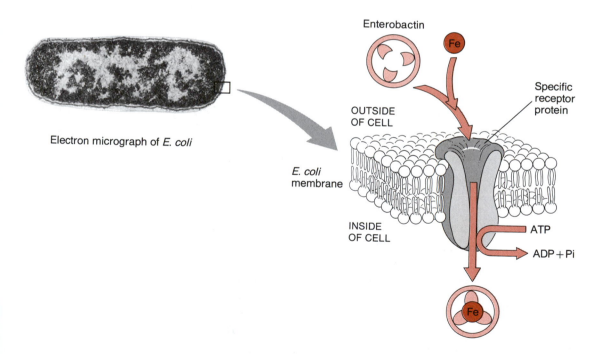

Figure 5.4
Scheme showing the ATP-driven uptake of ferric enterobactin into *E. coli* cells through a specific receptor protein in the cell membrane.

radioisotopes of iron. Studies of the kinetics of $[Fe(ent)]^{3-}$ uptake revealed saturation behavior analogous to that observed for most enzymes and inter-pretable in terms of a Michaelis-Menten type model. For $[Fe(ent)]^{3-}$, the apparent binding constant (K_m) is approximately 0.3 micromolar, indicative of reasonably tight binding to a specific receptor. The availability of such uptake assays has allowed investigation of the features of the $[Fe(ent)]^{3-}$ structure that are recognized by the receptor and has facilitated tests of the mechanism of iron release. For example, simple catecholate complexes such as $[Rh(cat)_3]^{3-}$ do not inhibit $[Fe(ent)]^{3-}$ uptake. Rhodium was used in these studies, since it forms kinetically inert complexes. Modification of the ligand to incorporate the amide linkages of enterobactin, however, yielded complexes that are quite potent inhibitors. Thus, $[Rh(N,N\text{-dimethyl-2,3-}$ dihydroxybenzamide)$_3]$ effectively blocks $[Fe(ent)]^{3-}$ uptake at 100 micro-molar levels. These observations suggest that, whereas the tris(catecholato) and amide portions of the metal-enterobactin complex are required for recog-nition by the membrane receptor, the trilactone ring is not essential. This result indicates that synthetic analogs of enterobactin retaining the dihy-droxybenzamide groups but not the remainder of the structure should com-pete for the receptor. Such compounds had been previously prepared as stable analogs of enterobactin lacking the hydrolytically sensitive trilactone functionality. One such compound is MECAM, shown in Figure 5.5, in which the trilactone is replaced by a 1,3,5-trimethylenebenzene grouping. This ligand binds iron(III) nearly as tightly as enterobactin itself ($K_f = 10^{46}$). Subsequent work revealed that the rhodium complex of MECAM is, in fact, a good inhibitor of $[Fe(ent)]^{3-}$ uptake. Furthermore, the iron complex of this ligand is absorbed by cells and can function as a source of iron. These struc-ture-activity relations have thus told us a lot about the nature of the receptor-$[Fe(ent)]^{3-}$ interaction, even though the receptor itself has yet to be characterized.

Finally, we come to the question of the mechanism of iron release. The binding of iron(III) to enterobactin is so tight that unaided release would be too slow to be biologically relevant. Several mechanisms appear possible. First, the sensitive nature of the enterobactin structure suggests that ligand hydrolysis could be utilized for iron release. The effectiveness of Fe(III)-MECAM, which lacks any easily hydrolyzable groups, as an iron source indicates that this mechanism cannot be the only one. A second possibility is that iron is reduced from Fe(III) to Fe(II), affording a more-labile and less-stable form of the complex that would release the metal in the cytoplasm. The reduction potentials for the tris(catecholato)iron(III) complexes at physi-ological pH are estimated to be approximately -750 mV versus NHE. This

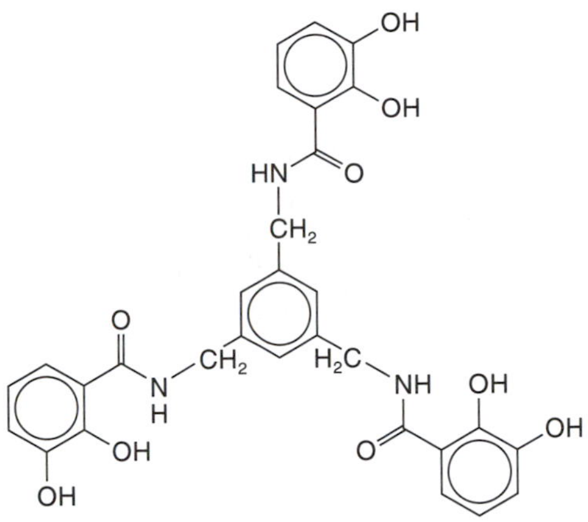

Figure 5.5
Chemical structure of MECAM, a synthetic model of enterobactin.

very negative value would appear to preclude reduction at neutral pH. Nonetheless, since six protons are released upon metal binding to enterobactin and since ferric enterobactin itself can be protonated with a pK_a slightly below 5, reduction at lower pH seems feasible. Finally, there could be a specific intracellular iron-binding ligand that, possibly in conjunction with redox chemistry, strips the metal from enterobactin in cells. Further studies are necessary to discover what mechanisms of iron release are actually used *in vivo*. Additional information about the control and utilization of iron in cells is provided in Chapter 6.

5.2. Enrichment Strategies and Intracellular Chemistry of Low-Abundance Metals

There are several mechanisms by which a cell can take up and concentrate a specific metal ion from its environment. In the dramatic case of vanadium mentioned earlier in this chapter, ascidians and tunicates must accumulate the element from sea water, where it exists almost exclusively in the form of VO_4^{3-} ion. This negatively charged species enters the cell by means of a specific system known to transport anions, such as sulfate ion, through the outer membrane. Once inside the cell, vanadium is reduced, binds to intracellular components, and so can no longer diffuse out into the medium. Other elements also gain access to the cytoplasm by means of the same anion-transport system. An interesting example is chromium, which in the form of the chromate anion, CrO_4^{2-}, is carcinogenic. Although full details of how this element causes cancer are unknown, its mechanism of entering cells and

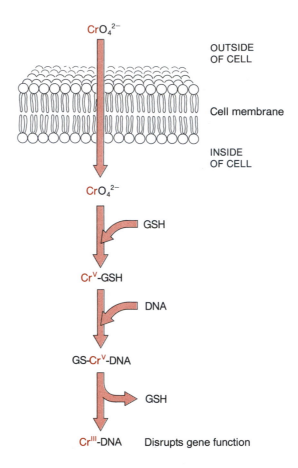

Figure 5.6
Model for cellular uptake and genotoxic action of chromate ion. The CrO_4^{2-} ion enters the cell through specific channels, is reduced by glutathione (GSH), and binds covalently to DNA.

becoming immobilized has recently been worked out, as indicated in Figure 5.6. Like vanadate ion, chromate is carried into cells by the anion-transport system. In the cytoplasm, it reacts with glutathione (GSH), an intracellular tripeptide present in ~ 5 mM concentration that contains a cysteinyl sulfhydryl group. In its reaction with GSH, CrO_4^{2-} becomes reduced to Cr(V) and Cr(IV), and a Cr–S bond forms. The metal ion is thus trapped in the cell and, like reduced forms of vanadium, cannot diffuse back out into plasma. Binding to DNA and subsequent reduction ultimately to Cr(III) are thought to constitute additional steps in the carcinogenic action of this ion. Interestingly, chromium in the $+2$ and $+3$ oxidations states is known not to be carcinogenic because it cannot readily enter cells. Thus, facilitated transport is an important aspect of its mechanism of toxicity.

For neutral metal complexes, cellular uptake can occur by passive diffusion across the cell membrane. This mechanism is the one that apparently operates for the antitumor drug *cis*-diamminedichloroplatinum(II) (*cis*-DDP), or cisplatin, briefly mentioned in Chapter 1. Cisplatin is administered by intravenous injection as an aqueous saline solution. Approximately half the platinum binds to serum proteins and is excreted. The rest is distributed among various tissues. In serum, the drug remains largely as *cis*-$[Pt(NH_3)_2Cl_2]$, owing to the relatively high chloride-ion concentration, ~ 0.1 M. As a neutral molecule, cisplatin diffuses passively across cell membranes into the cytoplasm, where it encounters a substantially lower chloride-ion concentration (~ 3 mM). Hydrolysis produces cationic complexes such as *cis*-$[Pt(NH_3)_2(OH_2)Cl]^+$ that diffuse to DNA, itself a polyanion, where they bind to form cytotoxic lesions. The hydrolysis reactions of *cis*-DDP are an important aspect of its biological activity. As shown in Figure 5.7, the chloride ions are displaced in a stepwise manner to form aqua and hydroxo species. These hydrolyzed forms of the drug react more rapidly with DNA than *cis*-DDP does. Studies of the rate of formation of *cis*-DDP complexes with calf-thymus DNA revealed the kinetics of DNA binding to be identical to those of hydrolysis of the first chloride ion from the platinum coordination sphere. Furthermore, ^{195}Pt ($I = 1/2$) NMR studies of the binding of *cis*- and *trans*-DDP to ~ 40 base-pair fragments of DNA from chicken erythrocytes revealed that platinum reacts first to form monofunctional adducts, in which only one bond is made to the nucleic acid. They then slowly close to bifunctional adducts, where there are two platinum-nucleobase bonds. Both chemical processes occur with half-lives of two or three hours, the rate-determining steps being loss of the chloride ion. What is important here is that, by changing its chemical form, in particular by transforming from a neutral molecule to a cation, the drug becomes trapped within cells, where it is able to damage their genomes and exhibit antitumor activity. This aspect of its mechanism is essentially the same as utilized by the other low-abundance metals, V and Cr, discussed previously.

Although binding to DNA is generally accepted to be an integral part of the antitumor mechanism of cisplatin, *cis*-DDP hydrolysis products also interact with other intracellular components. As with chromium, one such reaction is that with glutathione, which, as expected from hard-soft acid-base theory (Section 2.1.a), binds readily to cisplatin and its hydrolysis products. This chemistry probably ameliorates the toxic side effects of the drug without affecting its anticancer activity, which correlates best with the formation and persistence of *cis*-$\{Pt(NH_3)_2\}^{2+}$/DNA adducts. Another intracellular compo-

114

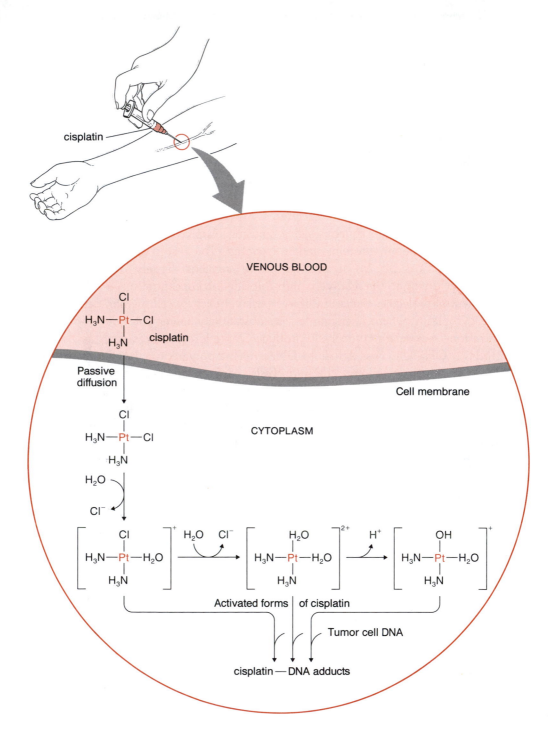

Figure 5.7
Administration and *in vivo* chemistry of the anticancer drug cisplatin.

nent postulated to react with platinum anticancer drugs is ascorbic acid, which is known to reduce Pt(IV) to Pt(II). This chemistry could be important in the activation of a new class of orally ingested Pt(IV) anticancer compounds that have recently entered clinical trials. Such reactions are therefore important subjects for investigation by the bioinorganic chemist.

5.3. Spontaneous Self-Assembly of Metal Clusters

Many, if not most, metals coordinate to their binding sites in proteins and nucleic acids as simple ions. Thus, Zn^{2+} binds to carboxypeptidase and to zinc-finger domains, copper binds plastocyanin as either Cu^+ or Cu^{2+}, and so forth. Sometimes, however, a more complex metal-containing unit is required. Two such examples are described here to show how nature can use inorganic cluster chemistry to achieve specific functions in metalloproteins. Although the detailed mechanisms by which these clusters are assembled and inserted into proteins *in vivo* are unknown, it is likely that the process involves spontaneous self-assembly, as described for the synthesis of model compounds in Chapter 2.

5.3.a. Iron-Sulfur Clusters. The discovery that iron-sulfur clusters occur as an integral component of electron-transfer proteins is an exciting chapter in bioinorganic chemistry. The chemical characteristics and nomenclature of these clusters are presented in Table 5.2. Although ultimate proof of the existence of the nFe-nS clusters came from protein X-ray crystal-structure determinations, values for the high-resolution metrical parameters of the clusters and, especially, their overall charge have often relied heavily on the synthesis and characterization of replicative model complexes. Moreover, the application of electronic absorption, EPR, and Mössbauer spectroscopy to proteins of unknown composition and structure, once the major cluster classes had been defined, enabled the various cores to be identified rapidly and reliably without recourse to protein X-ray crystal-structure determinations, which often take years to complete. Finally, several of the structures were unprecedented, not only in biology but in inorganic chemistry as well, further contributing to the enthusiasm with which each new result was met and raising important challenges for the synthetic-model chemist. Here we describe the basic structure types, charges on the protein clusters, and physical properties. The synthetic-model work is treated separately in the following section.

Table 5.2
Typical chemical and spectroscopic characteristics of iron-sulfur proteins

Protein	Cluster	Structure	Oxidation state	Formal valence	EPR g values (temp)	Mössbauer isomer shift (mm/sec)	λ_{max} (nm), extinction coeff. (X10^{-3}, per Fe)
Rubredoxin	1Fe-0S		Oxidized	Fe^{3+}	4.3, 9 (<20 K)	0.25	390(10.8), 490(8.8)
			Reduced	Fe^{2+}	None	0.65	310(10.8), 335(6.3)
2-Iron ferredoxin	2Fe-2S		Oxidized	$2Fe^{3+}$	None	0.26	325(6.4), 420(4.8), 465(4.9)
			Reduced	$1Fe^{3+}$, $1Fe^{2+}$	1.89, 1.95, 2.05 (<100 K)	0.25, 0.55	Absorbance declines 50 percent upon reduction
3-Iron ferredoxin	3Fe-4S		Oxidized	$3Fe^{3+}$	1.97, 2.00, 2.02 (<20 K)	0.27	305(7.7), 415(5.2), 455(4.4)
			Reduced	$2Fe^{3+}$, $1Fe^{2+}$	None	0.30, 0.46	425(3.2)
4-Iron ferredoxin	4Fe-4S		Oxidized	$3Fe^{3+}$, $1Fe^{2+}$	2.04, 2.04, 2.12 (<100 K)	0.31	325(8.1), 385(5.0), 450(4.6)
			Intermed.	$2Fe^{3+}$, $2Fe^{2+}$	None	0.42	305(4.9), 390(3.8)
			Reduced	$1Fe^{3+}$, $3Fe^{2+}$	1.88, 1.92, 2.06 (<20 K)	0.57	Unfeatured; absorbance declines upon reduction

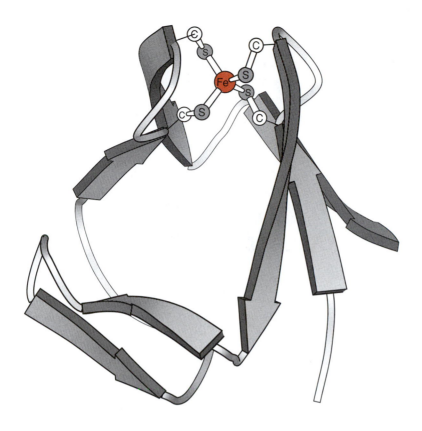

Figure 5.8
Structure of rubredoxin from *C. pasteurianum* depicting its iron core geometry.

(i) *[1Fe-0S]*. Figure 5.8 shows the result of a crystal-structure analysis of the protein *C. pasteurianum* rubredoxin (Rb). This protein contains a single iron atom tetrahedrally coordinated by four cysteinato sulfur atoms with an average Fe–S bond distance of 2.29 (2) Å (recall the discussion in Section 4.3.b). The cysteine ligands contributed from the protein chain occur in pairs, –Cys(6)–X–X–Cys(9)–Gly and –Cys(39)–X–X–Cys(42)–Gly, where –X– represents an intervening amino acid. Although crystallography cannot reveal the overall charge on the $\{Fe(SR)_4\}^{n-}$ unit in the protein, the iron was shown clearly to be high-spin ferric by EPR and Mössbauer spectroscopy (Table 5.2). The red color of the oxidized form of Rb derives from a S → Fe charge-transfer band at 490 nm. The measured redox potentials of several rubredoxins fall in the −50 to +50 mV range at pH 7.

(ii) *[2Fe-2S]*. The X-ray structure of a [2Fe-2S] ferredoxin (Fd) from *S. platensis* revealed the occurrence of an $\{Fe_2S_2(S\text{-cys})_4\}^{2-}$ cluster in which two

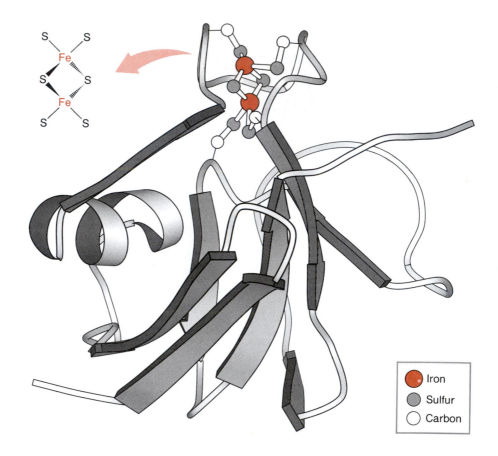

🔴	Iron
⚪	Sulfur
◯	Carbon

Figure 5.9
Structure of *S. platensis* ferredoxin and its 2Fe-2S cluster.

sulfide ligands bridge the two tetrahedrally coordinated iron atoms (Figure 5.9). The doubly bridged Fe_2S_2 unit has an $Fe \cdots Fe$ distance of 2.7 Å. Since the oxidation states readily available to iron in a tetrahedral sulfur environment are $+2$ and $+3$, the charge on the $\{Fe_2S_2\}^{n+}$ core can, in principle, vary from 0 ($Fe^{II}Fe^{II}$) to $+1$ ($Fe^{II}Fe^{III}$, mixed-valence form) to $+2$ ($Fe^{III}Fe^{III}$). These latter two forms exist in the proteins and are readily distinguishable by their physical properties (Table 5.2). In fact, the iron-sulfur core structure of the [2Fe-2S] protein class was correctly deduced from these physicochemical data before it was confirmed by protein crystallography.

Since the iron centers are four-coordinate, the crystal-field splitting is weak, and the metals are high-spin in both oxidized and reduced forms. The presence of the bridging sulfur atoms provides a pathway for antiferromagnetic exchange coupling, however, such that the ground state of the oxidized, diiron(III) form of the proteins is diamagnetic ($S_T = 0$). The para-

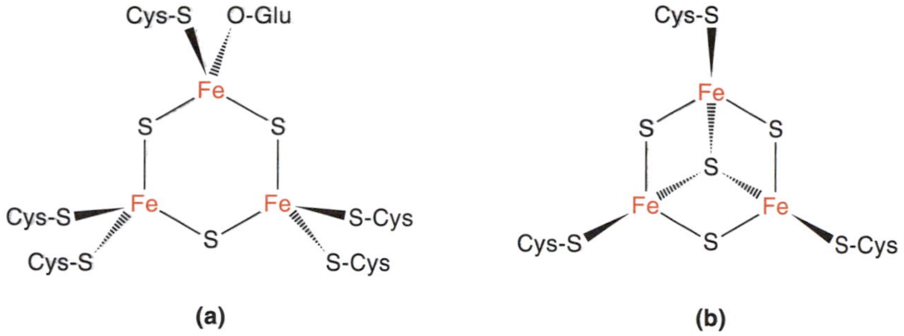

Figure 5.10
Scheme showing the original planar (left) and corrected non-planar (right) $[Fe_3S_4]$ cluster in *A. vinelandii* ferredoxin.

magnetic, mixed-valence $Fe^{II}Fe^{III}$ form of the [2Fe-2S] ferredoxins was a subject of much interest when they were first discovered. Superhyperfine coupling to both iron nuclei (^{57}Fe) and the bridging sulfide (^{33}S; also ^{77}Se in a selenide-substituted Fd) established electron delocalization over the cluster. The presence of g values below 2.00 (Table 5.2) was eventually explained by an exchange-coupling model. These characteristic g values thus became an important signature for identifying mixed-valence, antiferromagnetically coupled Fe(III)Fe(II) centers. From Mössbauer spectroscopy it was apparent that, whereas the spectra of the oxidized forms of the [2Fe-2S] proteins consist of a single quadrupole doublet characteristic of high-spin iron(III) in a single site, the reduced Fe(III)Fe(II) forms have two distinct quadrupole doublets indicative of trapped-valence states. In other words, electron exchange is slow on the Mössbauer time scale (see Table 4.1).

(iii) [3Fe-4S]. Originally the X-ray crystal structure of the 7-iron ferredoxin from *A. vinelandii* was interpreted as having a [4Fe-4S] and a [3Fe-3S] cluster. The latter was reported to have a nearly planar, six-membered Fe_3S_3 ring (Figure 5.10) and, as such, was unique not only among iron-sulfur proteins but among small-molecule iron-sulfur complexes as well. As was recently shown by two groups, including the original investigators, the structure originally reported was grossly in error, owing to the wrong choice of a space group from between the enantiomeric pairs $P4_12_12$ and $P4_32_12$. The correct structure has chain folding closely analogous to that of the 8Fe-8S ferredoxins, discussed below, a feature that was completely missed in the original analysis. An additional important change is reassignment of the tri-iron cluster as $[Fe_3S_4]$ having the geometry shown in Figure 5.10. This non-planar structure can be derived from the [4Fe-4S] core (see next section) by removing an Fe—S (cysteine) unit from one of the eight corners of the cube,

a formal transformation illustrated in Equation 5.2. EXAFS data on the *A. vinelandii* Fd, a 3Fe-4S Fd from *Desulfovibrio gigas*, and an inactive form

$$(5.2)$$

of aconitase all support the occurrence of the same [3Fe-4S] cluster. In *A. vinelandii* Fd, three cysteine thiolate groups anchor the cluster to the protein chain, as expected from Equation 5.2.

The sequence of historical events leading to the correct identity of the [3Fe-4S] cluster in *A. vinelandii* is of pedagogical interest. Mössbauer and EXAFS spectral studies on *D. gigas* Fd as well as a 3Fe form of aconitase revealed two distinct iron sites, in a 2:1 ratio and Fe⋯Fe distances of 2.7 Å. Both pieces of data disagreed with the original structure report for *A. vinelandii* Fd, which showed three equivalent iron sites and Fe⋯Fe distances of 4.1 Å. Moreover, addition of ferrous iron and no inorganic sulfide to beef-heart aconitase led to active enzyme. This result and a careful measurement of the inorganic sulfide content of the protein gave a [3Fe-4S] stoichiometry, in contrast to the [3Fe-3S] result from the earlier crystal structure. Thus either nature had two very different kinds of 3Fe clusters, or one of the structural determinations was in error, as proved to be the case. One lesson for the student of bioinorganic chemistry is that protein crystal structures should not be considered as credible as small-molecule X-ray structures. Even when errors in space-group assignment, as happened here, are not made, interpretation of electron-density maps from protein-crystallographic data is at least partly subjective. Figure 5.11 illustrates this point by displaying a stereoview of the actual electron density in the vicinity of the [3Fe-4S] cluster in *A. vinelandii* Fd for the *correctly* phased data to 2.7 Å resolution. Readers unfamiliar with protein crystal-structure determinations will probably not be able to recognize the cluster geometry of Figure 5.10 in the map of Figure 5.11. Moreover, it is *not* always the protein crystallographer who might overinterpret electron-density maps. The crystallographically inexperienced researcher interested in using the results in spectroscopic or mechanistic studies must be equally cautious.

Table 5.2 includes physical data on [3Fe-4S] proteins, from which it can be seen that this cluster has a unique spectroscopic signature. Except for *D.*

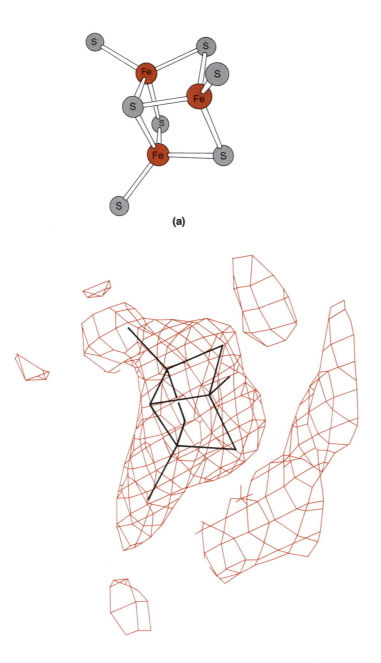

(a)

Figure 5.11

(a) 3Fe-4S cluster in *A. vinelandii* ferredoxin and (b) view of electron density at 2.7 Å resolution on which the structure shown in (a) was based. (Adapted from C. D. Stout, *Journal of Biological Chemistry* **263**, 9256–9260 (1988)).

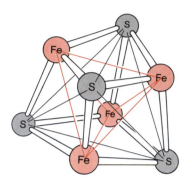

Figure 5.12
Distorted cube structure of a 4Fe-4S cluster. The cube can be visualized as two interpenetrating Fe_4 and S_4 tetrahedra.

gigas, the only protein that does not contain a second Fe_nS_n unit, procurement of the spectroscopic parameters for the [3Fe-4S] cluster has been a significant achievement. Even more difficult was the extraction of redox potential values for proteins such as *A. vinelandii* and *A. chroococcum* ferredoxin (called Fd I) that contain both 3Fe and 4Fe clusters. This latter information is especially important to know accurately because of the biological role of these proteins as electron-transfer agents.

(iv) [4Fe-4S]. The most ubiquitous iron-sulfur clusters in biology fall into this class. The basic structure is a distorted cube with alternating Fe and S atoms at the corners. The result is two interpenetrating concentric tetrahedra of four iron and four sulfide atoms, with mean $Fe \cdots Fe$ and $S \cdots S$ distances of ~ 2.75 Å and 3.55 Å, respectively (Figure 5.12). The cube is anchored to the protein by four cysteinyl sulfur atoms at the four Fe corners, resulting in a distorted tetrahedral coordination geometry (Figure 5.13). Protein X-ray structural results for these [4Fe-4S] clusters are available for both oxidized and reduced forms of HiPIP from *Chromatium*, *A. vinelandii* ferredoxin I, the 7Fe/8S protein discussed previously, and *P. aerogenes* ferredoxin, an 8Fe/8S protein that contains two distinct $\{Fe_4S_4\}^{2+}$ clusters. The relationship between the positions in the sequence of the eight cysteine residues that bind the two [4Fe-4S] centers in *P. aerogenes* Fd and their coordination to the clusters is interesting. As can be seen from Figure 3.6, and as has already been discussed in Chapter 3, the sequence runs $-X-Cys(8)-X-X-Cys(11)-X-X-Cys(14)-X-X-X-Cys(18)-\cdots-X-Cys(35)-X-X-Cys(38)-X-X-Cys-(41)-X-X-X-Cys(45)-\cdots$. Despite the fact that the eight cysteines fall in two clusters of four each, only three of the four bind to an individual [4Fe-4S] unit, corresponding to Cys–X–X–Cys but not to Cys–X–X–X–Cys. This preference has been explained as a means of providing $NH \cdots S$ hydrogen bonding from an amide hydrogen of the peptide backbone to the iron sulfur cluster.

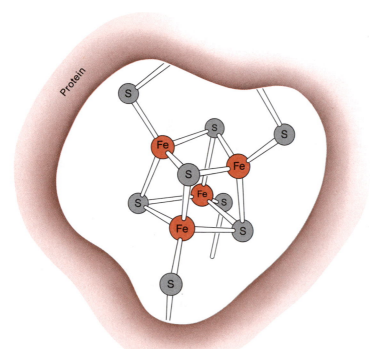

Figure 5.13
A 4Fe-4S cluster bound to a protein through four cysteine thiolate residues.

Finally, a recent crystal-structure determination of the MoFe protein of nitrogenase has revealed that 2 4Fe–4S cubes can be directly linked by means of a S–S bond.

(v) Synthetic models for iron-sulfur proteins. The beautiful core structures within the iron-sulfur proteins offered a formidable challenge to the bioinorganic chemist interested in preparing replicative models, for although the 4Fe-4S cube had been known in organometallic chemistry, stabilized by coordination to the cyclopentadienyl anion, iron complexes of thiolate ligands were difficult to prepare, because redox (Equation 5.3) or polymerization (Equation 5.4) reactions tend to occur. A major breakthrough came with the

$$2Fe^{3+} + 2RS^- \rightarrow 2Fe^{2+} + RSSR \qquad (5.3)$$

$$nFe^{2+} + 2nRS^- \rightarrow [Fe(SR)_2]_n \qquad (5.4)$$

discovery that reaction of $FeCl_3$, NaHS, and three equivalents of $Na(SCH_2Ph)$ in methanol yielded the $[Fe_4S_4(SCH_2Ph)_4]^{2-}$ cluster (Equation 5.5). In this reaction, some of the thiolate ligand is consumed to reduce half the iron

$$4FeCl_3 + 12RS^- \rightarrow 4Fe(SR)_3 + 12Cl^-$$
$$\xrightarrow[4HS^-]{4OMe^-} [Fe_4S_4(SR)_4]^{2-} + RSSR \atop + 6RS^- + 4MeOH \qquad (5.5)$$

formally to the $+2$ oxidation level required for the $\{4Fe\text{-}4S\}^{2+}$ unit. In subsequent work it was found that elemental sulfur could be substituted for sulfide in the presence of enough thiolate to serve as a reducing agent (Equation 5.6). Synthetic routes to the [2Fe-2S] analog family were discovered

$$4FeCl_3 + 14RS^- + 4S \rightarrow [Fe_4S_4(SR)_4]^{2-} + 5RSSR + 12Cl^- \quad (5.6)$$

shortly after the preparation of the first tetranuclear cluster, and the preparation and characterization of mononuclear rubredoxin models subsequently followed. Figure 5.14 pictorially summarizes the assembly of these complexes together with some of the identified intermediates.

(vi) Cluster chemistry and core extrusion. From studies of chemical reactions of iron-sulfur model complexes were derived further strategies to probe the properties of analogous clusters in the proteins. One reaction that proved to be of practical significance is ligand substitution, in which the terminal thiolates are replaced in a stepwise fashion (Equations 5.7 and 5.8). Arylthiols, for example, PhSH, are more effective than alkylthiols in promoting these reactions, owing to the weaker acidity of the latter. From

$$[Fe_4S_4(SR)_4]^{2-} + nR'SH \rightleftharpoons [Fe_4S_4(SR)_{4-n}(SR')_n]^{2-} + nRSH \quad (5.7)$$

$$[Fe_2S_2(SR)_2]^{2-} + 4R'SH \rightleftharpoons [Fe_2S_2(SR')_4]^{2-} + 4RSH \quad (5.8)$$

knowledge of this chemistry evolved parallel reactions with the proteins in which [nFe-nS], $n = 2$ or 4, clusters could be removed intact from the polypeptide chain. These reactions, which have been termed core extrusions, have proved to be valuable for identifying iron-sulfur cluster types in proteins of previously uncharacterized cores. They are usually carried out in an unfolding solvent, such as 80 percent v/v DMSO/H_2O, that denatures the protein enough to permit the substitution reaction to take place (Equation 5.9). Core-extrusion reactions have been applied successfully to identify, inter alia,

$$\text{holoprotein} + \underset{\text{excess}}{RSH} \xrightarrow[\text{solvent}]{\text{unfolding}} \underset{n = 2 \text{ or } 4}{Fe_nS_n(SR)_4]^{2-}} + \text{apoprotein} \quad (5.9)$$

[4Fe-4S] clusters in C. pasteurianum hydrogenase, the Fe protein of its nitrogenase, and [2Fe-2S] clusters in Fd II from A. vinelandii and milk xanthine oxidase. Although two-Fe and four-Fe clusters are known to interconvert in the presence of some thiols, sufficient care and control experiments usually enable one to avoid these potentially misleading reactions. A notable excep-

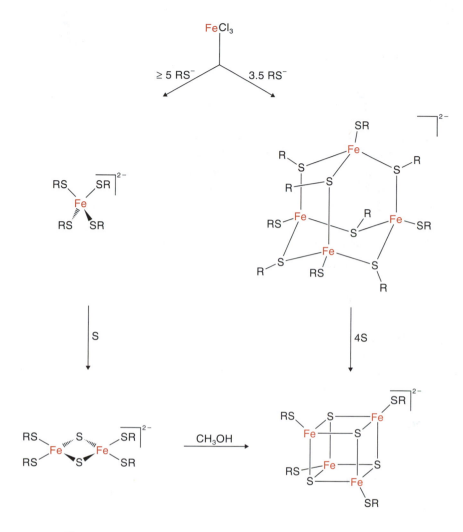

Figure 5.14
Routes to synthetic models of iron-sulfur clusters from proteins.

tion, however, was the extrusion of [2Fe-2S] clusters from the [3Fe-4S] proteins. The clusters extruded by the chemistry shown in Equation 5.9 can be identified by their characteristic absorption spectra (Table 5.2) and even by ^{19}F NMR spectroscopy when fluorinated thiols (HSR$_F$) such as *p*-trifluoromethylbenzenethiol are employed. The latter method works well because of the different paramagnetic contact shifts for $[Fe_2S_2(SR_F)_4]^{2-}$ and $[Fe_4S_4(SR_F)_4]^{2-}$ clusters.

5.3.b. Polyiron Oxo Clusters and Biomineralization. A recent development in bioinorganic chemistry has been the discovery of a class of proteins containing the dinuclear iron oxo core, $\{Fe_2O\}^{2+}$, and related units in which hydroxide or another monoatomic bridging ligand links the two iron atoms. Such species are well known in the hydrolytic chemistry of iron(III), and also form when iron(II) compounds are oxidized by dioxygen. Typical reactions are shown in Equations 5.10 and 5.11. In the metalloproteins hemerythrin (Hr) and ribonucleotide reductase (RR), the μ-oxodiiron(III) moiety is linked by one or two additional bridging carboxylate groups, which bring the iron atoms closer to one another and bend the Fe−O−Fe bridge

$$2Fe^{3+} + 6OH^- \rightarrow Fe_2O_3 \cdot 3H_2O \qquad (5.10)$$

$$12Fe^{2+} + 3O_2 + 12H_2O \rightarrow 6Fe_2O_3 + 24H^+ \qquad (5.11)$$

angle to $\sim 120-130°$. The μ-oxobis(μ-carboxylato)diiron(III) moiety found in these proteins is not unprecedented in inorganic chemistry, having been known for over a century in the basic iron carboxylates. These complexes have the general formula $[Fe_3O(O_2CR)_6L_3]^+$, in which a triply bridging oxygen atom links three iron(III) ions at the corners of an equilateral triangle. Two carboxylate ligands bridge pairs of iron atoms along each edge of the triangle, forming the $\{Fe_2O(O_2CR)_2\}^{2+}$ unit that occurs in hemerythrin. The core structures of Hr, RR, and the basic carboxylates are depicted in Figure 5.15.

A more dramatic example of the occurrence of a polyiron oxo cluster in biology is the iron-storage protein ferritin. Mammalian ferritins consist of a protein shell of 24 subunits surrounding a core of hydrated ferric oxide containing different amounts of phosphate. Each protein subunit consists of ~ 175 amino acids. Apo, or metal-free, ferritin can accommodate up to 4,500 iron atoms, giving a ratio of approximately one iron atom per amino acid! The overall structure of the protein is highly symmetric, as revealed by the X-ray crystal-structure determination of apoferritin (Figure 5.16). The asymmetric unit in this structure is one polypeptide chain; the symmetry of the crystallographic space group generates the entire 24-peptide assembly, which has cubic 432 (also called O) point group symmetry. The structure is roughly spherical, with an outside diameter of 130 Å and an inside diameter of 75 Å. Channels connecting the interior of the assembly with the outside lie along the three-fold and four-fold axes. The four-fold channels are lined with hydrophobic residues, whereas the three-fold channels are lined with hydrophilic aspartate and glutamate residues. These three-fold channels appear to be the most plausible sites of metal entry. The inside of the protein shell is also lined with hydrophilic residues.

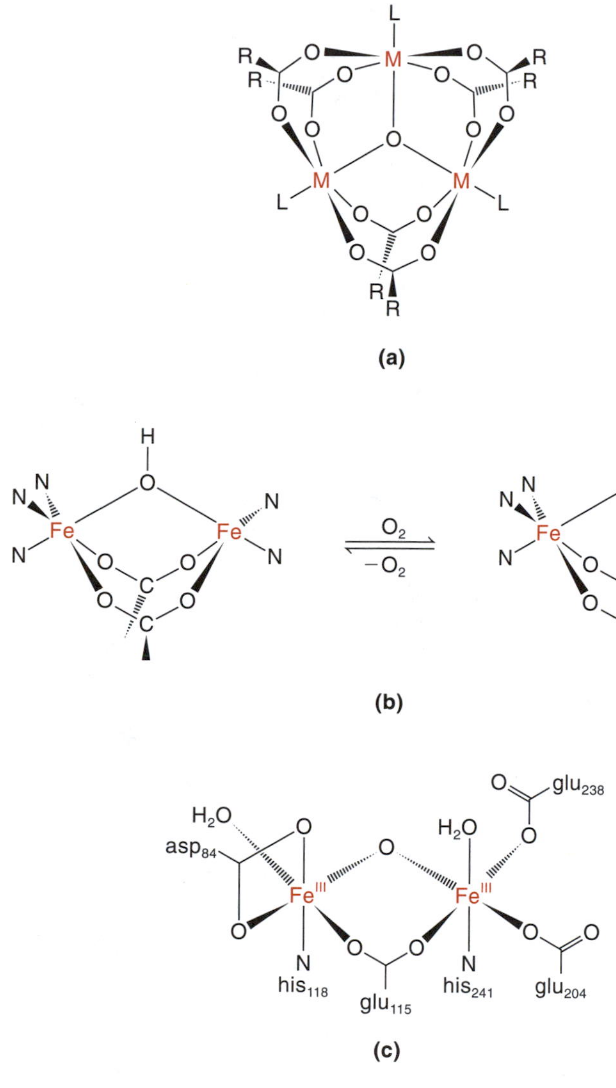

Figure 5.15
Polyiron carboxylate structures found in nature. (a) The basic metal carboxylate, not yet known in biology. (b) Carboxylate-bridged diiron unit in hemerythrin, illustrating the O_2-binding reaction. (c) Carboxylate-bridged diiron unit in oxidized R2 protein of ribonucleotide reductase.

128

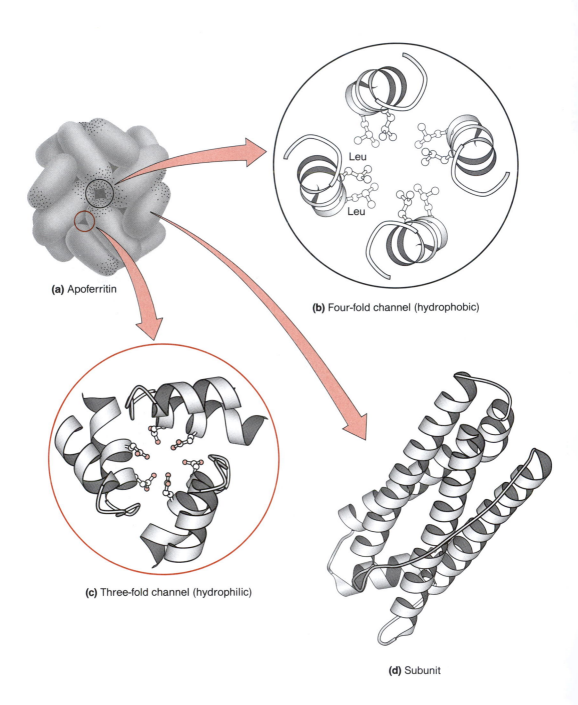

(a) Apoferritin

(b) Four-fold channel (hydrophobic)

Leu

Leu

(c) Three-fold channel (hydrophilic)

(d) Subunit

Figure 5.16
Structural features of the iron storage protein ferritin, including (a) the overall organization of 24 subunits in the apo protein, (b) hydrophobic channel of four-fold symmetry, (c) hydrophilic channel of three-fold symmetry, and (d) fold of an individual subunit.

The structure of the iron core of ferritin has been probed by a variety of spectroscopic techniques, including EXAFS, Mössbauer, and optical spectroscopy. The results are consistent with octahedrally coordinated iron(III) ions joined by bridging oxide and/or hydroxide ions. The EXAFS spectra could be fit to yield iron-oxygen distances of 1.95 Å and iron-iron distances of 3.3 Å, with approximately seven iron neighbors at this distance. The structure that is most consistent with these metric parameters consists of a close-packed array of oxygen atoms with iron atoms partially occupying the octahedral interstices. The vacant octahedral sites may be arranged irregularly, or may be in layers, so that the structure could be described as two-dimensional sheets built up of interconnected FeO_6 units that fold back on themselves to fill the three-dimensional core. A mineral termed ferrihydrite has a similar postulated structure. Ferrihydrite exhibits an X-ray diffraction pattern that resembles the electron diffraction pattern arising from the iron core of isolated ferritin molecules. These patterns do not contain enough information to define the structures completely, however, and there is need for more definitive work on the core geometry as well as the mechanism of iron entry and export.

The formation of the ferritin core is one example of a process that has been termed *biomineralization*. Organisms synthesize minerals from simple compounds for a variety of purposes. Examples include the calcium phosphate minerals that lend mechanical strength to bone, and the magnetite (Fe_3O_4) particles produced by magnetotactic bacteria for orientation purposes. Information on the mechanism of ferritin core assembly has been obtained by direct studies and by examination of model systems. The pathway appears to involve entrance of Fe(II) into the hydrophilic channels or to the edges of the interior, followed by aerial oxidation to Fe(III) with the formation of a few initiation complexes. These complexes grow through hydrolytic polymerization of Fe(II), with concomitant oxidation to Fe(III) to produce the core. Thus the inorganic reactions in both of Equations 5.10 and 5.11 appear to be involved.

Studies of the products generated by controlled hydrolysis of iron(III) complexes have been carried out in aqueous or nonaqueous solvents in the presence of appropriate ligands. A variety of complexes have been synthesized and characterized that contain dinuclear, trinuclear, tetranuclear, and even decanuclear, undecanuclear, dodecanuclear, heptadecanuclear, and octadecanuclear Fe(III) core structures. The structure of one recently characterized complex, $[Fe^{III}_4Fe^{II}_8(O)_2(OCH_3)_{18}(O_2CCH_3)_6(CH_3OH)_{4.67}]$, is shown in

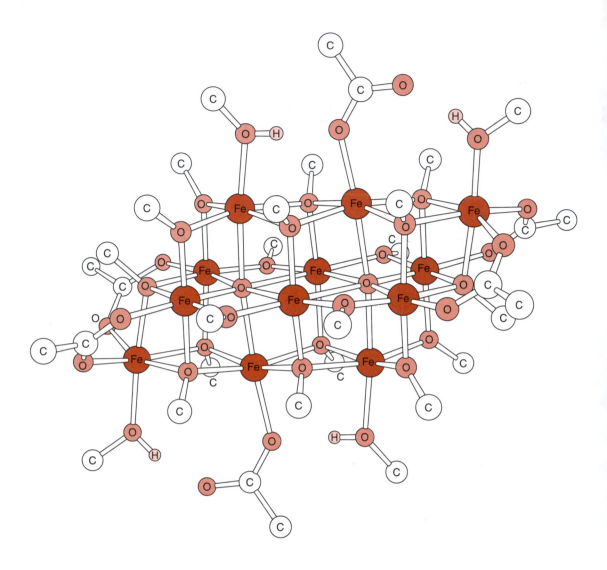

Figure 5.17
A mixed-valent polyiron oxo complex prepared as a model for intermediates in ferritin core formation. The overall formula for this complex is $[Fe_{12}O_2(OCH_3)_{18}(O_2CCH_3)_6(CH_3OH)_{4.67}]$.

Figure 5.17. This compound, prepared by the slow addition of dioxygen to ferrous acetate and lithium methoxide in methanol, has a distorted sodium-chloride structure type and resembles the three-dimensional lattice structure of iron and oxygen atoms proposed for ferritin. Further characterization of such complexes will undoubtedly shed light on the details of the structure of the ferritin core and the routes by which it is assembled and disassembled.

5.4. Specialized Units

Apart from the simple metal aqua ions and polymetallic clusters discussed thus far, there are some other specialized metal-binding molecules in biology that deserve mention. These units differ from ionophores such as enterobactin, which has a single functional purpose. Like the iron-sulfur clusters, they can be inserted into several different metalloproteins and their properties attenuated to meet a particular metabolic need. We encountered them previously in Chapter 3 in our discussion of specialized ligands and return to them here because of their role in the assembly of key bioinorganic components of proteins and nucleic acids.

5.4.a. Porphyrins.   Porphyrins are tetrapyrrole macrocycles that bind iron, magnesium, and other metal ions, forming a square-planar arrangement of four nitrogen atoms. The basic porphyrin structure was shown in Figure 1.3. Various substituents on the periphery of the porphyrin ring attenuate its electronic properties, further contributing to the utility of this important class of metal-binding ligand. Metalloporphyrins can be anchored to proteins either directly, through covalent bonds to the substituents on the pyrrole rings, or through ligand binding to the two available axial coordination positions above and below the ring of the macrocycle. Both kinds of bonds occur in porphyrin-containing proteins.

The tight structure at the center of the porphyrin ring, in both the free base and the metalloporphyrin, makes it difficult to metallate and demetallate the unit. Iron is inserted into porphyrins to form heme by means of an enzyme called *ferrochelatase*, discussed further in Chapter 8, that requires Fe^{2+} ion. The enzyme probably ruffles the porphyrin in order to direct one of the pyrrole nitrogen lone pairs out of the plane of the ring for iron binding. In support of this notion is the fact that metal ions are chelated by N-alkyl porphyrins about 10^4 times more rapidly than by nonalkylated analogs. Alkylation is also known to distort the porphyrin ring in a manner that facilitates metal binding. This observation has been used to great advantage in the design of antibodies to N-alkyl porphyrins that catalyze insertion of metals into porphyrins (see Section 8.1.b).

5.4.b. Corrins and Hydroporphyrins.   Another specialized unit is the corrin, the structure of which may be found in Figure 1.3. Like porphyrins, corrins are comprised of four pyrrole rings in the form of a macrocycle and utilize the four nitrogen atoms to coordinate a metal in an approximately square-planar fashion. The major difference, however, is the absence of a methine carbon atom linking the A and D pyrrole rings, a feature that distorts the geometry at the metal center and has been held responsible for some of the interesting properties of the resulting complex. Cobalt forms the major metal-corrinoid complex of interest in bioinorganic chemistry, the resulting vitamin-B-12 molecule and its coenzyme forms being important functional units for effecting 1,2-isomerization and radical-promoted redox reactions. Binding to the enzyme activates the coenzyme in ways not yet fully delineated. Corrins are also found in nature as their nickel complexes, which are used by hydrogenases of thermophilic organisms that evolve methane.

Another class of specialized ligands related to the porphyrins are the hydroporphyrins, including siroheme, chlorins, bacteriochlorins, isobacteriochlorins, corphins, and dioneheme. Corphin, for example, is a hydroporphyrin that is reduced from the porphyrin level by ten electrons. This tetrapyrrole occurs naturally as factor F_{430}, a nickel-containing prosthetic group of S-methyl coenzyme-M reductase found in methanogenic bacteria.

5.4.c. Metal-Nucleotide Complexes.   One of the most important and ubiquitous molecules in cells is adenosine triphosphate, or ATP. In particular, the Mg^{2+} complex of this polyanion is frequently required as a substrate for phosphoryl and nucleotidyl transferase enzymes. The Mg^{2+}, Ca^{2+}, Na^+, and K^+ complexes of ATP are important substrates for ATPases, enzymes that catalyze the hydrolysis of ATP and use the energy derived from this reaction for energy transduction. As will be shown in Chapter 7, this chemistry is of fundamental importance for information transfer and storage in cells.

The structures of ATP and several of its more important metal complexes are given in Figure 5.18. Metal ions can bind to the α-, β-, and γ-phosphate ester oxygen atoms as well as to the nitrogen heteroatoms of the base. For the Mg^{2+} ion, coordination occurs exclusively at the oxygen atoms, with water occupying the remaining octahedral positions about the metal.

5.4.d. Molybdenum-Binding Cofactors.   As already mentioned, Mo is the only essential second-row transition metal in biology. Its special value, as will be shown, is its ability to undergo two-electron transfer reactions, especially between the Mo(VI) and Mo(IV) oxidation levels. In addition, it is able to transfer an oxo atom to substrate(s), as illustrated in Equation 5.12. There are at least two specific cofactors that contain molybdenum. One of these is

$$Mo^{VI}(O)_2 + S \rightarrow S = O + Mo^{IV}(O) \tag{5.12}$$

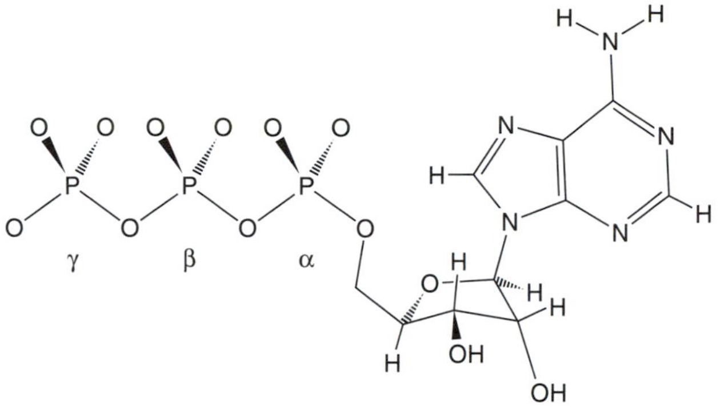

Adenosine triphosphate (ATP)

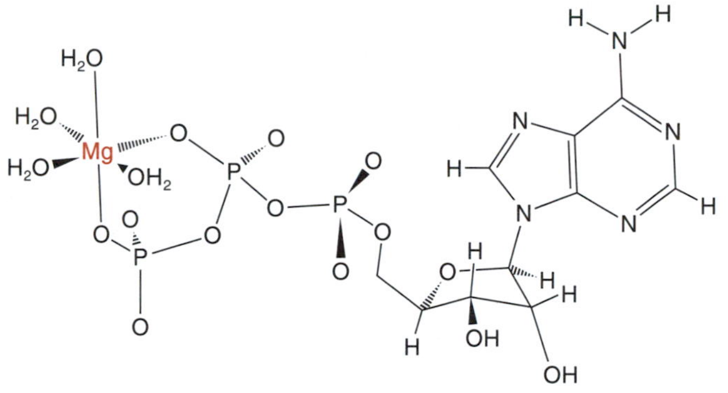

β, γ-Mg-ATP

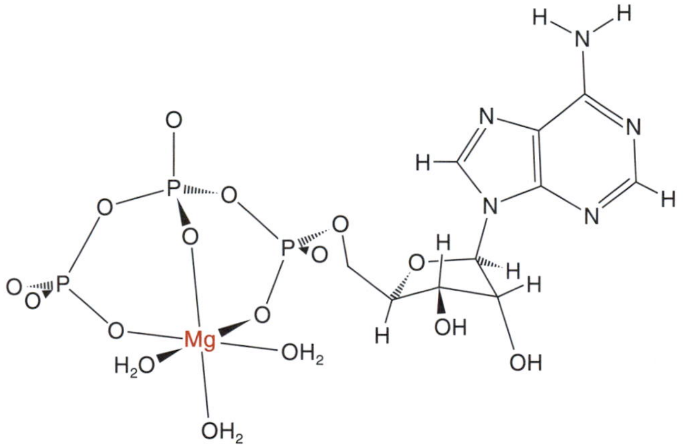

α, β, γ-Mg-ATP

Figure 5.18
Chemical structures of (top) ATP, (middle) ATP complexed to Mg^{2+} by the β and γ phosphates, and (bottom) ATP complexed to Mg^{2+} by the α, β, and γ phosphates.

the so-called iron-molybdenum cofactor, or FeMoco, that occurs in the enzyme nitrogenase. Recently the structure of the molybdenum-iron protein of nitrogenase has been crystallographically determined at 2.2 Å resolution (Figure 5.19) and a model for FeMoco has been proposed. The cofactor consists of two cuboidal fragments, one containing four iron atoms, and the other three iron atoms and the molybdenum atom. The Mo is octahedrally coordinated, whereas iron atoms at the interface of the two cube fragments appear to have open coordination sites. This feature has led to the speculation that dinitrogen binds and is activated for reduction by two or more iron atoms at the center of the cluster and may not coordinate directly to molybdenum. Such a hypothesis is appealing, since alternative nitrogenases have been discovered that do not contain molybdenum, but have other metals, specifically vanadium or iron, instead. Much more work is required to test such an idea.

The other molybdenum cofactor is present in reductases, oxidases, and dehydrogenases and functions in oxo transfer chemistry (Equation 5.12). As shown in Figure 5.20, the molybdenum-binding moiety is a substituted pterin that coordinates the metal via two sulfur atoms. This modified pterin was identified after the discovery that a compound called *urothione*, a molecule of known structure, is a major urinary metabolite of the molybdenum cofactor. Although all the known enzymes that require a molybdenum cofactor, except nitrogenase, contain this pterin derivative, the other metal-coordinating atoms differ significantly. This enzyme-specific property is similar to that previously encountered for the better-understood metalloporphyrins.

The role of the reduced pterin in the function of the molybdenum oxo-transferases remains to be determined. The organic moiety can undergo redox reactions that could be integral components of the electron-transfer pathways in the enzymes. Alternatively, the pterin could serve merely to store and transfer molybdenum from protein to protein. The aqueous chemistry of molybdenum is dominated by oxo anions such as molybdate, MoO_4^{2-}, and by polymolybdates such as $Mo_7O_{24}^{6-}$. Chelatable cationic species are unavailable at pH 7. This aspect of molybdenum aqueous chemistry may have required the development of special ligands for handling this metal and for incorporating it into the active sites of enzymes.

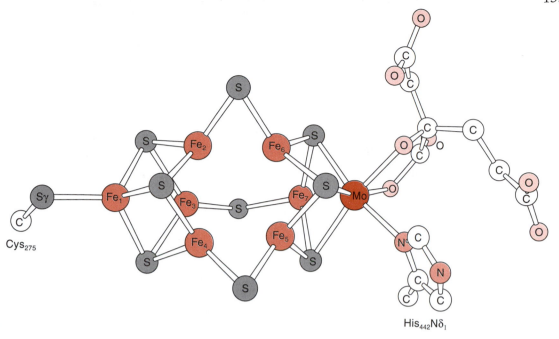

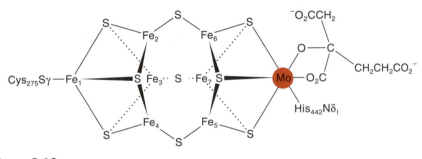

Figure 5.19
Structure of the iron-molybdenum cofactor of the molybdenum-iron protein of nitrogenase deduced from X-ray crystallography. The cavity in the center is a possible site of binding and activation of dinitrogen.

Molybdenum cofactor
(proposed structure)

Figure 5.20
Probable structure of the molybdenum cofactor found in proteins that promote oxo transfer reactions.

Study Problems

1. Given that the dissociation constant for $[Fe(ent)]^{3-}$ at high pH is 10^{-49} M, calculate the concentration of free Fe^{3+} in a 1 M solution of $[Fe(ent)]^{3-}$. How many free iron ions are present in one liter of such a solution? Repeat these calculations at neutral pH, where the dissociation constant is 10^{-25} M.

2. Would the enantiomer of enterobactin be expected to be a good chelator for iron? What value would you expect for its dissociation constant at neutral pH? Do you think that the enantiomer would be useful for iron uptake by bacteria that make enterobactin? Explain.

3. Model complexes for the [2Fe-2S] sites in ferredoxins have been prepared. The most well-characterized models are in the fully oxidized form $[Fe_2S_2(SR)_4]^{2-}$, where both irons are in the $+3$ oxidation state. One-electron reduction of these complexes produces species of the form $[Fe_2S_2(SR)_4]^{3-}$, which are unstable because two such complexes react to produce one $[Fe_4S_4(SR)_4]^{2-}$ complex. Write a balanced equation for this reaction. Why doesn't a similar reaction occur for the oxidized form? Can you think of a synthetic strategy to decrease the likelihood that this dimerization reaction will occur?

4. Magnetotactic bacteria produce approximately 10 new magnetite particles per generation. Each particle is approximately cubic, with an edge dimension of 500 Å. The density of magnetite (Fe_3O_4) is 5.18 gm/cm^3. Calculate how much iron each bacterium must take up to accomplish this feat.

5. Using the data in Table 5.2, make a plot of the iron isomer shift as a function of average oxidation state. You should observe a linear relationship. Given that the iron-molybdenum cofactor from nitrogenase has the metal-ion stoichiometry $MoFe_6$ and exhibits a Mössbauer spectrum with an isomer shift of 0.37 mm/sec, estimate the mean oxidation state for the iron atoms in the cofactor. Note that this analysis assumes tetrahedral sulfur coordination environments for iron.

Bibliography

Bioavailability of Metal Ions
E. I. Ochiai. 1987. *General Principles of Biochemistry of the Elements*. Plenum Press, New York.

K. N. Raymond. 1984. "Complexation of Iron by Siderophores. A Review of Their Solution and Structural Chemistry and Biological Function." *Top. Curr. Chem.* 123, 49–102.

Enrichment Strategies and Intracellular Chemistry of Low-Abundance Metals

D. P. Bancroft, C. A. Lepre, and S. J. Lippard. 1990. "^{195}Pt NMR Kinetic and Mechanistic Studies of *cis-* and *trans-*Diamminedichloroplatinum(II) Binding." *J. Am. Chem. Soc.* 112, 6860−6871.

P. H. Connett and K. E. Wetterhahn. 1983. "Metabolism of the Carcinogen Chromate by Cellular Constituents." *Structure and Bonding* 54, 93−124.

Spontaneous Self-Assembly of Metal Clusters

W. H. Armstrong, A. Spool, G. C. Papaefthymiou, R. B. Frankel, and S. J. Lippard. 1984. "Assembly and Characterization of an Accurate Model for the Diiron Center in Hemerythrin." *J. Am. Chem. Soc.* 106, 3653−3667.

J. M. Berg and R. H. Holm. 1982. "Structures and Reactions of Iron-Sulfur Protein Clusters and Their Synthetic Analogs," in T. G. Spiro, ed., *Iron-Sulfur Proteins.* Wiley, New York, 3−66.

D. Coucouvanis. 1991. "Use of Preassembled Fe/S and Fe/Mo/S Clusters in the Stepwise Synthesis of Potential Analogues for the Fe/Mo/S Site in Nitrogenase." *Acc. Chem. Res.* 24, 1−8.

K. S. Hagen, J. G. Reynolds, and R. H. Holm. 1981. "Definition of Reaction Sequences Resulting in Self-Assembly of $[Fe_4S_4(SR)_4]^{2-}$ Clusters from Simple Reactants." *J. Am. Chem. Soc.* 103, 4054−4063.

D. M. Kurtz, Jr. 1990. "Oxo- and Hydroxo-Bridged Diiron Complexes: A Chemical Perspective on a Biological Unit." *Chem. Rev.* 90, 585−606.

E. C. Theil. 1990. "The Ferritin Family of Iron Storage Proteins." *Adv. Enzymol. Relat. Areas Mol. Biol.* 63, 421−449.

Specialized Units

S. P. Cramer and E. I. Stiefel. 1985. "Chemistry and Biology of the Molybdenum Cofactor," in T. G. Spiro, ed., *Molybdenum Enzymes.* Wiley, New York, 411−442.

D. Dolphin, ed. 1978−1979. *The Porphyrins,* Volumes 1−7. Academic Press, New York.

D. Dolphin. 1982. B_{12}, Volumes 1−2. Wiley, New York.

D. Dunaway-Mariano and W. W. Cleland. 1980. "Preparation and Properties of Chromium(III) Adenosine 5′-Triphosphate, Chromium(III) Adenosine 5′-Diphosphate, and Related Chromium(III) Complexes." *Biochemistry* 19, 1496−1505.

J. Kim and D. C. Rees. 1992. "Structural Models for the Metal Centers in the Nitrogenase Molybenum-Iron Protein." *Science* 257, 1677−1682.

S. J. Lippard. 1993. "Bioinorganic Chemistry: A Maturing Frontier." *Science* 261, 699−700.

A. M. Stolzenberg, S. H. Strauss, and R. H. Holm. 1981. "Iron(II,III)-Chlorin and− Isobacteriochlorin Complexes. Models of the Heme Prosthetic Groups in Nitrate and Sulfite Reductases: Means of Formation and Spectroscopic and Redox Properties." *J. Am. Chem. Soc.* 103, 4763−4778.

CHAPTER 6

Control and Utilization of Metal-Ion Concentration in Cells

Principles: *Metal-ion concentrations must be maintained within certain ranges in cells. Metal-ion toxicity is most often associated with binding to inappropriate sites and, sometimes, subsequent reaction chemistry. Metal-ion homeostasis and detoxification are achieved by a variety of mechanisms, including extracellular metal carriers, metal-mediated protein structural changes that control passive transport through the cell membrane, ion channels and pumps, and metal-regulated transcription and translation. Changes in pH and redox state are used by the cell to effect the binding and release of metal ions to their receptor molecules. Ion-concentration gradients are used to store and transmit energy and information.*

6.1. Beneficial and Toxic Effects of Metal Ions

As is true for many biochemicals, metal-ion concentrations must be maintained within proper ranges. If the concentration of a given essential metal ion is too low, processes that need to use that ion will be adversely affected, and the organism can suffer from metal-ion deficiency. Once the concentration of a given metal ion is above a lower threshold, there will be enough of that ion to fulfill its biological functions. The concentration cannot be increased indefinitely without adverse consequences, however. Above an upper threshold, the effects of metal-ion toxicity will arise. For example, if a metal ion binds at an inappropriate site, it might compete with other beneficial metal ions for that site; or there might be undesirable reactivity of the metal ion when it is not properly controlled in its normal binding sites. This discussion assumes that the given metal ion has some biological function. Some metal ions have no known or presumed biological function; when present in cells, they may be rather innocuous or quite toxic. A graphical representation of these phenomena is presented in Figure 6.1.

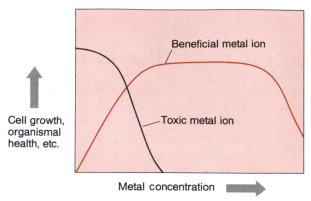

Figure 6.1

Representation of the concentration dependence of the toxic and beneficial effects of metal ions.

Homeostasis, the maintenance of the concentration of beneficial metal ions in the correct range, and detoxification, the removal of toxic concentrations of nonbeneficial metal ions, require a balance between the processes of metal-ion uptake, utilization, storage, and excretion. As noted above, metal ions not utilized in biological systems can be quite toxic, often because they tend to bind nonspecifically, but with high affinity, to certain types of sites. Because of this tight binding, which is often a consequence of kinetic inertness, these metals may bind to sites where they inhibit some normal process in such a manner that they are not easily removed and excreted. Other possible causes of metal-ion toxicity include the formation of insoluble salts in biological fluids, participation in hydrolytic reactions that degrade biopolymers, or redox chemistry that produces damaging byproducts, such as the hydroxyl radical.

These effects can be illustrated with two examples. Consider first iron, which is an essential metal ion for all organisms, including humans. In humans it is important for dioxygen transport and metabolism (Section 11.1), and it participates in a variety of electron-transport pathways (Section 9.1), among other functions. Iron deficiency resulting from diminished supply or uptake of the metal, or from its loss, produces anemia because of inadequate quantities required for hemoglobin synthesis. On the other hand, iron overload can also occur by accidental ingestion (iron is a common household poison) and through the treatment of diseases such as thalassemia, incorrect hemoglobin formation, with extensive blood transfusions. The excess iron can accumulate and is not easily excreted. Once the iron-storage mechanisms are saturated (Section 6.2.d.), excess amounts of the metal are released into the system, where they can catalyze the formation of various oxygen-based free radicals and extensively damage tissues. To avoid these effects, drugs such as desferrioxamine (Figure 3.24) are given to chelate the iron (Figure 6.2) and facilitate its excretion.

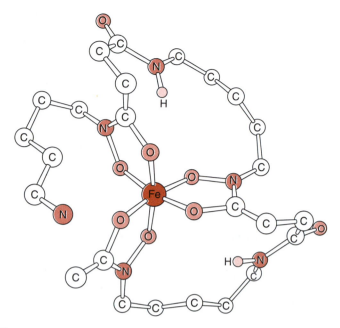

Figure 6.2
Iron complex of desferrioxamine, a drug given to chelate iron
and facilitate its removal from the body.

The second example is provided by platinum-based antitumor drugs such as cisplatin, shown in Figure 1.8. Platinum(II) complexes are rather inert to substitution and, in general, quite toxic. These properties are used to advantage in the chemotherapeutic applications of cisplatin. Its slow ligand substitution rate allows the drug to remain relatively unreactive in the presence of the high chloride-ion concentration (0.1 M) in the serum. As discussed in the previous chapter, substitution reactions of the chloride ions do occur once the platinum atom diffuses into the cell, and the resulting cationic aqua complexes eventually form stable adducts with DNA (Sections 7.3 and 8.3). These adducts are harmful, but more so to tumor than to normal tissue, leading to beneficial effects at the organismal level. By contrast, the corresponding palladium(II) compound, cis-[Pd(NH$_3$)$_2$Cl$_2$], undergoes ligand-substitution reactions at a rate 10^5 times more rapidly than cisplatin and is generally too toxic to be used in cancer chemotherapy. Presumably, the palladium complex reacts rapidly with many biological targets that it can damage or inactivate before irreversibly damaging DNA in the nucleus of cells.

6.2. A Beneficial Metal Under Tight Regulation: Iron

We focus here on the systems involved in the regulation of iron mobilization and storage, specifically the mammalian iron-transport protein transferrin and the storage protein ferritin. These systems illustrate the chemical and biological problems that must be solved in order to have effective metal-ion homeostatic control and introduce the emerging area of metal-ion effects on gene expression.

6.2.a. Solubilization, Uptake and Transport.

As discussed in Chapter 5, the hydrolytic chemistry of iron makes the levels of this element particularly challenging to regulate within organisms. Most importantly, iron(III) is quite insoluble at neutral pH. As a consequence, even though iron may be abundant in the environment, very low levels of the hydrated ion will be present in solution, and very little is available for uptake. The use of siderophores to sequester and facilitate iron transport into bacterial cells was described previously. We now turn to the mobilization of iron, insofar as it is understood, in mammalian systems.

In mammals, iron absorbed from food must be transported throughout the body. This function is accomplished by proteins of the transferrin family, glycoproteins with molecular masses of approximately 80 kilodaltons that tightly bind to Fe(III) ions. These proteins allow solubilization and transport of iron under conditions where the free ion would be hopelessly insoluble. The proteins bind Fe(III) with $K_{apparent}$ of approximately 10^{20} M^{-1}. Interestingly, these proteins also bind an anion, physiologically available carbonate ion, in a manner that is synergistic with the binding of iron; that is, the binding of carbonate facilitates iron binding and *vice versa*. This synergism is probably most important for iron release within cells.

The X-ray crystal structures of several proteins in the transferrin family have now been determined. As had been known from a variety of earlier studies, the proteins consist of two similar domains, each of which binds one iron atom. The structure is illustrated in Figure 6.3. Each domain, in turn, consists of two subdomains that bend like a hinge when bound to iron; that is, the subdomains clamp down on the iron-carbonate unit. This phenomenon is illustrated for human lactoferrin, a member of this protein family found in milk. The apo and iron-bound forms of one domain of lactoferrin are shown in Figure 6.4.

The distorted octahedral iron coordination sphere consists of two phenolate oxygens from tyrosine, one imidazole nitrogen from histidine, one carboxylate oxygen from aspartic acid, and two oxygens from the bound carbonate. The carbonate is also hydrogen-bonded to side chains from arginine and threonine and to two peptide NH groups. This structure is de-

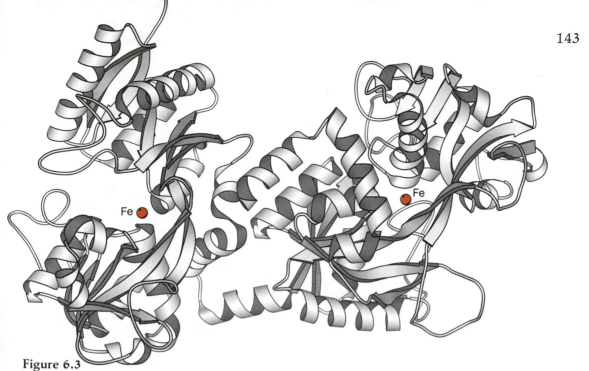

Figure 6.3
Structure of the iron-bound form of a transferrin, lactoferrin, as determined by X-ray crystallography.

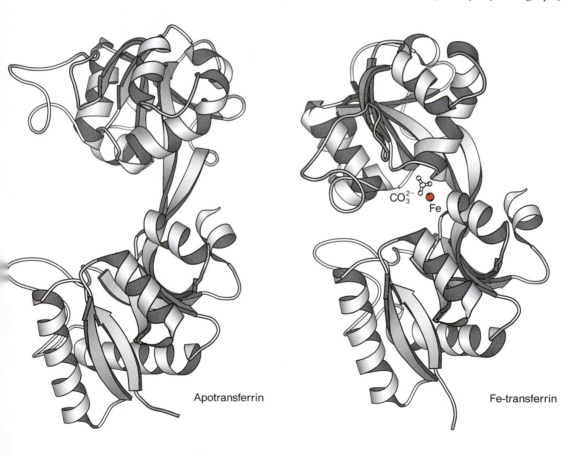

Apotransferrin

CO_3^{2-} Fe

Fe-transferrin

Figure 6.4
Comparison of the apo (left) and iron-bound (right) forms of lactoferrin, showing a hinge motion associated with iron/carbonate complexation.

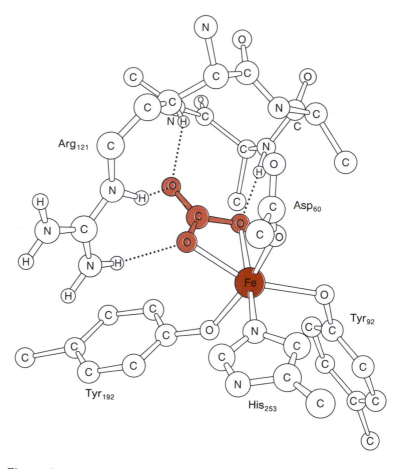

Figure 6.5
Active site structure of lactoferrin, showing the coordination of the iron
and the bound carbonate ion.

picted in Figure 6.5. The synergistic binding of the carbonate and iron atoms
can be explained by these geometric revelations. The protein apparently has
a binding pocket for carbonate that consists of the hydrogen-bonding units
contributed by side chains and the polypeptide backbone noted above. Once
carbonate binds in this site, the metal-binding position is fully organized,
so that iron(III) can now coordinate in an octahedral, oxygen-rich environment of ligands favorable for such a relatively hard metal ion.

Much like enterobactin, transferrin also binds to a membrane surface
receptor for uptake into cells. This process is one example of the general
phenomenon of *receptor-mediated endocytosis*. The transferrin receptor is a
dimeric glycoprotein that binds iron-containing but not apotransferrin. The
large conformational changes in transferrin that occur upon binding carbonate and iron noted above are essential to this differentiation between loaded
and unloaded transferrin. Once the receptor has bound transferrin, a portion

of the membrane containing the receptor pinches off to form a so-called "coated" vesicle. The vesicle coat is a protein called clathrin, which facilitates vesicle formation. Inside the cell, this vesicle uncoats to form an endosome. The membrane of the endosome contains ATP-driven proton pumps (Section 6.4.d). These molecules pump protons into the endosome to lower the pH to between 5 and 6. At this low pH, the iron is released as a result of protonation of the carbonate and tyrosinate ligands. The released iron is then available for use or for storage in ferritin. A portion of the vesicle containing the receptors that have apotransferrin still bound then fuses with the plasma membrane. Upon being reexposed to the outside of the cell with pH near 7.4, the apotransferrin is released, and the process can repeat itself. This scheme is summarized in Figure 6.6. The whole cycle appears to take about 15 minutes to complete under physiological conditions. Note that differences in pH are used to control both iron binding to transferrin and apotransferrin binding to the transferrin receptor.

6.2.b. Metalloregulation of Uptake and Storage.

Organisms have also developed methods for accumulating and storing metal ions for later use. The best-characterized metal-ion storage system is ferritin. The structure of ferritin and the formation and constitution of its polyiron core were previously introduced in Chapter 5. Ferritin serves as the repository for iron transported into the cell by transferrin and can release large quantities of the metal ion when required to do so. The molecular details of how the entry and departure of iron from ferritin are modulated by cells are not yet completely understood. Much progress has recently been made, however, on the mechanisms by which iron controls the genes that produce the proteins required for siderophore production as well as for the synthesis of the transferrin receptor and ferritin apoprotein.

The genes involved in siderophore biosynthesis in bacteria are regulated by iron at the transcriptional level. A single protein called Fur (iron uptake regulator) regulates a large number of genes in a coordinative manner. This protein is a dimer, with subunits of molecular mass 17 kilodaltons, that appears to bind to specific segments of DNA in the presence of divalent metal ions such as Fe^{2+} but not in their absence. Thus, under conditions where iron concentrations are reasonably high, the Fur protein contains bound metal ions, binds to specific DNA regulatory sequences, and represses the transcription of relevant genes. At lower metal-ion concentrations, where the Fur protein does not bind metal ions and does not bind DNA specifically, gene transcription can occur.

The expression of ferritin and the transferrin receptor in mammals is also regulated in a metal-dependent manner, but protein expression is regulated at the translational level. Stored ferritin mRNA is translated relatively slowly if cellular iron levels are low. If the cellular iron level increases, however,

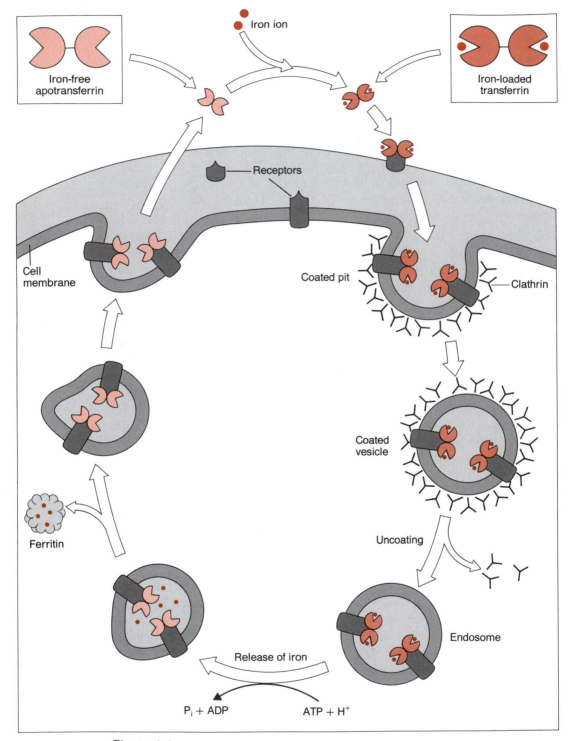

Figure 6.6
Scheme depicting the extracellular binding of iron to apotransferrin,
receptor-mediated endocytosis, ATP-driven release of iron into the endosome,
and loading of the metal ion into ferritin.

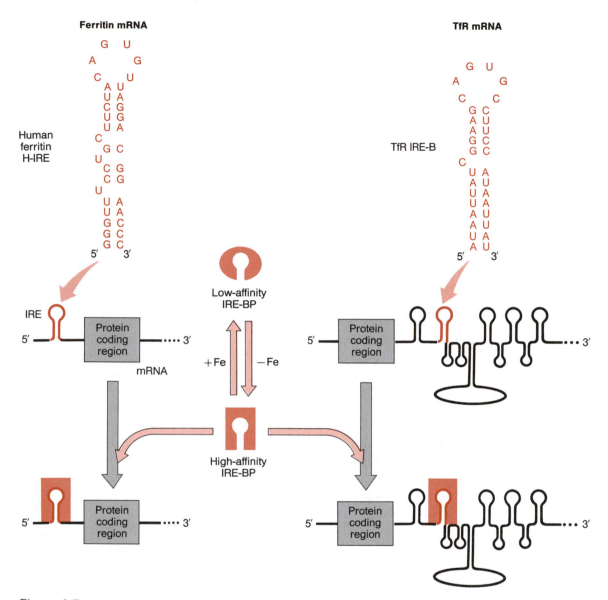

Figure 6.7
Regulation of ferritin (Ft) and transferrin receptor (TfR) expression by the
iron-responsive element binding protein (IRE-BP). At high iron levels, the IRE-BP
has bound iron and a low affinity for IREs, which facilitates Ft mRNA translation
and TfR mRNA degradation. At lower iron levels, the affinity of the IRE-BP for
IREs is increased, blocking Ft mRNA translation and stabilizing TfR mRNA.

the rate of ferritin mRNA translation increases dramatically. Conversely, the
mRNA for the transferrin receptor is relatively unstable under iron-rich conditions, but it becomes more stable under iron-poor conditions. These two
phenomena are mediated by similar sequences/structures in the mRNA molecules, structures that have been termed iron-responsive elements or IREs.
These IREs form stem-loop structures shown in Figure 6.7. An IRE occurs at

147

the 5′ end of ferritin mRNA, whereas several IREs are found in the opposite (that is, 3′) end of the transferrin receptor mRNAs. The IREs act as binding sites for a common protein called the *IRE-binding protein*, which has a molecular mass of around 90 kilodaltons. When the IRE-binding protein binds to the ferritin mRNA, it prevents translation of the message into protein; when it binds to the transferrin receptor mRNA, it inhibits degradation of the RNA but allows translation. The affinity of this protein for IRE-containing RNAs is regulated by iron. When isolated from cells treated with an iron chelator, the IRE-binding protein binds such RNAs with a dissociation constant near 10^{-11} M. In contrast, when isolated from cells treated with iron, the dissociation constant is closer to 5×10^{-9} M.

This protein has recently been purified, and a cDNA encoding it has been cloned and sequenced. Analysis of the deduced amino-acid sequence revealed a remarkable result. The IRE-binding protein is homologous to the mitochondrial enzyme aconitase, which contains an iron-sulfur cluster (Sections 5.3.a, 9.1.a, and 12.2.b). Aconitase undergoes interconversion between Fe_3S_4 and Fe_4S_4 cluster units under appropriate conditions. Further studies revealed that the IRE-binding protein itself has aconitase activity and is, in fact, equivalent to cytosolic aconitase. Although appreciation of the full functional and evolutionary significance of these observations must await further study, the results immediately suggest a model for understanding the ability of the IRE-binding protein to respond to changes in the iron concentrations in cells. At high levels of iron, the cluster in the protein has the Fe_4S_4 form and a relatively low affinity for IRE-containing RNA molecules. Under low iron conditions, the cluster is dissociated. This change increases the affinity for RNA. The Fur protein, acting at the transcriptional level, and the IRE-binding protein, acting at the translational level, are but early examples of the proteins that regulate gene expression in response to changes in metal-ion concentration.

6.3. An Example of a Toxic Metal: Mercury

We have just examined some of the strategies used by nature to regulate the concentration of the biologically essential element iron. A variety of similar mechanisms has also evolved for the effective removal of toxic elements. One of the best understood detoxification processes at present is the plasmid-encoded mercury detoxification system from bacteria. Remarkably, this system also manages aspects of the uptake and transport of mercury, as well as its detoxification, which has the effect of cleaning up the external environment of the bacterium. Another fascinating feature of the system is that the expression of genes that encode for proteins involved in detoxification is also regulated by metal-ion concentration.

6.3.a. The Enzymes Involved in Mercury Detoxification.   Mercuric ion is highly toxic to many organisms, partly because of its extremely high affinity for thiols. This interaction and the concomitant stability of the adducts formed presumably inactivate essential thiols in a variety of proteins and enzymes. Furthermore, normal metabolic processes can inadvertently produce species such as organomercurials that are even more toxic than mercury(II) itself. In order to detoxify such species effectively, the bacterial system utilizes two enzymes, organomercurial lyase and mercuric ion reductase. These enzymes catalyze Reactions 6.1 and 6.2, respectively. In addition, a typical operon that encodes these proteins includes genes that encode a

$$RHgX + H^+ + X^- \rightleftharpoons RH + HgX_2 \tag{6.1}$$

$$Hg(SR)_2 + NADPH + H^+ \rightleftharpoons Hg(0) + NADP^+ + 2RSH \tag{6.2}$$

periplasmic binding protein and a membrane-associated transport protein. The binding protein picks up mercury compounds from the environment surrounding the organism and delivers them to the transport protein, which then carries them into the cytoplasm, where the actual detoxification takes place. The operon also includes a transcriptional regulatory protein, MerR, that controls its own transcription as well as that of the other genes required for response to elevated mercury levels. We shall first consider the mechanism of detoxification and then the metalloregulation.

Most metal-carbon bonds are quite readily hydrolyzed. In contrast, mercury-carbon bonds are generally quite stable and are not easily cleaved. Even at the modest turnover rate of 1–100 mol per minute per mol of enzyme, organomercurial lyase accelerates these hydrolysis rates by approximately one million-fold. This 22-kilodalton enzyme effects this catalysis without cofactors or metal ions other than the substrate. A mechanism for its hydrolytic reaction has been developed based on the manner in which substrate structure influences product formation. The results of such experiments revealed essentially complete maintenance of the structural and stereochemical integrity of substrates upon replacement of the mercury by a proton, ruling out radicals and other intermediates. A concerted S_E2 mechanism,

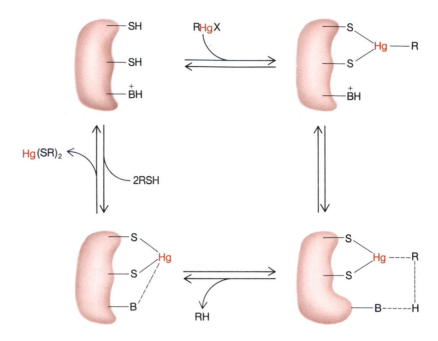

Figure 6.8
Postulated mechanism for cleavage of Hg-C bonds at the active site of organomercurial lyase.

shown in Figure 6.8, is favored. Note that the active site of this enzyme, like those of essentially all the other proteins in this system, consists chiefly of cysteine sulfhydryl residues as ligands for mercury.

The product of organomercurial lyase is a mercury thiolate adduct, $Hg(SR)_2$, produced by exchange of endogenous thiols such as glutathione for the cysteinate ligands of the enzyme. This complex, in turn, is a substrate for mercuric-ion reductase, a dimeric enzyme having a subunit molecular mass of approximately 60 kilodaltons and containing one flavin per subunit. Like organomercurial lyase, the enzyme also contains several essential cysteine residues, some of which are involved in binding the Hg(II) substrate. Use of cysteine thiolates to coordinate the mercuric ion, however, presents a chemical dilemma. Reduction of free Hg(II) by NADPH is highly favorable thermodynamically. When linked to two thiolate ligands, however, the redox potential of the mercuric ion is dramatically reduced, so that the driving force for reduction is nearly zero or, perhaps, even unfavorable. Thus, a compromise had to be struck in designing the active site in order to achieve enough affinity for binding Hg(II) to remove this ion from other thiolate ligands in solution, but at the same time affording the appropriate redox potential. A recent crystal-structure analysis of the Cd(II) complex of a mercuric-ion reductase suggests that the enzyme might accomplish this goal by using accessory nonthiolate ligands for substrate binding. Once the mercury ion is bound, it is reduced to Hg(0). This process appears to involve hydride transfer from NADPH to flavin, followed by attack of the reduced flavin at one of the Hg(II)-coordinated thiolates. Reduction of the mercury then ensues, with concomitant formation of a flavin-thiolate adduct. Mercury-atom release and binding of the next Hg(II) with accompanying carbon-sulfur bond cleavage completes the cycle. It should be noted that, once mercury atoms are produced, they simply diffuse through the cell membrane and out into the environment. The ability of mercuric-ion reductase to tune its active center to achieve the required balance between substrate-binding affinity and reduction potential is an early example of the general principles discussed in Chapter 12.

6.3.b. Metalloregulation of the Mercury Detoxification Genes.
The foregoing machinery of active mercury uptake and binding is required only when mercury levels are relatively high. Under other conditions, the synthesis of these proteins would waste energy and precursors. In addition, the system might take up from the environment other ions that could be potentially harmful. These concerns would be obviated if the system were expressed or activated only in the presence of threatening concentrations of mercury compounds. This objective is achieved via the MerR protein, an intracellular mercury sensor that transcriptionally controls the expression of the mercury detoxification genes.

MerR is a dimeric DNA-binding protein. It binds site specifically to DNA in the presence or absence of mercury with only subtle differences. In the absence of bound mercury, it represses the transcription of the mer genes. Once mercury is bound, it activates transcription of these genes. This effect is mediated through changes in the structure of the protein-DNA complex. Binding of mercury to the protein results in an unwinding of the DNA by 33 degrees. Thus, mercury acts on the MerR-DNA complex and, in so doing, regulates the ability of RNA polymerase to initiate transcription. Quantitative examination of the effects of mercury on transcription reveals a remarkable result; the system shows ultrasensitivity, in that the effect on transcription is more sharply dependent on mercury concentration than can be accounted for by the simple binding of a single mercury ion to the protein-DNA complex. This phenomenon is illustrated in Figure 6.9. Once the mercury concentration inside the cell reaches a threshold limit, the detoxification-apparatus genes are rapidly transcribed. The concentration required to produce half the maximal rate of transcription is approximately 10 nM. Other metal ions such as Cd(II), Au(I), Zn(II), and Au(III) will also induce

152

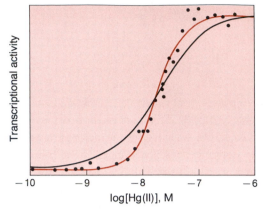

Figure 6.9
Dependence of the rate of transcription from a MerR-dependent promotor on mercury concentration. The dependence is steeper than that expected for a single Hg(II) ion binding to the MerR protein. (Adapted from D. M. Ralston and T. V. O'Halloran, *Proceedings of the National Academy of Sciences USA* **87**, 3846–3850 (1990)).

transcriptional activation, but require two to three orders of magnitude higher concentrations.

The structure of the MerR-Hg(II) complex has been probed by a variety of methods. Direct-binding experiments revealed that one mercuric ion binds per dimer of MerR. Spectroscopic and mutagenesis studies suggest that each mercury is bound by three thiolates, probably one from one subunit and two from another. The binding of the mercury ion at the subunit interface probably induces a change in the dimer conformation that is transmitted to the DNA. Further work on the structural characterization of this fascinating protein and its complexes with DNA should reveal its mechanism of action in greater detail. In addition, as more systems such as this and the Fur protein discussed above are characterized, it should be possible to reach some general conclusions about how metal ions control gene expression and, hence, influence metal-ion homeostasis.

6.4. The Generation and Uses of Metal-Ion-Concentration Gradients

Gradients of ions across membranes are used to store energy and information. The ions of particular importance are sodium, potassium, calcium, and hydrogen, and their ionic gradients are of fundamental significance in bioenergetics and in the function of nervous systems. Two types of proteins are essential for the functioning of these systems, namely, those involved in pumps and those comprising channels. Pumps function to couple the transport of ions across membranes with concomitant interconversion of ATP and ADP. Since these proteins catalytically hydrolyze ATP or synthesize ATP from ADP and inorganic phosphate, they are alternatively referred to as ATPases. They can generate and maintain ionic gradients or use existing ionic gradients to synthesize ATP. Channels transport ions across membranes, driven by ionic gradients. They differ from pumps in that the transport of ions is not directly coupled to ATP hydrolysis or synthesis. Channels are not simply passive ion-transport systems, however, because they are *gated*: their ability to transport ions is affected by other conditions, either the membrane potential or the concentration of exogenous ligands such as neurotransmitters. Gating allows channels to transduce information from one form to another. Techniques that allow observation of the behavior of single molecules in channels have been developed. The ability to monitor these systems at such a high level of resolution presents a unique opportunity to the chemist. Before discussing specific examples of pumps and channels in detail, we shall present several concepts that are useful in understanding these processes.

The first concept is that of *membrane potential*, the difference in potential between the two sides of a membrane. The membrane potential can be considered as the sum of the potentials due to each ion as determined by the differences in ion concentrations across the membrane. Because of such differences, these equilibrium potentials can be considered as the values necessary to produce the net driving force for transport of a particular ion. The concentration differences determine the potentials according to the Nernst equation (Equation 6.3), where R = gas constant, T = absolute temperature, z = charge on the ion, F = Faraday's constant, c_{out} is the ion concentration

$$V = (RT/zF) \ln(c^{out}/c^{in}) \tag{6.3}$$

outside the cell, and c_{in} is the concentration inside the cell. The ionic compositions inside and outside a typical mammalian cell are given in Table 6.1. Thus, the equilibrium potential for Na^+ is $((8.314 \text{ J K}^{-1} \text{ mol}^{-1})(310 \text{ K}))/(+1)(96,485 \text{ C mol}^{-1}) \ln(145 \text{ mM}/12 \text{ mM}) = +68 \text{ mV}$. Similarly, for K^+ the equilibrium potential is -99 mV.

The second concept is that of *conductance*. Channels do not allow transport of all ions with equal efficiency. Instead, they show varying degrees of ion selectivity, ranging from the simplest differentiation between cations and anions to a high degree of selectivity for a particular ion such as Na^+

Table 6.1
Free ionic concentrations and equilibrium potentials for mammalian skeletal muscle

Ion	Extracellular concentration (mM)	Intracellular concentration (mM)	Ratios of $[ion]_o$ to $[ion]_i$	Equilibrium potential (mV)
Na^+	145	12	12	$+68$
K^+	4	155	0.026	-99
Ca^{2+}	1.5	$< 10^{-7}$ M	$> 15,000$	$> +128$
Cl^-	123	4.2	30	-90

with the strong discrimination against even quite similar ions. This ability to discriminate between closely related ions is of great importance for function, although understanding the details of its chemical basis remains an important challenge for the future. The voltage dependence of the current conducted by a channel is governed by Ohm's law, Equation 6.4, where i is the current, g is the conductance, and V is the voltage or net difference in electrochemical

$$i = g \times V \tag{6.4}$$

potential across the membrane. The conductance, g, is in units of reciprocal ohms or siemens, S. As we shall see in Section 6.4.d, the conductance of an ion channel is a molecular property.

6.4.a. Generation of Ionic Gradients. The generation of ionic gradients is an energy-dependent process, where the energy utilized is supplied by the hydrolysis of ATP to ADP and inorganic phosphate. The mechanism most responsible for generation of the Na^+ and K^+ gradients noted above is the enzyme Na^+-K^+ ATPase. This intrinsic membrane protein consists of two components, a 100-kilodalton catalytic subunit and a 45-kilodalton associated glycoprotein, organized into an $\alpha_2\beta_2$ tetramer. The enzyme catalyzes the process shown in Equation 6.5. Several points about this reaction deserve

$$3Na^+_{in} + 2K^+_{out} + ATP + H_2O \rightleftharpoons 3Na^+_{out} + 2K^+_{in} + ADP + P_i \tag{6.5}$$

comment. First, it involves the net transport of charge across the membrane; three sodium ions are transported out for every two potassium ions pumped in. Second, the hydrolysis of ATP is tightly coupled to the ion-transport processes. This coupling is crucial for the function of the enzyme, since uncoupling could result in unproductive hydrolysis of ATP. Even with essen-

OUTSIDE OF CELL

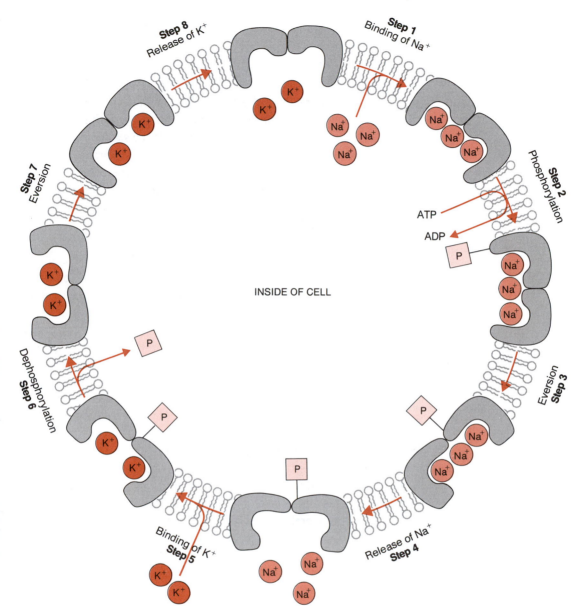

Figure 6.10
Scheme depicting the steps in the ATP-dependent export of three sodium ions and the concomitant import of two potassium ions by the Na$^+$-K$^+$ ATPase.

tially perfect coupling, it is estimated that a typical mammalian cell consumes 30 percent of its ATP through this enzyme.

The detailed molecular structure and mechanism of the Na$^+$-K$^+$ ATPase are not yet known. An outline of the mechanism is shown in Figure 6.10.

Sodium binding to the inner side of the membrane facilitates phosphorylation of a specific aspartic-acid residue on the enzyme. This process then induces a conformational change that leads to transport of the three sodium ions across the membrane. Next, potassium binding on the outer membrane catalyzes dephosphorylation of the enzyme, followed by the inverse conformational change to transport the two potassium ions into the cell. The molecular details, including the coordination chemistry of the sodium and potassium ions, remain to be elucidated.

6.4.b. Ion Transport by Ion Channels.

The Na^+-K^+ ATPase can turn over at a rate of approximately 100 times per second. In contrast, ion channels are proteins that are capable of transporting 10^6–10^7 ions per second. This fast rate of ion transport is essential for rapid signaling events, such as the generation of the action potentials that are responsible for signal transmission in nerves. Channels can exist in a variety of conformational states, some that are "open" and capable of transporting ions, others that are "closed" or "inactivated" and not capable of ion transport. We shall discuss two ion channels in some detail: the acetylcholine receptor, a channel gated by a neurotransmitter, and the sodium channel, a voltage-gated channel.

6.4.c. The Acetylcholine Receptor.

The acetylcholine receptor is a large and complex molecule, the structure of which has not yet been determined in atomic detail. Nonetheless, studies using a wide variety of biochemical, biophysical, and molecular biological techniques have produced a coherent picture of the overall structure of the molecule. Furthermore, electrophysiological methods, including those that allow visualization of single acetylcholine receptor molecules in action, have produced much information on the mechanism of action of this protein. The biological role of the acetylcholine receptor is illustrated in Figure 6.11.

The acetylcholine receptor is a large oligomeric glycoprotein. It contains a total of five polypeptide chains of four different types with an overall stoichiometry of $\alpha_2\beta\gamma\delta$. The total molecular mass is approximately 300 kilodaltons. It is found primarily in muscle, brain, and specialized organs such as the electroplax from the electric fish torpedo. The amino-acid sequences of all subunits from both calf muscle and torpedo have been deduced from cDNA clones. Analysis clearly shows that the four subunits are homologous, related to a common ancestral sequence. The chains range from 437 to 501 amino acids, the latter occurring in torpedo. Each glycoprotein binds up to two molecules of the neurotransmitter acetylcholine, one on each of the α subunits.

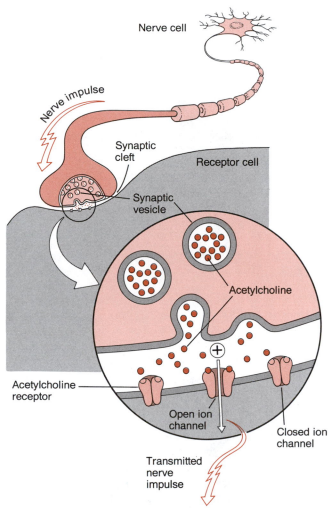

Nerve cell

Nerve impulse

Synaptic cleft

Receptor cell

Synaptic vesicle

Acetylcholine

Acetylcholine receptor

Open ion channel

Closed ion channel

Transmitted nerve impulse

Figure 6.11
Scheme depicting the release of acetylcholine from a nerve terminal leading to binding of acetylcholine to receptors in the opposing membrane and the concomitant ion channel opening.

The overall architecture of the receptor has been determined from electron microscopic studies. Schematic views of the structure are shown in Figure 6.12. The five subunits are arranged in a pentagonal array around a relatively open central core. A large domain extends to the extracellular side of the membrane, whereas a smaller one is present on the intracellular side. Analysis of the distribution of the hydrophobic residues in the amino-acid sequence of the subunits allows identification of the regions that probably span the membrane. The proposed schematic structure is shown in Figure 6.12. There are four stretches of 20–25 contiguous hydrophobic residues in

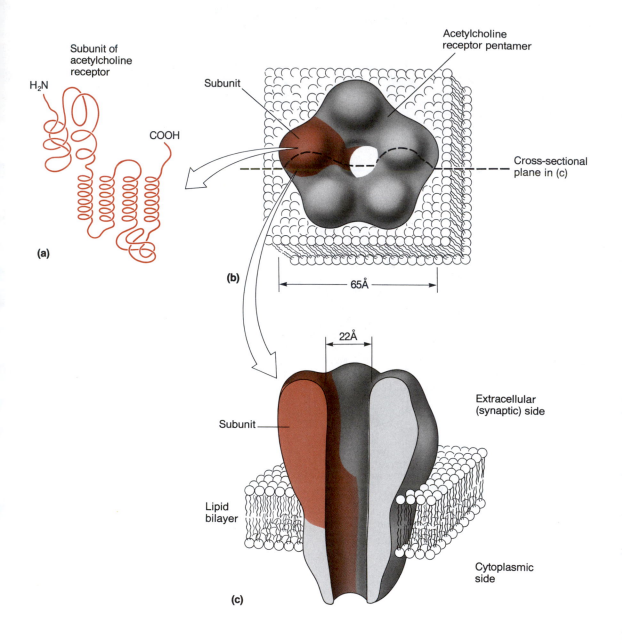

Subunit of acetylcholine receptor

H$_2$N

COOH

(a)

Acetylcholine receptor pentamer

Subunit

Cross-sectional plane in (c)

(b)

65Å

22Å

Extracellular (synaptic) side

Subunit

Lipid bilayer

Cytoplasmic side

(c)

Figure 6.12

Structure of the acetylcholine receptor. (a) A proposed schematic structure of one subunit based on analysis of the amino acid sequence. (b) A top view based on electron microscopic studies showing the pentameric arrangements of the five subunits. (c) A cross-sectional view revealing the membrane-spanning region that includes the ion channel.

corresponding positions in each subunit. These appear to be analogous to similar sequences in the bacterial photosynthetic reaction center, which have been shown by X-ray crystallographic studies to be membrane-spanning α-helices. These regions are of particular interest, for some of them must play a role in forming the channel in the membrane through which ions must pass. Site-directed mutagenesis studies, as we shall see, have revealed that the second of these putative transmembrane helices probably forms the inner lining of the channel.

Ions can be transported across membranes by molecules that act by one of two fundamentally different mechanisms. In one, an ion binds to a carrier, which then diffuses, translationally and/or rotationally, across the membrane, where the ion is released. In the other, a pore is formed across the membrane through which ions may pass without any large motions of the transporter. The acetylcholine receptor is an example of the latter class. Under normal circumstances, the channel is closed, and no significant number of ions passes through the membrane. In the presence of acetylcholine, however, the receptor molecule has a greatly increased probability of undergoing a transition to an open state through which ions may freely pass. The acetylcholine receptor is a nonspecific cation channel. Only positively charged ions may pass through, but there is little discrimination between various cations. Under physiological conditions, sodium ions are the major species that are transported. As noted above, ions may pass through such channels at quite rapid rates. The muscle acetylcholine receptor typically remains open for 10 msec per transition, but during this time, approximately 10^4 to 10^5 ions move across the membrane. This flux corresponds to a current of several picoamps.

Because of the large amplification factor resulting from the movement of 10^5 ions across the membrane per molecular conformational change, and because the signal produced is in an easily measurable form (current), it has been possible to develop a method to monitor individual channels in action. The method is called *patch-clamping* and is illustrated in Figure 6.13. A micropipette with a very smooth surface is moved up to the cell membrane. Slight suction produces a very tight seal of the pipette against the cell such that any current that flows between two electrodes, one inside the pipette and one inside the cell, must pass through the patch of membrane covered by the seal. The advent of suitably sensitive methods for measuring picoamp currents allowed single-ion channel currents to be recorded with this method in the presence of acetylcholine. Such a recording is shown in Figure 6.13. The jumps in current correspond to single acetylcholine receptor molecules undergoing conformational changes from closed to open states. These and older bulk methods that monitor current flowing through more than one channel at

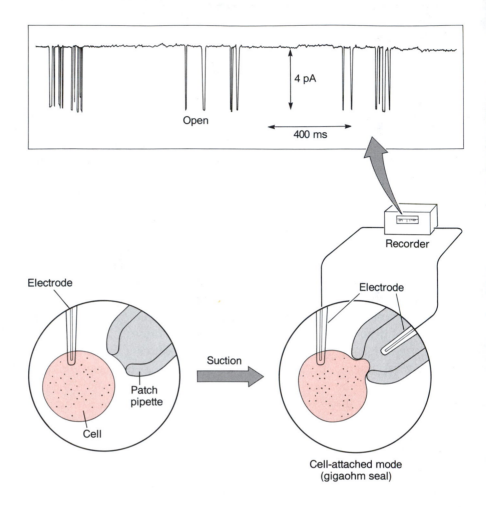

Figure 6.13
The patch-clamp method for measuring conductance through ion channels. The downward spikes in the trace correspond to the opening of single ion channel molecules.

a time allow detailed measurements to be made of the permeability of channels to various ions. As noted above, the acetylcholine channel is quite nonspecific in the ions that can pass through it. In addition to biologically relevant ions, cations as large as diethylammonium can pass through these channels nearly as well as sodium. From these and other considerations, it is clear that ions flow through the acetylcholine receptor channel in their solvated form without the necessity of removing tightly bound water molecules.

The behavior of a region of membrane containing a large number of acetylcholine receptors can be reproduced by adding together the traces from a number of individual channels or, alternatively, of one channel measured

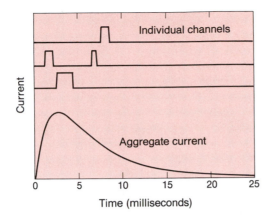

Figure 6.14
Scheme illustrating that the aggregate current through a membrane can be dissected into components corresponding to the opening and closing of individual ion channel molecules.

repeatedly over time. This aggregate current is illustrated in Figure 6.14 and reveals that the bulk behavior can be explained in terms of the properties of a large number of individual molecules acting independently. The ability to monitor single molecules allows direct investigation of questions that are normally only *gedanken* experiments in statistical thermodynamics. Moreover, these methods have allowed models for explaining the gating by acetylcholine to be developed and tested. One such scheme is shown in Figure 6.15. In the absence of high concentrations of acetylcholine, the channel is closed. Acetylcholine is then released from vesicles (Figure 6.11), causing a sudden increase in its concentration. One molecule of acetylcholine binds to the receptor, but the receptor remains closed. Upon binding of a second acetylcholine molecule, the probability of undergoing a conformational change to an open state increases significantly, so that open-channel records can now be seen. Eventually, the acetylcholine concentration is reduced by diffusion and hydrolysis by the enzyme acetylcholine esterase. In addition, upon prolonged exposure to acetylcholine, the receptor can undergo a transition to an inactivated state that is distinct from the other closed states. This mechanism predicts that channel openings should occur in bursts, followed by longer periods in which only one molecule of the neurotransmitter is bound. Behavior of this kind is observed.

These recording techniques have been combined with some powerful molecular-biological methods to allow the region of the receptor actually involved in channel formation to be probed. It is possible to express acetylcholine receptors by preparing messenger RNAs for the four subunits and microinjecting them into *Xenopus* oocytes (frog eggs). Patch-clamp studies reveal that functional receptors can be found on the surfaces of the oocytes

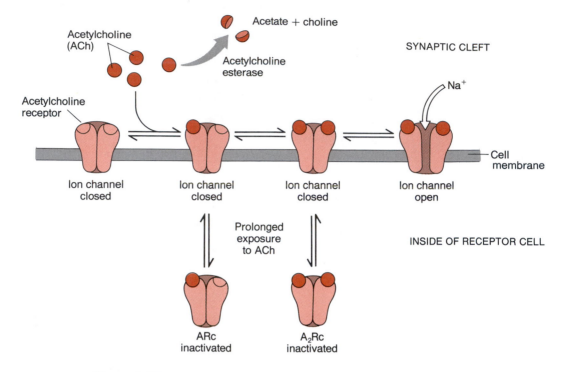

Figure 6.15

A scheme that rationalizes the dependence of acetylcholine receptor ion channel opening on acetylcholine concentration and on time.

and that these have properties indistinguishable from the receptor isolated from tissue. This expression system allows receptors, the amino-acid sequences of which have been modified by site-directed mutagenesis, to be investigated. In one study, serine residues that lie in equivalent positions just before the second transmembrane helix in the four subunits were individually mutated to alanine. Patch-clamp experiments revealed that the affinity of a compound that binds specifically to the open-channel conformation of the receptor decreased as the number of serines decreased, suggesting that these residues may form part of the channel. Further studies of this type will be crucial for testing more detailed models of the structure and functional properties of this important prototype for ion-channel-forming macromolecules.

6.4.d. The Voltage-Gated Sodium Channel. The acetylcholine receptor opens in response to a sudden increase in acetylcholine concentration, as discussed above. This process allows sodium ions to move through the channel, so that the membrane potential in the region near high concentration of the receptors changes from its resting value near the potassium equilibrium potential of ~ -100 mV toward the sodium equilibrium potential of

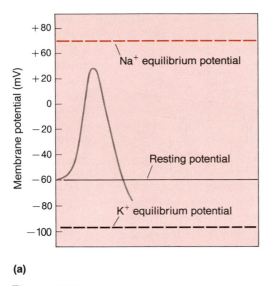

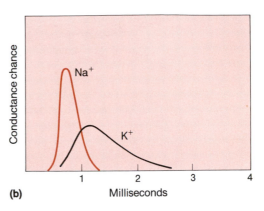

(a)　　**(b)**

Figure 6.16
An action potential (a) and its components (b) resulting from the conductance of sodium ions into the cell followed by conductance of potassium ions out of the cell.

$\sim +70$ mV. This local depolarization of the membrane causes an *action potential* to be produced. An action potential is a large depolarization of the membrane followed by a rapid repolarization, as shown in Figure 6.16. Movement of the potential down the membrane at a rate of 50 meters per second is responsible for nerve conduction. The action potential is largely due to the presence of voltage-sensitive sodium channels in the membrane. These channels are closed when the membrane potential is near the resting value, but they undergo transitions to an open state with increasing probability as the potential becomes less negative. Thus the opening of the acetylcholine receptors and the concomitant membrane polarization cause the sodium channels to open, which further depolarizes the membrane. As more sodium ions flow across the membrane, the depolarization spreads to nearby regions of membrane, causing sodium channels nearby to open. A pulse is produced since the sodium channels remain open for only a short time, a few milliseconds, and then undergo transitions to an inactivated closed state. The membrane then repolarizes with the aid of specific potassium channels that open in response to the initial depolarization but at a slower rate than the sodium channels. The propagation of an action potential is illustrated in Figure 6.17.

The voltage-sensitive sodium channel has also been purified, cloned, and found to consist of a major protein and two smaller accessory proteins. The large subunit has a molecular mass of ~ 270 kilodaltons. Analysis of the deduced amino-acid sequence revealed that it consists of four internally repeated domains, each of which has ~ 600 amino acids, including six putative

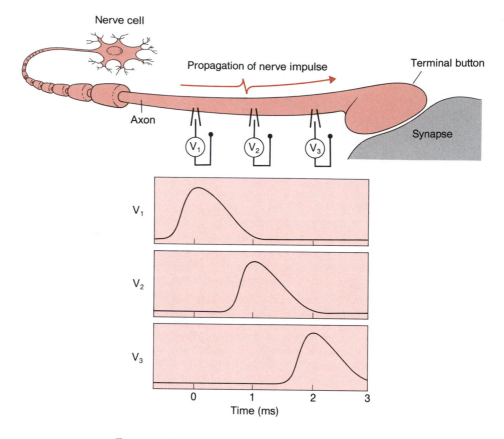

Figure 6.17
Propagation of an action potential along a nerve axon.

transmembrane helices. A schematic structure is shown in Figure 6.18. The internal repeats of this protein appear to be analogous in some sense to the subunits of the acetylcholine receptor. Little additional structural information is available about these channels; however, several lines of evidence bear on their ion-transport properties.

Unlike the acetylcholine receptor, the sodium channel is quite selective for sodium over other ions, including potassium. Examination of the permeability of a variety of ions through this channel strongly supports the theory that the selectivity for sodium over potassium results from the smaller size of the former, which enables it to pass through a narrow part of the channel. The permeability for sodium is 11 times greater than that for potassium. Other ions such as lithium and hydroxylammonium are readily transported through the channel. These observations are most easily reconciled by the hypothesis that the channel has a restriction that allows a sodium ion with one bound water to pass easily, but a potassium ion, with an ionic radius 0.3 Å larger, is much more strongly inhibited. The second line of evidence involves the pH

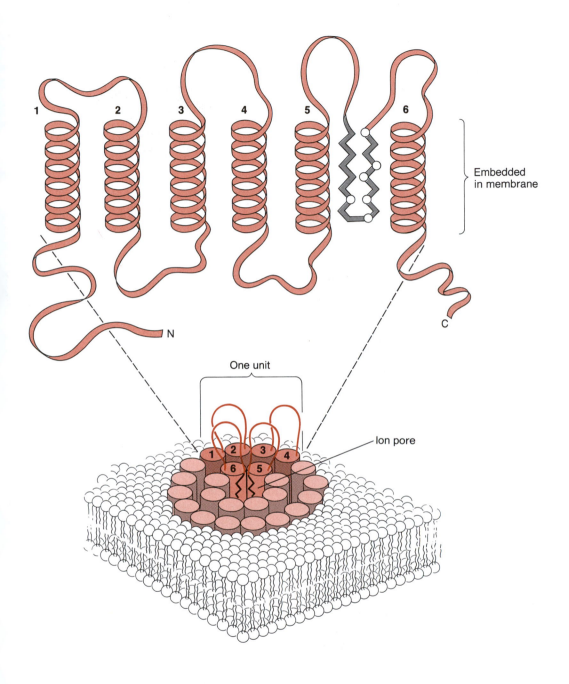

Figure 6.18
Proposed structure for a voltage-sensitive sodium channel. The molecule is
comprised of four similar units, each of which is proposed to include six
membrane-spanning helices and a beta strand structure that appears to form at least
part of the ion pore.

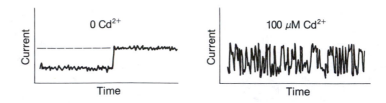

Figure 6.19
Flickering of the sodium channel from cardiac muscle as a result of transient blocking by cadmium ion. (Adapted from P. H. Backx et al., *Science* **257**, 248–251 (1992)).

dependence of the sodium-ion conductance. The conductance decreases as the pH is lowered, and the data can be fit with an apparent single-proton-binding constant, $pK_a = 5.5$. This result has been used to propose that a single carboxylic acid residue is close to, or interacting with, the ion as it passes through the rate-limiting region of the channel. Finally, the voltage dependence of the ion transport has been investigated by using purified sodium channels reconstituted into planar lipid bilayers. The conductance of a single channel is not dependent on the membrane potential; the current carried through each open channel depends on Ohm's law with a fixed conductance. Instead, the probability of a channel being open is strongly dependent on the potential. At a potential of -100 mV, a given channel spends less than 5 percent of its time in the open state. If the membrane is depolarized to -70 mV, each channel is open approximately 90 percent of the time. This behavior can be fit by assuming that n charges move across the membrane as the channel changes from a closed to an open conformation. For the reconstituted channels, this *gating charge* is estimated to be four electronic charges, compared to approximately six for the sodium channel in native membranes observed with patch-clamp methods.

Site-directed mutagenesis and expression have also been used to probe the structure and function of the sodium channel. One set of results reveals the exquisite details available via single-channel recordings. The sodium channel from cardiac muscle is transiently blocked by Cd^{2+} ions, resulting in flickering in the channel conductance, as shown in Figure 6.19. No such effect is seen for the corresponding channel from skeletal muscle. Examination of the amino-acid sequences of the channel proteins revealed the presence of a unique cysteine in the cardiac channel, located in the loop between the fifth and sixth putative membrane-spanning helices in the first of the internal repeats. This region had been implicated as being part of the pore itself in studies of other channels. A mutant of the skeletal-muscle channel was pro-

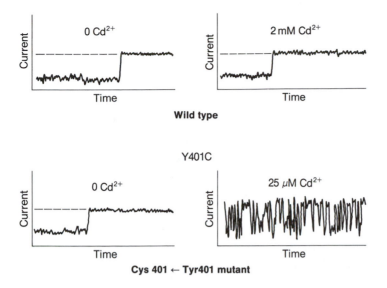

Figure 6.20
The absence of cadmium-induced flickering for wild-type sodium channels from skeletal muscle and the induction of such flickering by the introduction of a unique cysteine residue by site-directed mutagenesis. (Adapted from P. H. Backx et al., *Science* **257**, 248–251 (1992)).

duced in which a cysteine was substituted for the naturally occurring tyrosine residue that normally occupies the same position in this channel as does the cysteine in the cardiac muscle channel. Remarkably, this substitution produced a protein with virtually the same sensitivity to Cd^{2+} as the cardiac channel, as shown in Figure 6.20. Each of the interruptions in conductance, which last a few milliseconds, appears to correspond to binding of a Cd^{2+} ion to the cysteine residue, and perhaps other nearby groups. The physiological significance of these findings remains unclear, but they illustrate the rich bioinorganic chemistry of these fascinating and important molecules.

These examples represent several of the better-characterized ion-transport systems at present. Systems involved in signal transduction and storage in such diverse processes as vision, olfaction, and memory have now been at least partially characterized. In upcoming years, additional functional, molecular, and, hopefully, detailed three-dimensional structural characterization of these systems will provide deeper insights into the fundamental roles that gradients of inorganic ions, especially those of the group I and II metals, play in biology.

6.4.e. Synthetic Models for Ion Channels. Ion channels appear to be extremely complex and sophisticated molecules. Despite this complexity, a remarkably simple approach for generating active peptide-based models for ion channels has proved to be successful. A 21-amino-acid peptide with the sequence NH_2-(Leu-Ser-Ser-Leu-Leu-Ser-Leu)$_3$-CONH$_2$ was designed, with the goal of producing a molecule that would self-assemble into an aggregate of alpha-helical units that can span a membrane and, potentially, produce a pore that could transport ions. Specifically, molecular modeling suggested that this peptide could form a hexamer, with the hydrophobic leucine residues packed against one another and exposed to the outside, where they could contact with alkyl groups of the lipid bilayer, and an inner pore lined with the more hydrophilic serine residues. This structure is shown in Figure 6.21. The peptide was synthesized, purified, and incorporated into bilayer membranes in such a way that ion transport across the membrane could be readily detected. Step increases in conductance lasting a few milliseconds were observed that closely resembled the single-channel currents discussed above. Additional studies revealed that this behavior was the result of channel formation and conductance of cations such as potassium across the bilayer. No such behavior was observed for a 14-amino-acid peptide of similar design, presumably because it was not long enough to span the membrane. Studies of ion selectivity showed a pattern similar to that for the acetylcholine receptor with little discrimination among monovalent cations. In addition, the ion-channel model could transport organic cations as large as Tris, trihydroxymethylammonium ion, but not larger ones. These studies indicate that ion-channel activity can be produced with relatively simple, appropriately conceived peptides. They may offer a clue about the types of molecules that formed ion channels early in evolutionary history.

6.5. Tissue Selectivity in Metal Drug Distribution

Control of the distribution of metal ions or complexes within the body is also of importance for metal-based pharmaceuticals. Drugs or imaging agents must achieve useful concentrations within appropriate tissues with as little nonspecific accumulation in other tissues as possible. An important example that illustrates this point is that of radiometal-based imaging agents, of which those based on technetium are the most important. Specific complexes that accumulate in different tissues have been developed. For example, $[Tc(RNC)_6]^+$, shown in Figure 1.8, is now widely used as a heart-imaging agent, taking advantage of radioactivity emitted from the isotope ^{99m}Tc, which has a six-hour half-life. The basis for the tissue-specific accumulation of this and other technetium radiopharmaceuticals is not well understood at

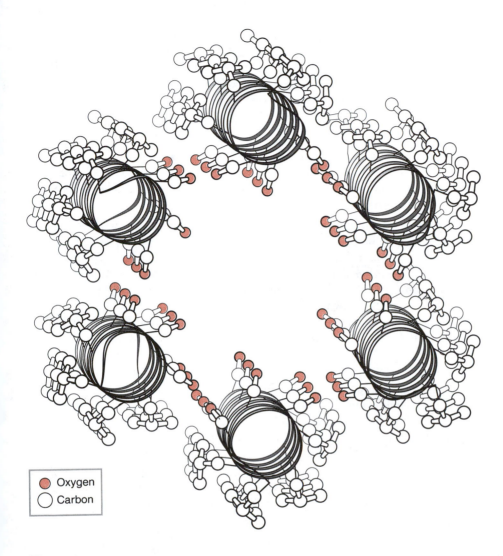

Figure 6.21
A proposed structure for ion channels formed by aggregation of 21-amino acid synthetic peptides into a hexamer.

present. Attempts are being made, however, to target radiometals in a rational manner to specific tissues or tumors by using radio-metal-labeled monoclonal antibodies and other elaborate ligands. In addition, with the development of magnetic resonance imaging methods, new reagents that allow targeting of highly paramagnetic metal complexes to specific tissues are in the process of being developed.

Study Problems

1. Suppose you wanted to design a biological transport protein for copper. Propose a reasonable active site, exogenous cofactor, and chemical strategy, different from those used for transferrin, to bind and release copper from the protein. Be sure to specify oxidation states.

2. Metallothionein in yeast binds copper tightly and is used to detoxify the organism in the presence of high concentrations of the metal ion. The copper is bound as Cu(I) in the form of a cluster with the approximate stoichiometry $Cu_6(Cys)_8$. The expression of the metallothionein gene is regulated at the transcriptional level by the concentration of copper. This regulation is accomplished via the action of a protein termed ACE1 (alternatively named CUP2). This protein also binds 6–7 Cu(I) ions in the form of a copper-cysteinate cluster. Propose several reasons why formation of such a polynuclear cluster might be advantageous for a selective copper-ion-concentration switch.

3. As noted in Chapter 5, there are very few known examples of biological systems that utilize Co(II), despite the fact that many synthetic Co(II) complexes exhibit very favorable reactivity properties. What difficulties would be encountered in developing systems for the transport of Co(II) and for regulation of Co(II) levels? For what other first-row transition metals would these problems also arise?

4. Calculate the Ca^{2+} ion-equilibrium potential for a cell with an intracellular Ca^{2+} concentration of 0.1 μM and an extracellular concentration of 1 mM. Application of a hormone to the surface of the cell results in the release of Ca^{2+} from internal stores, so that the intracellular concentration increases to 10 μM. How does this change affect the equilibrium potential?

5. A variety of neurotoxins act by binding tightly to ion-channel-forming molecules. For example, compounds such as tetrodotoxin from the puffer fish and saxitoxin from a marine dinoflagellate block voltage-sensitive sodium channels. From the structures below, speculate about what portions of these molecules are the most important for their activity.

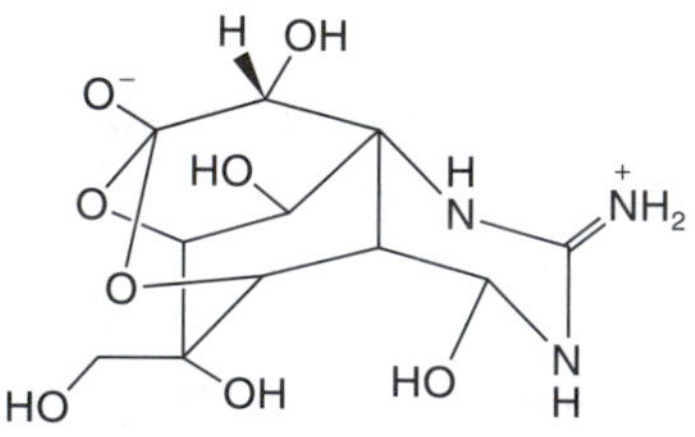

Tetrodotoxin

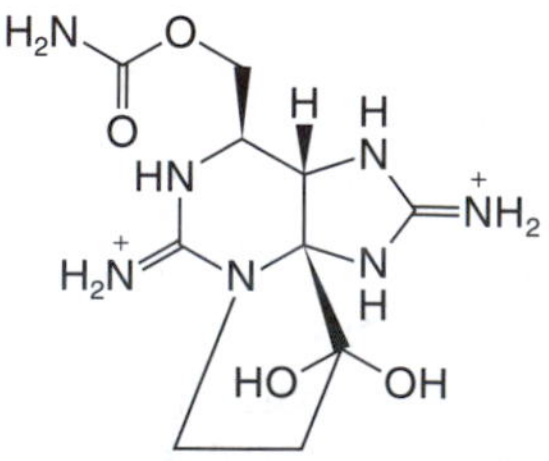

Saxitoxin

6. Many local anesthetics and related drugs act at ion channels. One such compound is QX-314:

Interestingly, if this compound is applied to a membrane containing voltage-sensitive sodium channels, little binding or blockage of the channel current is observed. If the membrane is briefly depolarized, however, the binding becomes much stronger, and the current is much more effectively blocked. Propose an explanation for these observations.

Bibliography

Beneficial and Toxic Effects of Metal Ions

P. S. Dobbin and R. C. Hider. 1990. "Iron Chelation Therapy." *Chem. Brit.* 26, 565–568.

D. H. Hamer and D. R. Winge, eds. 1989. *Metal-Ion Homeostasis: Molecular Biology and Chemistry: UCLA Symposia on Molecular and Cellular Biology,* Volume 98. Alan R. Liss, New York.

W. Mertz. 1981. "The Essential Trace Elements." *Science* 213, 1332–1338.

A Beneficial Metal Under Tight Regulation: Iron

E. N. Baker and P. F. Lindley. 1992. "New Perspectives on the Structure and Function of Transferrins." *J. Inorg. Biochem.* 47, 147–160.

S. H. Banyard, D. K. Stammers, and P. M. Harrison. 1978. "Electron Density Map of Apoferritin at 2.8 Å Resolution." *Nature* 271, 282–284.

J. L. Casey, D. M. Koeller, V. C. Ramin, R. D. Klausner, and J. B. Harford. 1989. "Iron Regulation of Transferrin Receptor mRNA Levels Requires Iron-Responsive Elements and a Rapid Turnover Determinant in the 3' Untranslated Region of the mRNA." *EMBO J.* 8, 3693–3699.

R. R. Crichton. 1991. *Inorganic Biochemistry of Iron Metabolism.* Ellis Horwood, New York.

R. R. Crichton and R. J. Ward. 1992. "Iron Metabolism: New Perspectives in View." *Biochemistry* 31, 11255–11264.

S. Kaptain, W. E. Downey, C. Tang, C. Philpott, D. Haile, D. G. Orloff, J. B. Harford, T. A. Rouault, and R. D. Klausner. 1991. "A Regulated RNA Binding Protein Also Possesses Aconitase Activity." *Proc. Natl. Acad. Sci. USA* 88, 10109–10113.

E. A. Leibold and H. N. Munro. 1988. "Cytoplasmic Protein Binds *in vitro* to a Highly Conserved Sequence in the 5' Untranslated Region of Ferritin Heavy- and Light-Subunit mRNAs." *Proc. Natl. Acad. Sci. USA* 85, 2171–2175.

E. W. Müllner, B. Neupert, and L. C. Kühn. 1989. "A Specific mRNA Binding Factor Regulates the Iron-Dependent Stability of Cytoplasmic Transferrin Receptor mRNA." *Cell* 58, 373–382.

T. A. Rouault, C. David Stout, S. Kaptain, J. B. Harford, and R. D. Klausner. 1991. "Structural Relationship Between an Iron-Regulated RNA Binding Protein (IRE-BP) and Aconitase: Functional Implications." *Cell* 64, 881–883.

P. Saltman and J. Hegenauer, eds. 1982. *The Biochemistry and Physiology of Iron*. Elsevier Biomedical, New York.

S. Welch. 1992. *Transferrin: The Iron Carrier*. CRC Press, New York.

An Example of a Toxic Metal: Mercury

N. L. Brown. 1985. "Bacterial Resistance to Mercury: Reductio ad Absurdum?" *TIBS* 10, 400–403.

M. J. Moore, M. D. Distefano, L. D. Zydowsky, R. T. Cummings, and C. T. Walsh. 1990. "Organomercurial Lyase and Mercuric Ion Reductase: Nature's Mercury Detoxification Catalysts." *Acc. Chem. Res.* 23, 301–308.

D. M. Ralston and T. V. O'Halloran. 1990. "Ultrasensitivity and Heavy-Metal Selectivity of the Allosterically Modulated MerR Transcription Complex." *Proc. Natl. Acad. Sci. USA* 87, 3846–3850.

J. G. Wright, M. J. Natan, F. M. MacDonnell, D. M. Ralston, and T. V. O'Halloran. 1990 "Mercury(II)-Thiolate Chemistry and the Mechanism of the Heavy Metal Biosensor MerR," in S. J. Lippard, ed., *Progress in Inorganic Chemistry*, Volume 38. Wiley-Interscience, New York, 323–412.

The Generation and Uses of Metal Ion Concentration Gradients

P. H. Backx, D. T. Yue, J. H. Lawrence, E. Marban, and G. F. Tomaselli. 1992. "Molecular Localization of an Ion-Binding Site Within the Pore of Mammalian Sodium Channels." *Science* 257, 248–251.

A. Brisson and P. N. T. Unwin. 1985. "Quaternary Structure of the Acetylcholine Receptor." *Nature* 315, 474–477.

B. Hille. 1992. *Ionic Channels of Excitable Membranes*, 2nd ed. Sinauer, Sunderland, MA.

J. D. Lear, Z. R. Wasserman, and W. F. DeGrado. 1988. "Synthetic Amphiphilic Peptide Models for Protein Ion Channels." *Science* 240, 1177–1181.

R. J. Miller. 1987. "Multiple Calcium Channels and Neuronal Function." *Science* 235, 46–52.

E. Neher. 1992. "Ion Channels for Communication Between and Within Cells." *Science* 256, 498–502.

E. Neher and B. Sakmann. 1992. "The Patch Clamp Technique." *Sci. Amer.* 266, 44–51.

B. Sakmann. 1992. "Elementary Steps in Synaptic Transmission Revealed by Currents Through Single Ion Channels." *Science* 256, 503–512.

J. C. Skou and J. G. Norby, eds. 1979. *Na⁺-K⁺ ATPase: Structure and Kinetics*. Academic Press, New York.

R. Y. Tsien and M. Poenie. 1986. "Fluorescence Ratio Imaging: A New Window into Intracellular Ionic Signalling." *TIBS* 11, 450–455.

Tissue Selectivity in Metal Drug Distribution

L. G. Marzilli, R. F. Dannals, and H. D. Burns. 1980. "Technetium-99m Radiopharmaceuticals: Problems, Potential Applications in Nuclear Medicine," in A. E. Martell, ed., *Inorganic Chemistry in Biology and Medicine*. ACS Symposium Series, Volume 140. American Chemical Society, Washington, DC, 91–102.

U. Mazzi, M. Nicolini, and G. Bandoli. 1991. "How a Technetium Complex Becomes a Radiopharmaceutical," in M. Nicolini and L. Sindellari, eds., *Lectures in Bioinorganic Chemistry*. Raven Press, New York, 91–102.

CHAPTER 7

Metal Ion Folding and Cross-Linking of Biomolecules

Principles: *Metals can act as templates for organizing three-dimensional biological structures. In binding to protein sites, metal ions discard most, if not all, of their coordinated water molecules. When interacting with nucleotides and nucleic acids, many of the water ligands remain and help organize the structure. The tertiary structures induced by metal binding can facilitate interactions between macromolecules. The formation of cross-links is important for the mechanism of action of some inorganic-element-based drugs. The large ionic radii of metal ions (versus the proton) and the ability to afford high coordination numbers are used to advantage.*

7.1. Metal-Ion Stabilization of Protein Structure

Although in many metalloproteins metal ions can facilitate a variety of chemical reactions (Chapters 9−11), in others, the metal appears to play a purely structural role. Because the metal ions, usually calcium and zinc, that organize the structures of biomolecules are spectroscopically unspectacular, they have been somewhat underappreciated by inorganic chemists. Nonetheless, recent discoveries have highlighted the importance of this type of metal center in biology.

7.1.a. Aspartate Transcarbamoylase.
The enzyme aspartate transcarbamoylase (ATCase) catalyzes the condensation of aspartate and N-carbamoyl phosphate to form carbamoyl aspartate plus inorganic phosphate. This reaction is the first step of pyrimidine biosynthesis. The enzyme is allosterically regulated; addition of CTP, an end product of the biosynthetic pathway, decreases enzyme activity, whereas addition of ATP, a measure of the cellular purine pool, increases activity. This *allosterism*, a cooperative or

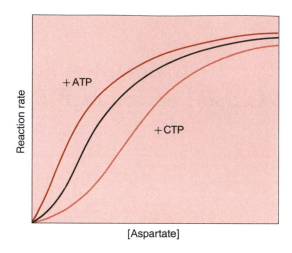

Figure 7.1
Plot of the dependence of aspartate transcarbamoylase enzymatic activity on the allosteric effectors CTP and ATP. At a given aspartate concentration, CTP inhibits enzymatic activity whereas ATP stimulates it.

anticooperative interaction between protein subunits triggered by binding substrate or other exogenous molecules, is illustrated in Figure 7.1.

The holoenzyme is quite large. It contains a total of twelve polypeptides, six copies of the catalytic subunit that is a chain of 302 amino acids, and six copies of the regulatory subunit that has 154 amino acids. Partial disruption of the quaternary structure of the enzyme results in separation of the catalytic and regulatory units. The catalytic chains form trimers that have all the enzymatic activity, whereas the regulatory chains form dimers that contain binding sites for inhibitors and activators such as CTP and ATP. The holoenzyme contains a total of six zinc ions, one in each of the regulatory chains.

The allosteric behavior of ATCase can be interpreted in terms of a relatively simple model first introduced by Monod, Wyman, and Changeux. In this model, the enzyme can exist in two quaternary forms. One form is designated the T (tense) state and has a low affinity for substrate; the other, R (relaxed), state has a higher affinity. In such a scheme, the effect of an inhibitor is to shift the equilibrium toward the T state, whereas the effect of an activator is to shift the equilibrium toward the R state.

The three-dimensional structure of ATCase has been determined by crystallographic methods in several forms. The overall structure of the native enzyme is shown in Figure 7.2. The structure consists of two catalytic trimers with their threefold axes aligned, bridged by three regulatory dimers arranged along three twofold rotation axes perpendicular to the threefold axis of the trimers. The overall molecule has D_3 symmetry. The regulatory chains

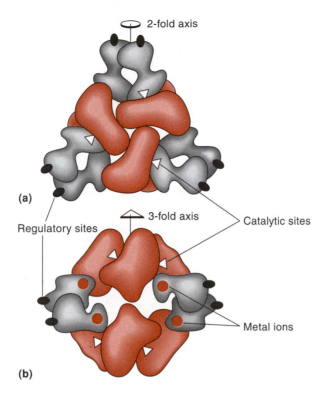

Figure 7.2
Structure of aspartate transcarbamoylase depicting the catalytic and regulatory
sites. (a) View down the three-fold symmetry axis. (b) View along a two-fold
symmetry axis. The bound zinc ions stabilize domains involved in interactions
between the regulatory and catalytic subunits.

consist of two domains. The amino-terminal 90 amino acids form a domain
that is responsible for dimerization and for interactions with the allosteric
effectors. The carboxyl-terminal 60 amino acids make up another domain that
is structurally organized around a zinc ion coordinated by four cysteinate
residues. The zinc domain is involved in direct contacts between the regulatory and catalytic subunits.

The importance of the zinc in allowing these interactions has been directly probed in several ways. Before the structure of the enzyme had been
elucidated, chemical modification studies revealed that removal of the zinc
resulted in dissociation of the regulatory and catalytic subunits. In particular,
the enzyme was treated with the cysteine-modifying agent *p*-hydroxymer-
curibenzoate in the presence of the zinc-sensitive dye 4-(2-pyridylazo)-
resorcinol. Cysteine modification was monitored by the increase in absorbance accompanying formation of the cysteinate-mercury adduct, and zinc
release was monitored by the visible absorbance of the zinc-dye complex.
Modification of approximately 24 cysteine residues resulted in the release of

the six zinc ions from the holoenzyme. This process also caused dissociation of the enzyme into catalytic trimers and regulatory dimers as determined by ultracentrifugation. Cleavage of the cysteinate-mercury adducts with excess thiol produced apo-regulatory subunits that would reassociate with the catalytic subunit only when zinc was added. These experiments strongly suggested the presence of the zinc-cysteinate complexes in the regulatory domains, a conclusion later confirmed crystallographically.

More recently the role of the zinc domain has been probed more directly by production of regulatory subunits in bacteria that have had the amino-terminal domain deleted, that is, subunits that consist of the zinc domain only. The resulting 9-kilodalton protein fragments form tight 3:1 complexes with catalytic trimers. Interestingly, the formation of these complexes is cooperative, with no intermediates being detectable during titration of the catalytic subunits with the zinc domain. The complex is easily disrupted with mercurials, but can be reformed by treatment with thiol and zinc as described above. The catalytic properties of the enzyme are changed by addition of the zinc domain. The affinity for aspartate becomes about twice that of the isolated catalytic domains, whereas the V_{max} is reduced. These properties are strikingly similar to those deduced for the R state of the enzyme. Thus, it appears that zinc allows formation of a small protein domain that is capable of binding to the catalytic subunit in a manner that produces conversion from a T-like to an R-like state.

7.1.b. Zinc-Binding Domains in Nucleic-Acid-Binding Proteins.

Structural zinc ions have recently been discovered to occur in a wide variety of nucleic-acid-binding and gene-regulatory proteins. The first such system to be characterized was that found in transcription factor IIIA (TFIIIA). TFIIIA is a site-specific DNA-binding protein that plays a central role in controlling the transcription of 5S ribosomal RNA genes in the African clawed toad *Xenopus*. A role for zinc in this protein was established via two lines of experiments. First, the protein as isolated from *Xenopus* oocytes, as part of a specific complex with 5S RNA, was found to contain two to three equivalents of zinc. These ions were resistant to removal by dialysis of the TFIIIA-5S RNA complex against solutions of the chelating agent EDTA. Destruction of the RNA by nuclease treatment, however, did allow the zinc ions to be removed by dialysis against EDTA solutions. Second, the bound zinc ions were found to be essential for specific binding to DNA. The protein as isolated from the 5S RNA complex, which contains zinc, binds to a sequence of approximately 45 base pairs and protects it from nuclease digestion. Removal of the zinc by EDTA treatment resulted in a loss of this specific binding activity. Addition

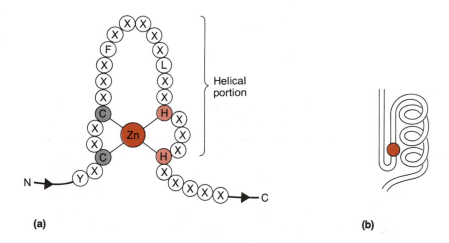

Figure 7.3
Schematic representations of (a) the amino acid sequence and features characteristic of zinc finger domains from transcription factor IIIA and related proteins, and (b) the three-dimensional structure adopted by these domains upon binding zinc.

of zinc but not similar concentrations (15 μM) of Co^{2+}, Ni^{2+}, Fe^{2+}, or Mn^{2+} restored the activity.

Analysis of the amino-acid sequence of TFIIIA, deduced from the nucleotide sequence of a cDNA clone, led to an exciting hypothesis for the zinc-dependent DNA binding. The amino-terminal three-quarters of the sequence was found to consist of nine tandem sequences that closely approached the consensus (Tyr, Phe)-X-Cys-$X_{2,4}$-Cys-X_3-Phe-X_5-Leu-X_2-His-$X_{3,4}$-His-X_{2-6}, where X represents relatively variable residues. Each of these sequences has two cysteines and two histidines, which were recognized to have potential zinc-binding side chains. Furthermore, it was found that the TFIIIA-5S RNA complex, when purified under conditions that avoided chelating agents such as EDTA, contained 7 to 11 zinc ions per protein. Given these observations, it was proposed that each of the nine sequences formed a structural domain organized around a zinc ion bound to the invariant cysteine and histidine residues. These domains were termed *zinc fingers*. A schematic representation of the proposed structure is shown in Figure 7.3.

A wide variety of data supports this proposal. The nature of the ligands coordinated to the zinc ions has been directly probed by EXAFS spectroscopy. The EXAFS spectra obtained from the TFIIIA-5S RNA complex could be fit with two sulfur atoms at 2.30 Å and two nitrogen atoms at 2.00 Å, consistent with the postulated structure. Further support for the structure of the metal-binding sites, as well as information indicating the independent nature of the structural units, came from studies of single zinc-finger peptides. A peptide corresponding to the second zinc-finger domain of TFIIIA was prepared and purified. This peptide was found to bind one equivalent of Zn^{2+} or Co^{2+}. The absorption spectrum of the cobalt complex included bands in the visible region with extinction coefficients greater than 500 $M^{-1}cm^{-1}$, suggestive of tetrahedral coordination, especially when compared with appropriate model complexes. Furthermore, examination of the circular dichroism spectra of the peptide in the absence of added metal ions indicated the lack of a well-defined structure. Addition of stoichiometric amounts of divalent zinc or cobalt ions resulted in significant changes of the CD spectrum indicative of structure formation. Finally, proteolysis studies demonstrated that the addition of zinc made the peptide much less susceptible to cleavage by trypsin.

Single-domain peptides have also been useful for investigating the metal-binding specificity of zinc fingers. As noted above, in addition to Zn^{2+}, such peptides will also bind Co^{2+} in a stoichiometric fashion. Additional studies have revealed that Fe^{2+} and Ni^{2+} also bind. Determination of the dissociation constants for a series of metal ions via metal-ion titration experiments, however, revealed that Zn^{2+} binds 10^3 to 10^5 times more tightly than the other ions. Thus, the specificity for zinc in reconstituting the DNA-binding activity of EDTA-treated TFIIIA preparations can be explained by the relative affinities of the zinc-finger domains for various metal ions.

Shortly after the discovery that the TFIIIA sequence could be understood in terms of zinc-finger domains, several other proteins were found to contain quite similar imperfectly repeated sequences. These included the yeast transcription factor ADR1, which has two tandem zinc-finger sequences, and the *Drosophila* developmental-control protein Krüppel, which has five. In subsequent years, both through independent discoveries and through intentional search procedures, a very large number of DNA sequences has been characterized that code for repeated zinc-finger domains. The human genome has been estimated to contain approximately 500 genes that encode domains of this class. The number of domains in a single protein ranges from as few as one to as many as 37. Some of these are summarized in Figure 7.4. Experiments performed on the proteins themselves have shown them to have sequence-specific DNA-binding activity.

The availability of single-domain peptides afforded an appealing approach to determining the three-dimensional structures of zinc-finger domains by using NMR methods. These peptides consist of only 25 to 30 amino acids, yet they fold in aqueous solution. Initial studies were reported for a single-domain peptide from ADR1. One-dimensional spectra clearly indicated that the peptide underwent a transition from an essentially unfolded form in the absence of zinc to a well-ordered structure in its presence. Two-

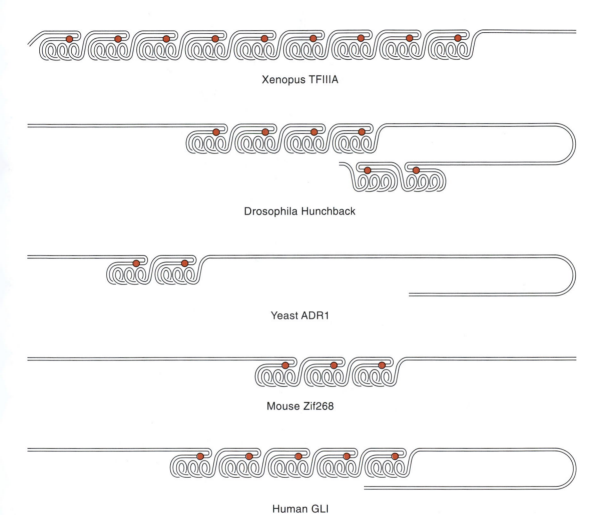

Figure 7.4
Schematic view of several representative zinc finger proteins indicating the number of zinc finger domains and their relative positions within the primary structures of the proteins.

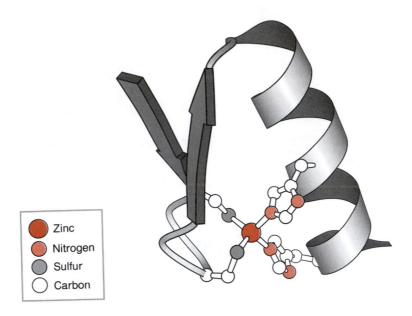

Zinc
Nitrogen
Sulfur
Carbon

Figure 7.5
View of the three-dimensional structure of a zinc finger domain deduced by NMR methods illustrating the overall fold and the coordination of zinc by two cysteinate and two histidine residues.

dimensional methods allowed complete assignment of the spectra as well as determination of the overall structure via measurement of short proton-proton distances. Subsequent studies have also been performed on other peptides. One deduced structure, shown in Figure 7.5, consists of a hairpin turn containing the two cysteinate residues followed by an alpha helix. An interesting aspect of the structure is that the amino and carboxyl termini of each domain are relatively far from one other. This result suggested that a protein containing tandem zinc-finger domains would wrap around the DNA when it bound. The structure of the individual zinc-finger domains as well as this mode of interaction with DNA have recently been confirmed by an X-ray crystallographic analysis of a fragment of the mouse transcription factor Zif268 consisting of three tandem zinc-finger domains bound to an oligonucleotide recognition site. This structure is shown in Figure 7.6.

In addition to the large family of proteins that contain sequences quite similar to those found for TFIIIA, several other classes of nucleic-acid-binding proteins have been identified that contain patterns of cysteine and histidine residues suggestive of metal-binding domains. Among them are the steroid-receptor superfamily of proteins containing an approximately 66 amino-acid domain with nine conserved cysteine residues. The cysteines are organized into two somewhat independent units, each of which is centered about a zinc

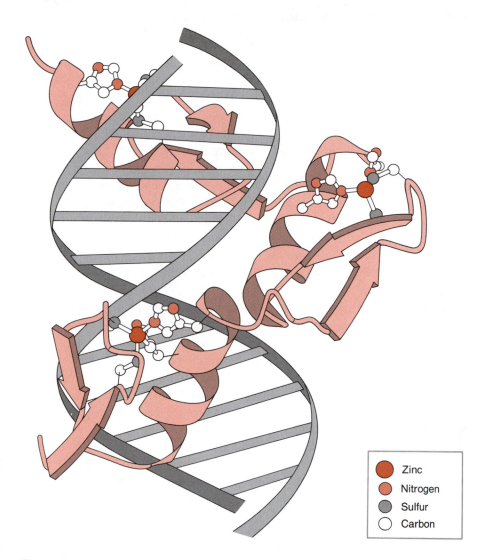

🔴	Zinc
🔴	Nitrogen
⚫	Sulfur
⚪	Carbon

Figure 7.6
Structure of the three zinc finger-DNA binding domain of Zif 268 complexed to an oligonucleotide binding site as determined by X-ray crystallography.

ion bound by four cysteines. The three-dimensional structures of these domains have been established, originally by NMR spectroscopy and, more recently, by a crystallographic study of a complex with DNA. One interesting aspect of the structure is that one of the zinc-based units does not play a significant role in DNA recognition, but instead mediates dimerization of the protein.

Another family is typified by the yeast gene-regulatory protein GAL4 and is defined by the presence of dinuclear $Zn_2(Cys)_6$ centers. Again, the structures have been elucidated by both NMR and X-ray crystallographic methods. The first evidence for the presence of zinc in GAL4 came from genetics studies. A strain of yeast with a mutant GAL4 gene that displayed regulatory activity only in the presence of additional zinc in the growth medium was discovered. The mutation converted a conserved proline residue into leucine, and structural studies revealed that this proline is involved in a *cis* rather than *trans* peptide bond. This unusual configuration is favored when the amino group of the peptide bond comes from proline. It seems likely that the affinity of the mutant GAL4 is reduced by the strain resulting from the requirement that a *cis* peptide bond occur at a position not involving proline.

One additional family is the retroviral nucleocapsid proteins that have either one or two metal-binding sequences with three cysteines and one histidine. NMR studies have revealed the three-dimensional structures of these domains. EXAFS experiments have demonstrated that the domains contain zinc inside intact retrovirus particles. Since each such virus particle capsid has a diameter of approximately 50 nm and contains $\sim 1,000$ copies of the nucleocapsid protein, the internal zinc concentration can be computed to be nearly 1 mM.

The structures of the zinc-based domains from these functionally diverse protein classes are quite distinct, as shown schematically in Figure 7.7. In all these families, it has been demonstrated that zinc binding induces structural changes and stabilization. Thus it is clear that nature uses the coordinating ability of metal ions to produce or stabilize structures important for interactions between macromolecules. An additional intriguing idea is that zinc binding might sometimes regulate nucleic-acid binding and, hence, gene expression. As discussed in Chapters 5 and 6, metal-responsive gene-regulatory proteins that respond to iron or mercury levels in the cell have been identified. Time will tell if zinc is merely a structural device in these proteins or whether, in fact, it is also an information carrier, or messenger, in these gene-regulatory networks.

7.1.c. Calcium-Binding Proteins: Transient Structural Stabilization.
Many processes of signal transduction involve the release of Ca^{2+} ions as one part of an interconnected set of pathways. Increases in intracellular Ca^{2+} are sensed by a family of calcium-binding proteins. A prototypical member of this family is calmodulin, which couples changes in the intracellular calcium concentration to the state of activation of a variety of enzymes, such as protein kinases, NAD kinase, and phosphodiesterases, as well as some Ca^{2+}-pumping ATPases.

Calmodulin is a monomeric protein consisting of a chain of 148 amino acids that is capable of binding up to four Ca^{2+} ions. The primary structure

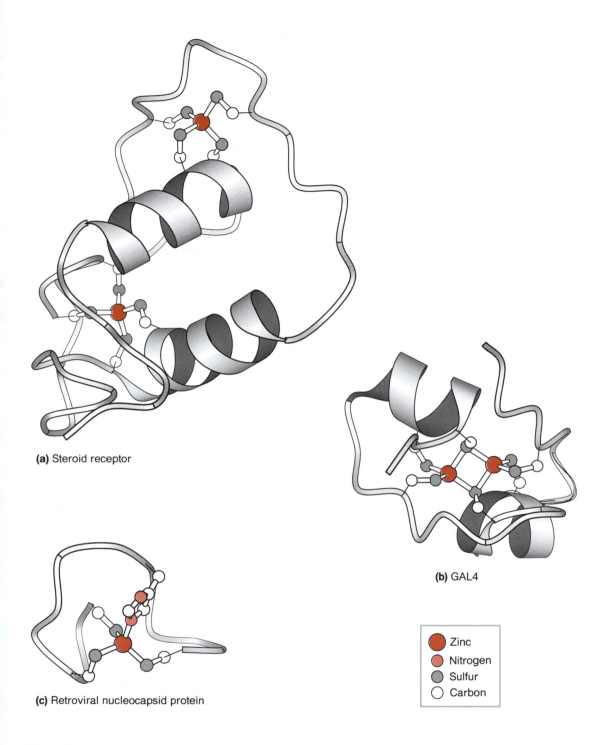

(a) Steroid receptor

(b) GAL4

(c) Retroviral nucleocapsid protein

🔴	Zinc
🔴	Nitrogen
⚫	Sulfur
⚪	Carbon

Figure 7.7
Structures of other classes of zinc-based domains from nucleic acid-binding proteins.

of calmodulin is remarkably conserved in species ranging from paramecia to humans. Examination of the primary structure revealed the presence of four sequences that match a pattern previously discovered in other calcium-binding proteins. The structure adopted by this sequence has also been observed crystallographically in a number of calcium-binding proteins. It was first observed in parvalbumin and termed the *EF hand*, where the letters *E* and *F* denote alpha helices. The structure consists of an alpha helix of about ten residues, a ten-amino-acid loop, and a second alpha helix. The sequences of the two calcium-binding EF hand sequences from parvalbumin as well as the four sequences from calmodulin are shown below:

```
ADDVKKAFAII D Q D K S GF I EED E LKLFLQNFKA
DGETKTFLKAG D S D G D GK I GVD E FTALVKA

IAEFKEAFSLF D K D G D GT I TTK E LGTVMRSLGQ
EAELQDMINEV D A D G N GT I DFP E FLTMMARKMK

EEEIREAFRVF D K D G N GY I SAA E LRHVMTNLGE
DEEVDEMIREA N I D G D GQ V NYE E FVQMMTAK
```

The positions marked in boldface type correspond to sites of calcium ligation. These residues bind to the calcium via side-chain carboxylate, alcohol, or carboxamide oxygens or via backbone carbonyl groups. The EF-hand domains come in pairs. Whereas the coordination of each calcium occurs within one EF-hand sequence, the two members of a pair pack against one another via nonpolar residues. The structure of one of the two pairs of EF-hand units from calmodulin is shown in Figure 7.8. A more detailed view of one EF-hand unit, including the coordination sphere around the calcium ion, is shown in Figure 7.9.

Each calcium is seven-coordinate, with three monodentate aspartate or asparagine residues, one bidentate glutamate residue, one coordinated peptide carbonyl group, and one bound water molecule. The nature of these coordination sites is important for metal-binding specificity. Calcium-responsive proteins must specifically recognize calcium in the presence of relatively high concentrations of other metal ions such as Mg^{2+}. The high coordination numbers produced by the structure of the EF-hand site favors Ca^{2+} binding over Mg^{2+} binding because the latter ion is smaller. For calmodulin, the dissociation constants for Ca^{2+} binding range from 10^{-6} to slightly less

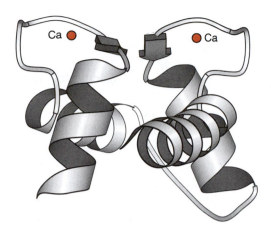

Figure 7.8
Structure of a pair of EF-hand domains showing the Ca-binding loops and packing of the two domains against one another.

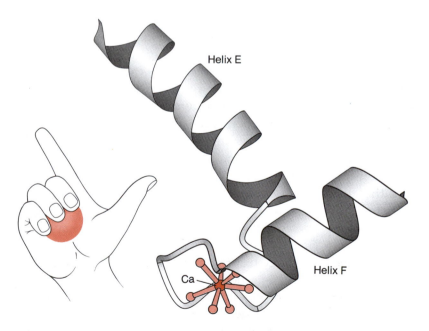

Figure 7.9
A more detailed view of one EF-hand domain illustrating the locations of the E (finger) and F (thumb) helices and the calcium-binding site. The seven calcium-bound oxygen atoms derived from amino acid side chains, the peptide backbone, and from water are also shown.

than 10^{-7} M, whereas the dissociation constants for Mg^{2+} binding are approximately 10^{-3} M.

The structure of the entire calmodulin molecule, as determined by X-ray crystallographic methods, is shown in Figure 7.10. The molecule is strikingly dumbbell-shaped with a long, exposed alpha helix connecting the two pairs of EF-hand domains. This helix is an extension of the F helix of domain 2 and leads directly into the E helix of domain 3. The occurrence of such long solvent-exposed alpha helices in calmodulin and in the related calcium-modulated protein troponin C was unprecedented and surprising. The helix appears to be stabilized by salt bridges between positively (lysine, arginine) and negatively (aspartate, glutamate) charged amino acids that are spaced with i to $i + 4$ relationships along the helix. Furthermore, as will be discussed in more detail later, this helix must be metastable, since it is disrupted when calmodulin binds to a target molecule. A final notable feature in the structure is the presence of a large hydrophobic patch on the surface of each of the two lobes of the dumbbell. These patches are flanked by regions of highly negative electrostatic potential.

As was noted earlier, calmodulin can modulate the activity of many different enzymes. Interestingly, however, calmodulin binds to its targets quite tightly, with dissociation constants generally 10–100 nM or less. These two observations appear almost contradictory, in that it is hard to conceive of many mechanisms that allow binding to many very different substrates, yet bind each with high affinity. A structural explanation that resolves this apparent paradox has recently been proffered. Calmodulin binds peptides that are capable of forming basic, amphipathic, alpha-helical structures with high affinity. Moreover, examination of the sequences of many calmodulin targets and mapping of the sites by mutagenesis and chemical methods has revealed that regions that are responsible for calmodulin binding are small, approximately 20 amino-acid, stretches that are positively charged and capable of forming amphipathic helical structures. Several examples of such sites are shown in Figure 7.11. Studies using fluorescence quenching and photoaffinity labeling indicate that the hydrophobic portions of the peptide are partially buried upon interaction with calmodulin. Furthermore, experiments with doubly labeled peptides reveal that both ends of such peptides are capable of interacting with the two domains of calmodulin simultaneously. This result is striking, in that interaction with the two hydrophobic patches simultaneously is not possible with the interconnecting alpha helix unperturbed. The combination of these results suggests that the connecting helix is structurally flexible and that the two lobes of calmodulin clamp down on a

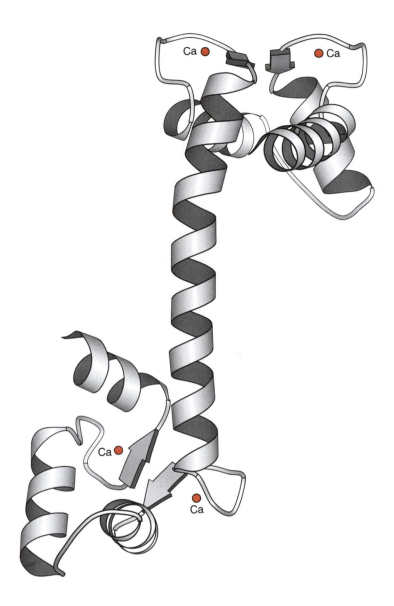

Figure 7.10
Overall structure of calmodulin determined by X-ray crystallography.

Skeletal muscle myosin light chain kinase	K R R W K K N F I A V S A A N R F K K I S S
Smooth muscle myosin light chain kinase	R R K W Q K T G H A V R A I G R L S S S
Calmodulin kinase	R R K L K A A V K A V V A S S R L G
Ca²⁺ pump	Q I L W F R G L N R I Q T W I K V V N F S S

Figure 7.11
Amino acid sequences of several calmodulin binding sites from proteins activated by calmodulin.

target amphipathic peptide. Examination of the sites of photo-cross-linking revealed an additional surprising result, namely, that methionine residues were almost always the site of modification. Analysis of the nature of the hydrophobic patches provided a simple explanation. The accessible surface of the two hydrophobic patches is nearly 50 percent contributed by methionine residues. The significance of this observation lies in the fact that methionine is hydrophobic, yet unbranched, and can be very flexible. Thus the resolution of the paradox of broad specificity yet high-affinity binding appears to be the presence of the structurally adaptable hydrophobic surfaces on the two calcium-binding lobes, which are, in turn, connected to one another by a relatively flexible tether.

These hypotheses have been confirmed and extended by structural studies of two calmodulin-target peptide complexes. The structure of one of these complexes is compared with that of free calmodulin in Figure 7.12. One of these analyses, carried out on a calmodulin complex of a 26-residue peptide from skeletal-muscle myosin light-chain kinase, applied newly developed NMR methods. Calmodulin was uniformly labeled with ^{13}C and ^{15}N, a procedure that facilitated assignments based on spin-spin coupling between these nuclei as well as differentiation between protein-protein and protein-peptide contacts. The resulting complex is one of the largest structures worked out to date by NMR methods. The other analysis, by X-ray crystallography, was carried out on a calmodulin complex with a 20-amino-acid peptide from smooth-muscle myosin light-chain kinase. The two peptides share a number of common features, including residues with spacing Trp-X₅-Ala-Val-X₅-(Phe, Leu). The Trp and some nearby residues are bound in the hydrophobic pocket of the carboxyl terminal lobe, whereas the (Phe, Leu) region is bound to the amino terminal lobe. The Ala-Val region interacts with the tether between lobes. Numerous contacts with methionine residues are also observed. Each of the complexes overall has approximate

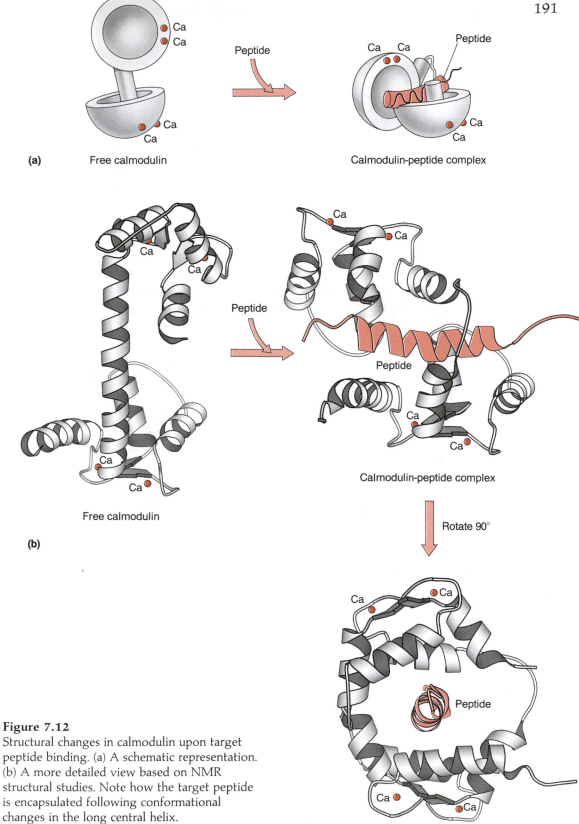

(a) Free calmodulin

Peptide

Calmodulin-peptide complex

(b)

Free calmodulin

Peptide

Calmodulin-peptide complex

Rotate 90°

Peptide

Figure 7.12
Structural changes in calmodulin upon target peptide binding. (a) A schematic representation. (b) A more detailed view based on NMR structural studies. Note how the target peptide is encapsulated following conformational changes in the long central helix.

twofold symmetry, with the central helix from free calmodulin unwound to form a bend.

We now turn to the role of calcium in mediating these processes. The binding of calcium to calmodulin induces distinctive conformational changes as determined by a variety of spectroscopic and chemical probes. Although the detailed structures of the apo or partially ligated forms of the protein are not known, it seems clear that the conformation and rigidity of the two lobes built up from the pairs of EF-hand units must be significantly dependent on bound calcium. Thus the hydrophobic surfaces discussed above are probably not well established prior to the binding of calcium, precluding their interactions with targets. In this way, an increase in the intracellular concentration of calcium from below the value necessary for adequate binding to calmodulin can lead to the binding of calcium to the protein, the concomitant conformational changes, the binding of the calcium-containing protein to target proteins, and, hence, to target activation.

There is evidence that the required changes in intracellular calcium concentration do, indeed, occur. The most direct information comes from the use of membrane-permeable fluorescent calcium-binding dyes, such as Fura-2. The basis of action of these dyes is illustrated in Figure 7.13. As an example of the use of these dyes, let us consider the effects of treatment of cells from a cell line derived from vascular smooth muscle of mouse brain with the drug phenylephrine. Before treatment, the intracellular calcium concentration was determined to be 150 nM by comparison of the Fura-2 fluorescence produced by excitation at 350 and 385 nanometers. Addition of 10 micromolar phenylephrine resulted in an increase in intracellular calcium levels to approximately 0.5 millimolar. Calcium is said to act as a second messenger; the initial message, an increase in drug concentration at the cell surface, is transduced into an increase in calcium concentration, which, in turn, affects specific cellular processes via the action of calmodulin and other calcium-responsive proteins.

7.2. Metal-Ion Stabilization of Nucleic-Acid Structure

Nucleic acids are polyanions and, as such, require counterions in order to adopt stable, compact structures. These counterions are required to neutralize partially or completely the negatively charged phosphate groups, so that electrostatic repulsions do not overwhelm other stabilizing effects. This charge-neutralization requirement is generally accomplished by relatively

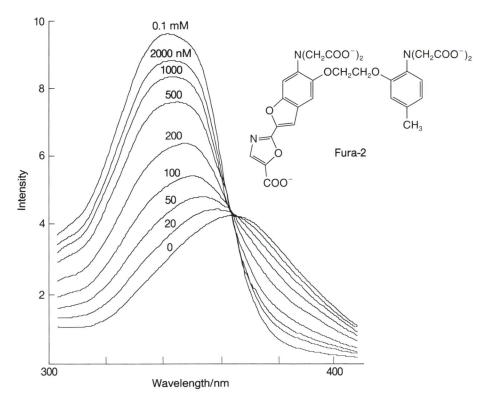

Figure 7.13
Fluorescence spectra of the calcium-binding dye Fura-2 as a function of calcium concentration. (Adapted from R. Y. Tsien and M. Poenie, *Trends in Biochemical Sciences* **11**, 450–455 (1986)).

nonspecific complexes with cations such as Na^+, K^+, and Mg^{2+} or by cationic polyamines such as spermine. In some important cases, however, more specific metal-ion-binding sites have been identified and at least partially characterized. Some key examples are discussed below.

7.2.a. Transfer RNA. The best structurally characterized RNA molecules are various types of transfer RNAs. These RNA molecules, previously discussed in Section 3.2.b, contain approximately eighty bases and fold up into the L-shaped structure shown in Figure 3.14. The binding of metal ions to transfer RNAs has been thoroughly investigated, both in an effort to understand the roles that metal ions play in stabilizing RNA structure and in the course of preparing heavy-atom derivatives for X-ray crystallographic studies. The positions of some of the metal-binding sites that have been identified on yeast phenylalanine tRNA are shown in Figure 7.14. The site most important for stability is one that can be occupied by Mg^{2+} or, nearly but not quite isostructurally, by Mn^{2+}. Structural stabilization measurements

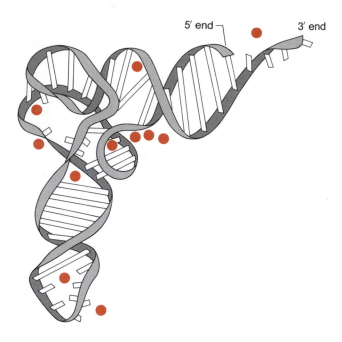

Figure 7.14
Positions of some of the metal binding sites on tRNA
deduced from X-ray crystallographic studies.

in solution have identified the locus of this site as near the D and T loops
(Figure 3.14). The structure of the Mg^{2+} site, as deduced from X-ray crystallography, is shown in Figure 7.15.

An important distinction exists between the structural metal-binding sites
for proteins discussed in Section 7.1.a. and the sites on nucleic acids. In
proteins, the metal ions are very nearly or completely dehydrated, with all
the water molecules bound in aqueous solution being removed and replaced
by protein-derived ligands. In contrast, for the sites on tRNA and probably
other nucleic-acid molecules, most of the waters of hydration remain coordinated, with only one or two nucleic-acid-derived ligands bound directly to
the metal ion. The coordination sites occupied by the waters are still important for the specificity of binding, however, since they mediate hydrogen-
bonding interactions with phosphates and other groups on the nucleic acid.

7.2.b. Catalytic RNA Molecules. One of the most exciting discoveries in recent years is that certain RNA molecules can act as catalysts for
biochemical reactions. A number of different catalytic RNA molecules, or
ribozymes, have been identified, and both functional and structural characterization is proceeding. In many instances the role(s) of metal ions in the
molecules has been investigated, and in all, it is clear that metal ions function
in two different ways: by stabilizing the folded structure of the RNA molecules and by actually facilitating the chemical reactions themselves. One

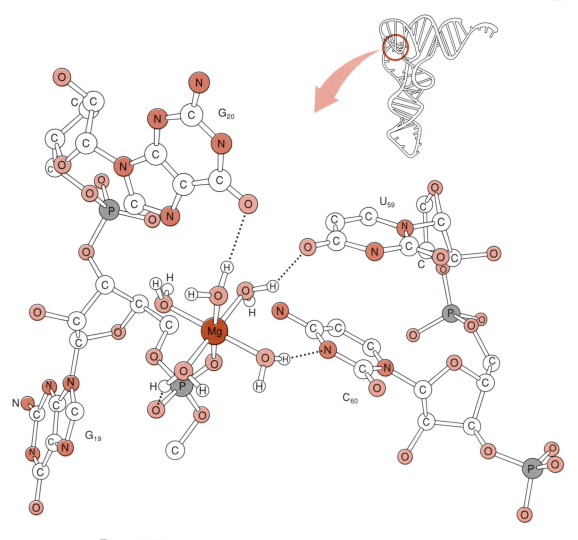

Figure 7.15
Structure of a Mg^{2+} binding site on yeast tRNA^{Phe}.

example that has been extensively studied in the self-splicing intron from *Tetrahymena thermophila*, the secondary structure of which is shown in Figure 7.16. This RNA molecule is inactive in the absence of Mg^{2+} or Mn^{2+}. Other cations, such as Ca^{2+}, Ba^{2+}, and Sr^{2+}, can reduce the requirement for Mg^{2+} or Mn^{2+} but will not alone produce active catalysts. The different roles for these metal ions have been more directly probed with the development of a technique that can probe the tertiary structure of the RNA, namely, hydroxyl radical footprinting. In this method, which had been previously developed for studying DNA (see Section 11.4.c.), hydroxyl radicals are generated via Fe^{II}(EDTA) and hydrogen peroxide in order to cleave the radiolabeled RNA.

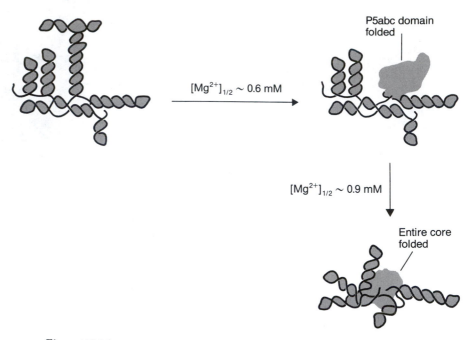

Figure 7.16
Schematic view of the proposed folding of the self-splicing intron from *Tetrahymena thermophila* upon the binding of Mg^{2+}.

The sites of cleavage can then be revealed by examining the lengths of the resultant fragments by gel electrophoresis using the radioactivity for detection. The rates of cleavage at different sites are dependent on the folded structure of the molecule. With the use of this method it was found that Mg^{2+} and Ca^{2+} at concentrations near 1 mM can produce a fully folded RNA that is not cleaved nearly as well as an RNA molecule that lacks the folded tertiary structure. The Ca^{2+} complex, although folded nearly as well as the Mg^{2+} complex as judged by protection from hydroxyl radical cleavage, does not show catalytic activity, but addition of Mg^{2+} to this complex produced active species. Thus Mg^{2+} ion uniquely folds and activates the ribozyme. Similar dual roles for metal ions have been observed for the RNA from RNAase P and from the "hammerhead" ribozymes.

7.2.c. Telomeres. In general, DNA structures are more regular than RNA structures, with the classical Watson-Crick double helix being the most widely occurring structure by far. Specialized DNA structures do exist, however, one example being *telomeres*, the structures that occur at the ends of chromosomes. Specialized structures are required here to avoid loss of genetic material from the ends of chromosomes during multiple rounds of replication.

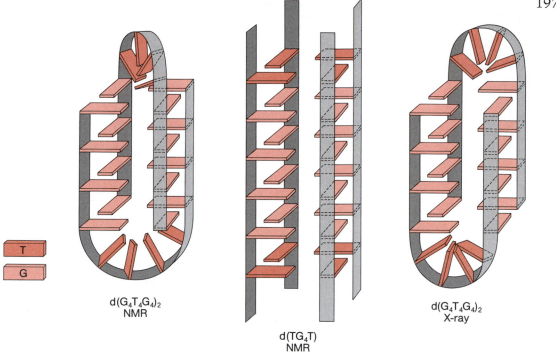

Figure 7.17
Structures of three guanine tetrad-containing oligonucleotide complexes deduced by NMR or X-ray crystallography.

The sequence of a telomere from tetrahymena is:

5'-GGGGTTGGGGTTGGGGTTGGGGTTGGGGTTGGGG-3'

3'-CCCCAACCCCAACCCCAACCCC-5'

It was discovered that oligonucleotides containing such sequences, or indeed other oligonucleotides containing runs of G residues, form structures having anomalous gel-electrophoretic mobilities. The formation of such species was dependent on the salt conditions used in the experiments. Potassium ions were found specifically to stabilize one of the structures, whereas sodium was more effective in stabilizing others. The potassium-stabilized structure is a dimer that involves the 3' overhanging tail folding back on itself, a specific geometry for which has been proposed and supported by chemical-protection studies. This structure has a "hole," formed by the O6 oxygens from eight guanine residues, that appears to be uniquely suited for the size of the potassium ion. This postulated structure is shown in Figure 1.7.

Recent investigations using NMR and X-ray crystallographic methods have confirmed the general features of these models. The three structures that have been determined are shown schematically in Figure 7.17. NMR data for a dimer of sequence d($G_4T_4G_4$), prepared in 50-mM NaCl solution, revealed a structure having a tetraplex of guanine bases. Detailed analyses were most consistent with a structure of the type shown in Figure 7.17, in which half the adjacent DNA strands are parallel, the other half being antiparallel. In contrast, NMR studies of a tetramer having the sequence d(TG_4T) in 100-mM NaCl revealed a different structure, with all the strands parallel to each other. Finally, a crystal structure has been reported for the dimeric d($G_4T_4G_4$) molecule. The crystals were grown in the presence of 60 mM K^+ ion. The results were interpreted in terms of yet a third structure, having a mixture of parallel and antiparallel strands as before, but different overall connectivity. One exciting observation in the crystal-structure analysis is the presence of a partially disordered potassium ion in the center of the guanine tetrad. Further studies are required to understand the relationships among the stabilities of these and perhaps other structures, in both chemical and biological terms. The possibility of metal-ion-dependent switching among the different structures has been raised from other studies and could be of considerable significance.

7.3. Protein Binding to Metallated DNA

The natural structures that arise in DNA as a result of specific sequences are used to control a variety of cellular functions, such as transcription. We have already seen how coordination of zinc to the zinc-finger domains folds them into structures that bind DNA and activate genes. It has recently been found that the modification of DNA by platinum antitumor drugs brings about structural changes that activate the binding of specific classes of cellular proteins to the platinated DNA. This discovery further illustrates the ability of metal ions to fold biopolymers into conformations that alter their physiological states. Moreover, as our understanding of this phenomenon grows, it may be possible to design metal complexes as chemotherapeutic agents based on the information that emerges.

7.3.a. Unwinding and Bending of DNA Modified by Platinum Anticancer Drugs.   When the aquated cisplatin (see Chapter 5) reacts with DNA, a variety of adducts forms. The major species are intrastrand cross-links of the kind cis-[Pt(NH_3)$_2$\{d(GpG)\}] and cis-[Pt(NH_3)$_2$\{d(ApG)\}], which together account for up to 90 percent of the bound platinum. The ability of cancer patients to form and maintain these adducts has been reported to correlate with their response to treatment with cisplatin. The exact binding

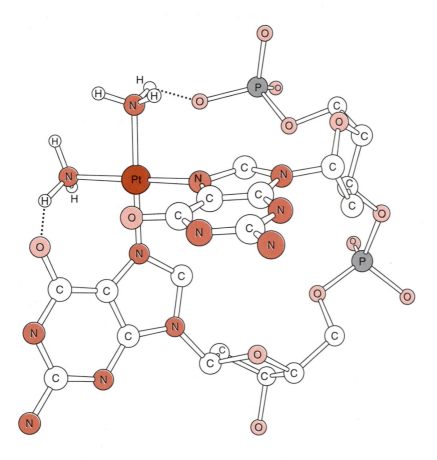

Figure 7.18
X-ray structure of the cisplatin adduct of d(pGpG).

positions of platinum on the DNA bases have been elucidated by NMR spectroscopic and single-crystal X-ray diffraction studies. In these experiments, short oligonucleotides are first synthesized and then platinated by cis-DDP. The products are separated and purified by HPLC. From NMR spectroscopic study of the major adduct, cis-[Pt(NH$_3$)$_2${d(pGpG)}], it was deduced that platinum binds to N7 positions of two guanine bases and that the 5′-nucleotide deoxyribose ring has a C3′-endo conformation. X-ray crystallography confirmed these conclusions and provided metric details about the structure; it also revealed a tendency for ammine ligands to form hydrogen bonds with the 5′-terminal phosphate group in d(pGpG). Figure 7.18 displays the structure. These features persist in longer oligonucleotides in which intrastrand d(GpG) or d(ApG) platinum adducts are embedded. The purine rings are completely destacked as a result of platination and have dihedral angles in the 75−90° range, compared to ∼ ± 10° in unplatinated B-DNA. This unstacking of the bases leads to a pronounced and directed bend in the DNA at the site of platination.

The bent structure has been established in gel-electrophoresis studies, where several duplexes containing either a single cis-[Pt(NH$_3$)$_2$\{d(GpG)\}] or a single cis-[Pt(NH$_3$)$_2$\{d(ApG)\}] adduct were oligomerized (Figure 7.19). Molecules of 88, 110, and 132 bp lengths, corresponding to tetramers, pentamers, and hexamers of the platinated 22-mers, respectively, formed circles as a consequence of intramolecular end-to-end ligation of kinked duplexes. Linear oligomers exhibited retarded electrophoretic gel mobilities consistent with a kink angle of 32° or 34° at the sites of platination for the d(GpG) and d(ApG) cross-links, respectively. These and related studies also established that the DNA is bent in the direction of the major groove, as expected for platination at guanine N7 atoms. The structure of a bent, cis-DDP-platinated DNA model is shown in Figure 7.20. The bending produces a remarkable difference in shape compared to normal B-DNA (see Figure 3.16).

By repeating the experiment and changing the positions of the platinum atoms with respect to one another in the ligated oligomers, it was possible to determine the change in the local twist, or unwinding, angle of DNA containing intrastrand d(GpG) and d(ApG) cross-links. As shown in Figure 7.19, the maximum shift in gel mobility, corresponding to optimal phasing of directed bends among platinum atoms in an oligomer, occurred when the platinum atoms were 21.38 base pairs apart. The normal helical repeat of DNA is 10.5 base pairs; so two turns of DNA would correspond to 21.0 base pairs. The fact that the platinated duplex is 0.38 base pairs longer per bound platinum atom indicates a net unwinding of $0.38/10.5 \times 360° = 13°$. When cis-DDP forms d(GpTpG) intrastrand cross-links, the DNA is unwound by 23° (Figure 7.19). Interestingly, when similar experiments were carried out on other, therapeutically inactive platinum complexes, they were found not to produce directed bends in DNA and/or to unwind the double helix by an amount significantly different from that produced by cisplatin.

7.3.b. Recognition and Binding of High-Mobility-Group Box Proteins to DNA Containing Cisplatin Intrastrand d(GpG) or d(ApG) Cross-Links.

A major goal in studies of the mechanism of action of cisplatin is to discover how the drug and its active analogs selectively destroy tumor cells at levels that healthy cells can tolerate. One reasonable hypothesis is that the platinated DNA is processed differently in the two different cell types. For example, if platinum lesions were preferentially repaired faster in nontumor tissue, they would be more toxic to the population of cancer cells. In pursuit of evidence to check this hypothesis, several laboratories have identified proteins in the nuclei and cytoplasm of human cells that bind preferentially to DNA containing the major adducts of cisplatin. The existence of such factors was proved by the gel-mobility-shift assay, outlined in Figure 7.21. DNA

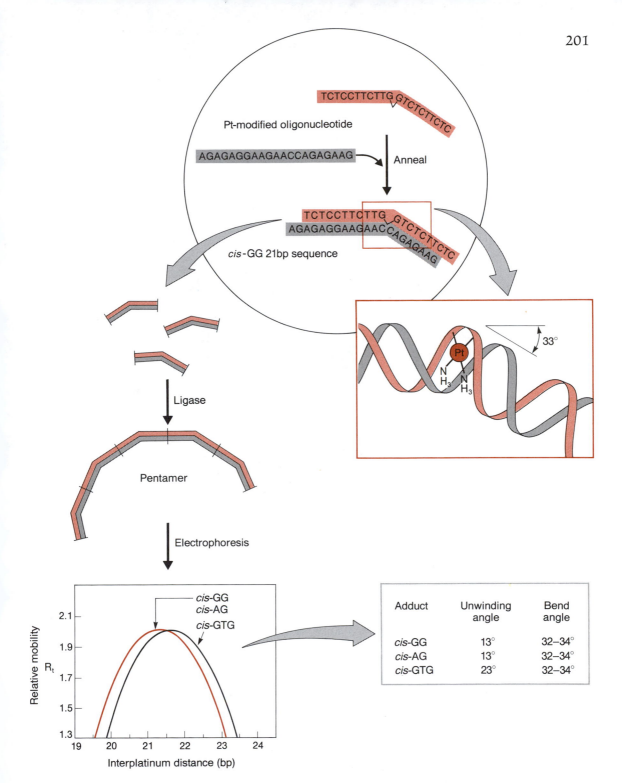

Figure 7.19

Scheme illustrating the method used to determine the unwinding of duplex DNA due to *cis*-diammineplatinum(II) adduct formation. A synthetic oligonucleotide is platinated and annealed to its complement. The resulting synthetic duplex is oligomerized via DNA ligase and the resulting multimers are isolated. The relative electrophoretic mobilities are then measured. The experiment is repeated with oligonucleotides of different interplatinum distances, and the results are compared to calculate the unwinding angle as described in the text.

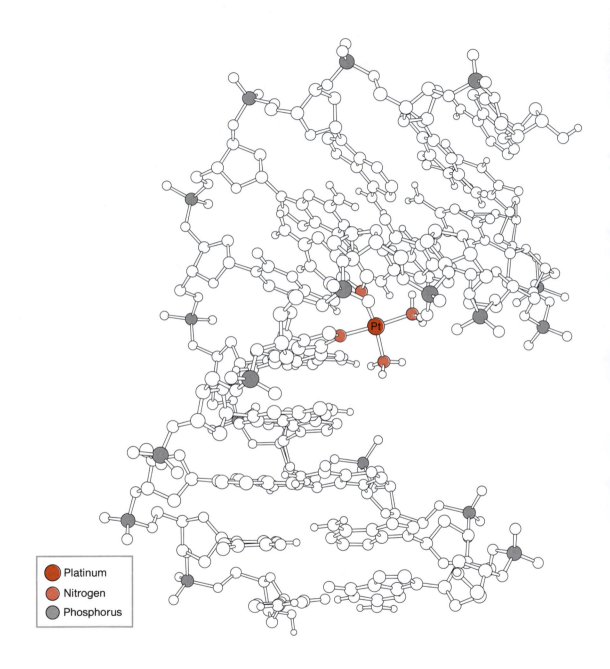

Figure 7.20
Model for the bent structure of a DNA duplex containing a *cis*-diammineplatinum-(II) intrastrand cross-link as deduced from molecular mechanics calculations.

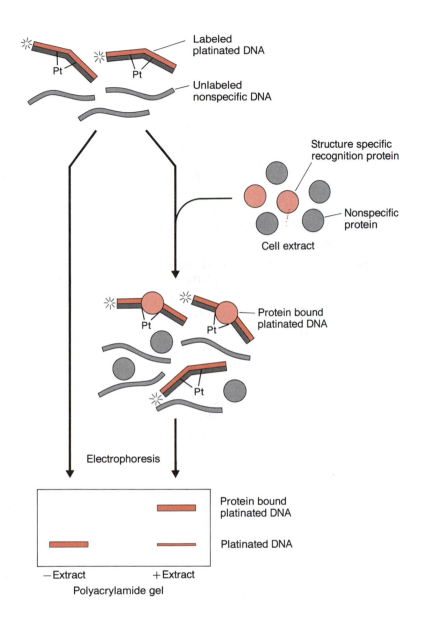

Figure 7.21
Scheme illustrating the gel-mobility-shift assay used to detect the presence of structure-specific recognition proteins in cell extracts. The negatively charged DNA molecules migrate from the top (negative pole) to the bottom (positive pole) of the polyacrylamide gel. Protein binding reduces the rate of migration.

restriction fragments, short stretches of duplex DNA cut from larger pieces with sequence-specific restriction enzymes, are platinated, radiolabeled, and run out in nondenaturing electrophoresis gels in the presence and absence of cellular extracts. A shift of the band corresponding to the labeled DNA in the presence of the extract is taken as evidence for protein binding, especially if the band shift can be reversed by pretreatment with proteases. In this manner, proteins were found that bind specifically to DNA containing *cis*-DDP, but not *trans*-DDP or [Pt(dien)Cl]$^+$, adducts. By using site specifically modified DNA molecules, researchers demonstrated that the recognition motif was afforded by the *cis*-[Pt(NH$_3$)$_2${d(GpG)}] or *cis*-[Pt(NH$_3$)$_2${d(ApG)}] intrastrand cross-links, but not the *cis*-[Pt(NH$_3$)$_2${d(GpTpG)}] 1,3-intrastrand cross-links. Two classes of proteins, termed structure specific recognition proteins (SSRPs), were identified according to their molecular masses, one ~28 kDa, the other 80–100 kDa, in size.

Further information about the nature of these proteins has recently been obtained from genetics studies. A cDNA expression library from human cells was screened by using cisplatin-modified DNA as the probe, and the full-length gene was ultimately obtained by further library screening. The sequence revealed an 81-kilodalton encoded protein, SSRP1, with several charged domains and a stretch of 75 amino acids having 47 percent identity with the nuclear protein HMG1 (Figure 7.22). This latter domain has been termed an HMG box. The high-mobility-group protein (so called because of its fast migration in electrophoresis gels of nuclear proteins) HMG1 binds to cruciform structures in DNA and has been postulated to have a role in transcription, but its true function in the cell is yet unknown. Its molecular mass is 24.5 kilodaltons. Subsequent work revealed that HMG1 itself binds specifically to DNA containing *cis*-[Pt(NH$_3$)$_2${d(GpG)}] or cis-[Pt(NH$_3$)$_2${d(ApG)}] intrastrand cross-links, but not the *cis*-[Pt(NH$_3$)$_2${d(GpTpG)}] 1,3-intrastrand cross-links or adducts made by *trans*-DDP or [Pt(dien)Cl]$^+$.

These results strongly suggest that HMG1 and SSRP1 recognize the unwinding and bending of DNA that occurs at the loci of *cis*-diammineplatinum(II) intrastrand d(GpG) and d(ApG) cross-links. Thus cisplatin binds to DNA to form structures that lead to the specific binding of HMG1 and HMG box proteins. This interaction could facilitate the anticancer activity of the drug in one of several ways, as outlined in Figure 7.23. These mechanisms suggest that, if the levels of the SSRPs could be controlled in cancer cells, they might be selectively sensitized to cisplatin.

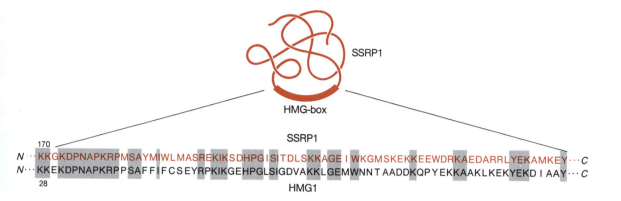

Figure 7.22
Comparison of HMG box sequences in SSRP1 and HMG1.

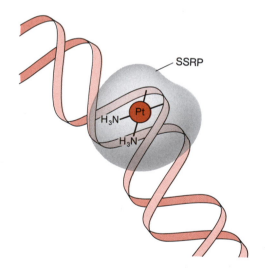

- Damage recognition as the first step in excision repair of the Pt/DNA lesion

- Titration of a tumor regulatory protein; removal from its normal functional role

- Protection of the cisplatin/DNA lesion from repair, thus assuring its ability to block replication and cell division

- Enhancement of the replication/ transcription inhibitory properties of cisplatin.

Figure 7.23
Proposals for how binding of a structure-specific recognition protein (SSRP) to cisplatin-modified DNA might alter gene function. Cells could be sensitized to the drug if levels of the SSRP were diminished (first proposal) or elevated (remaining three proposals).

7.4. Metal-Organized Structures as Probes for Conformation

Metal ions can also be used to organize small molecules into configurations that are appropriate for specifically binding to particular macromolecules. The relative ease (compared with organic chemistry) of producing species with elaborate three-dimensional structures based around central metal ions has been used to great advantage in recent years. This situation is particularly true for complexes that interact with nucleic acids.

7.4.a. Metallointercalators.   Intercalation is the mode by which planar aromatic heterocyclic cations such as ethidium (Figure 4.3) bind to double-stranded DNA. A variety of planar cationic metal complexes having nitrogen heterocyclic ligands also intercalates into DNA; one of these has been used to prove the neighbor exclusion binding model, as previously discussed in Section 4.3.a. In the complex $[Pt(terpy)(SCH_2CH_2OH)]^+$, the three rings of the terpyridine ligand are required to be coplanar, a conformation that facilitates intercalative binding to DNA (Figure 4.3). An interesting series that illustrates this point is the set of related complexes $[Pt(bpy)(en)]^{2+}$, $[Pt(o\text{-}phen)(en)]^{2+}$, and $[Pt(py)_2(en)]^{2+}$. As indicated in Figure 7.24, the X-ray fiber-diffraction diagrams of DNA containing the first two complexes are diagnostic of the intercalation mode. For the bis(pyridine) complex, however, in which the two pyridine rings cannot rotate to become coplanar with one another, the binding is purely ionic and has no intercalative component. Other planar metal complexes that bind to DNA by intercalation include metalloporphyrins. A different class of metallointercalators, in which the metal does not itself insert between the base pairs, includes complexes such as MPE-Fe(II) and $[Pt(AO\text{-}en)Cl_2]$ (Figure 7.25). Here the metal fragment can be used to generate hydroxyl radicals (MPE) or bind covalently to DNA ($[Pt(AO\text{-}en)Cl_2]$). Although metallointercalators are useful in probing the structures of nucleic acids, they have not yet been found to function in naturally occurring systems.

7.4.b. Conformational Recognition.   Although double-stranded DNA forms structures that are quite similar to those originally proposed from fiber-diffraction data, examination of oligonucleotide structures by higher-resolution single-crystal methods revealed much variability in DNA structure. Included are the further characterization of known differences, such as those between A- and B-forms, the discovery of an entirely new structure, Z-DNA, and the realization that more subtle but significant sequence-specific differences actually occur along the helix of a given DNA molecule. Physical

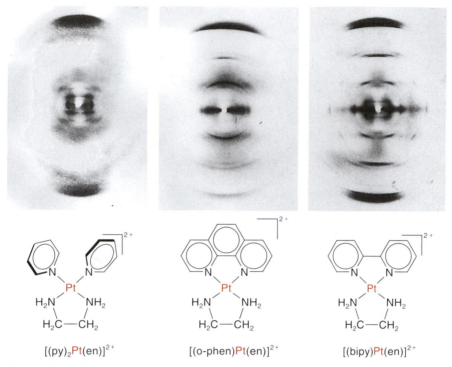

(a) X-ray fiber diffraction and structures of intercalating reagents

[(py)$_2$Pt(en)]$^{2+}$ [(o-phen)Pt(en)]$^{2+}$ [(bipy)Pt(en)]$^{2+}$

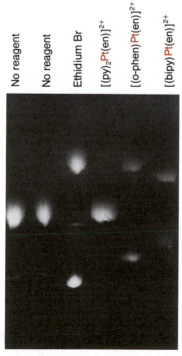

(b) Electrophoresis pattern

Figure 7.24
(a) X-ray fiber diffraction patterns for non-intercalating (left) and two intercalating (middle, right) platinum complexes. The strong reflections on the meridian at 10.2 Å and 5.1 Å are hallmarks of metallointercalation. (b) Gel electrophoretogram of a mixture of relaxed circular and nicked DNAs in the absence and presence of various reagents. Note that only the intercalators shift the mobility of the closed circular DNAs as a result of supercoil formation that accompanies intercalation. (Adapted from S. J. Lippard et al., *Science* **194**, 726–728 (1976)).

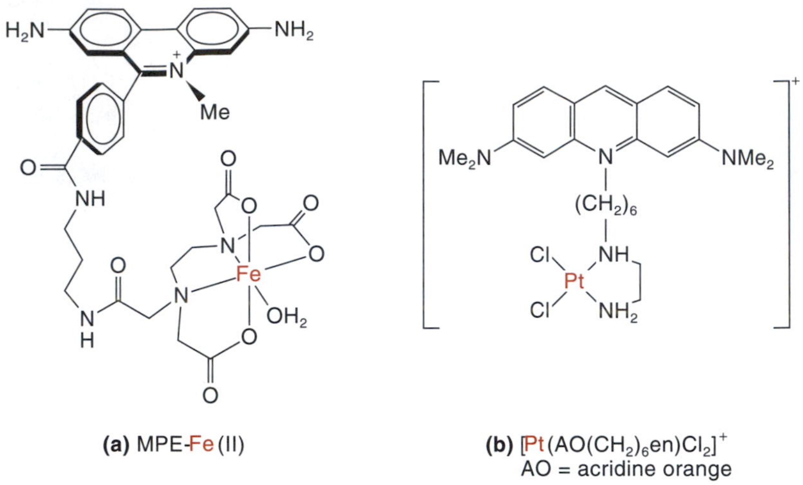

(a) MPE-Fe (II)

(b) [Pt(AO(CH$_2$)$_6$en)Cl$_2$]$^+$
AO = acridine orange

Figure 7.25
Structural diagrams of (a) MPE-Fe(II) and (b) [Pt(AO)(CH$_2$)$_6$en)Cl$_2$]$^+$, two metallointercalators in which the metal ion is not inserted between the base pairs upon DNA binding.

methods such as X-ray crystallography and NMR spectroscopy have been useful in extending these observations to other DNA molecules. These methods are time-consuming, difficult to apply to large DNA molecules, and subject to constraints imposed by the sample conditions (for example, crystal versus solution state). These obstacles have motivated the development of other, less-direct probes of DNA and RNA structures in solution.

An exciting early observation was that the two enantiomers of tris(phenanthroline) complexes of various metal ions bind to DNA with different affinities. This enantioselective binding laid the groundwork for the synthesis of more elaborate DNA-binding metal chelates that also have the advantage that they can be induced to cleave DNA via oxygen-based or photochemical reactions. These points are illustrated by tris(3,4,7,8-tetramethylphenanthroline)Ru(II), [Ru(TMP)$_3$]$^{2+}$ shown in Figure 7.26. Although the methyl groups preclude intercalative binding, studies reveal that the Λ enantiomer is preferentially bound at equilibrium and that nucleic acids that can adopt an A-like structure, such as DNA-RNA hybrids, are bound preferentially over B-form DNA. This and related Ru(II) complexes are capable of cleaving DNA in a light- and oxygen-dependent process that involves a diffusible species such as singlet oxygen. Based on this chemistry, [Ru(TMP)$_3$]$^{2+}$ can be used to probe large DNA species for sequences that are capable of forming A-like structures. More recently, additional Ru and Rh

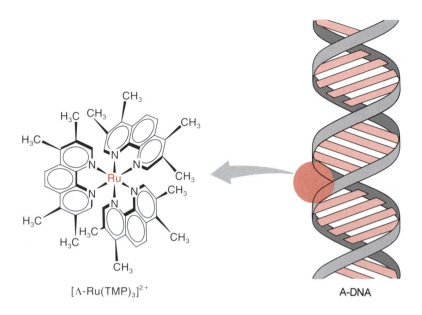

Figure 7.26
A chiral metal complex that preferentially binds to the A-form of duplex DNA.

tris(chelate) complexes in this class have been developed as probes for the tertiary structures of RNA molecules in solution. By studying the regions of protection against cleavage by compounds of known structure, such as tRNA, knowledge of the metal-binding-site preferences was obtained that could be used to deduce features of the tertiary structure of other RNA molecules, such as 5S RNA, from their cleavage patterns.

Study Problems

1. Many RNA molecules that are stabilized by added Mg^{2+} ions can also be stabilized by the addition of $[Co(NH_3)_6]^{3+}$. Account for this observation. Do you think that $[Co(NH_3)_6]^{3+}$ will be useful as a substitute for Zn^{2+} for the stabilization of protein structures? Explain.

2. One of the major mechanisms by which bound metal ions appear to stabilize protein structure is by introducing cross-links that reduce the conformational degrees of freedom of the unfolded protein. For a metal-binding sequence of the form Cys-X_i-Cys-X_j-Cys-X_k-Cys, the reduction in conformational entropy is estimated to be $R(-10.4 - 1.5 \ln[(i + 1)(j + 1)(k + 1)])$, where $R = 1.987$ cal/mol-K. Calculate this factor for one of the domains from a steroid receptor with $i = 2$, $j = 13$, $k = 2$. What factor does the cross-linking contribute to the equilibrium constant for protein folding at room temperature?

3. Some of the proteins that recognize cisplatin-DNA adducts have also been characterized by their ability to bind to cruciform structures that occur in double-stranded DNA having palindromic sequences. Comment on the implications of this observation for the likely structures of these cruciforms. Does this observation offer any potential insight into the possible mechanism(s) of action of cisplatin?

4. Human-growth hormone binds to two different receptors, one of which also binds the hormone prolactin. Studies revealed that the growth-hormone-prolactin receptor interaction is weak in the presence of EDTA, but increases 8,000-fold upon the addition of 50 μM zinc. Mutagenesis studies indicated that changing His18, His21, or Glu174 in growth hormone to alanine reduced the strength of this interaction by two orders of magnitude. Even larger effects were observed following mutation of His188 in the prolactin receptor to alanine. Propose a model that accounts for these observations [*Science* 250, 1709 (1990)].

Bibliography

Stabilization of Protein Structure
Y. S. Babu, C. E. Bugg, and W. J. Cook. 1988. "Structure of Calmodulin Refined at 2.2 Å Resolution." *J. Mol. Biol.* 204, 191–204.

J. M. Berg. 1990. "Zinc Fingers and Other Metal-Binding Domains." *J. Biol. Chem.* 265, 6513–6516.

M. J. Cormier. 1983. "Calmodulin: The Regulation of Cellular Function," in T. G. Spiro, ed., *Calcium in Biology*. Wiley-Interscience, New York, 53–106.

R. M. Evans and S. M. Hollenberg. 1988. "Zinc Fingers: Gilt by Association." *Cell* 52, 1–3.

M. Ikura, G. M. Clore, A. M. Gronenborn, G. Zhu, C. B. Klee, and A. Bax. 1992. "Solution Structure of a Calmodulin-Target Peptide Complex by Multidimensional NMR." *Science* 256, 632–638.

A. Klug and D. Rhodes. 1987. "'Zinc Fingers': A Novel Protein Motif for Nucleic Acid Recognition." *TIBS* 12, 464–469.

B. F. Luisi, W. X. Xu, Z. Otwinowski, L. P. Freedman, K. R. Yamamoto, and P. B. Sigler. 1991. "Crystallographic Analysis of the Interaction of the Glucocorticoid Receptor with DNA." *Nature* 352, 497–505.

R. Marmorstein, M. Carey, M. Ptashne, and S. C. Harrison. 1992. "DNA Recognition by GAL4: Structure of a Protein-DNA Complex." *Nature* 356, 408–414.

W. E. Meador, A. R. Means, and F. A. Quiocho. 1992. "Target Enzyme Recognition by Calmodulin: 2.4 Å Structure of a Calmodulin-Peptide Complex." *Science* 257, 1251–1255.

K. T. O'Neil and W. F. DeGrado. 1990. "How Calmodulin Binds Its Targets: Sequence-Independent Recognition of Amphiphilic α-Helices." *TIBS* 15, 59–64.

N. P. Pavletich and C. O. Pabo. 1991. "Zinc Finger–DNA Recognition: Crystal Structure of a Zif268-DNA Complex at 2.1 Å." *Science* 252, 809–817.

M. F. Summers, L. E. Henderson, M. R. Chance, J. W. Bess, Jr., T. L. South, P. R. Blake, I. Sagi, G. Peret Alvarado, R. C. Sowder III, D. R. Hare, and L. O. Arthur. 1992. "Nucleocapsid Zinc Finger Detector in Retroviruses: EXAFS Studies of Intact Viruses and the Solution-Structure of the Nucleocapsid Protein from HIV-1." *Pro. Sci.* 1, 563–574.

Aspartate Transcarbamoylase

E. R. Kantrowitz and W. N. Lipscomb. 1988. "*Escherichia coli* Aspartate Transcarbamylase: The Relation Between Structure and Function." *Science* 241, 669–674.

D. W. Markby, B.-B. Zhou, and H. K. Schachman 1991. "A 70-Amino Acid Zinc Binding Polypeptide from the Regulatory Chain of Aspartate Transcarbamoylase Forms a Stable Complex with the Catalytic Subunit Leading to Markedly Altered Enzyme Activity." *Proc. Natl. Acad. Sci. USA* 88, 10568–10572.

M. E. Nelbach, V. P. Pigiet, Jr., J. C. Gerhart, and H. K. Schachman. 1972. "A Role for Zinc in the Quaternary Structure of Aspartate Transcarbamoylase from *Escherichia coli*." *Biochemistry* 11, 315–327.

H. K. Schachman. 1988. "Can a Simple Model Account for the Allosteric Transition of Aspartate Transcarbamoylase?" *J. Biol. Chem.* 263, 18583–18586.

Metal-Ion Stabilization of Nucleic-Acid Structure

F. Aboul-ela, A. I. H. Murchie, and D. M. J. Lilley. 1992. "NMR Study of Parallel-Stranded Tetraplex Formation by the Hexadeoxynucleotide d(TG$_4$T)." *Nature* 360, 280–282.

E. H. Blackburn. 1990 "Telomeres: Structure and Synthesis." *J. Biol. Chem.* 265, 5919–5921.

D. W. Celander and T. R. Cech. 1991. "Visualizing the Higher Order Folding of a Catalytic RNA Molecule." *Science* 251, 401–407.

D. E. Draper. 1992. "The RNA-Folding Problem." *Acc. Chem. Res.* 25, 201–207.

A. Jack, J. E. Ladner, D. Rhodes, R. S. Brown, and A. Klug. 1977. "A Crystallographic Study of Metal-Binding to Yeast Phenylalanine Transfer RNA." *J. Mol. Biol.* 111, 315–328.

C. Kang, X. Zhang, R. Ratliff, R. Moyzis, and A. Rich. 1992. "Crystal Structure of Four-Stranded *Oxytricha* Telomeric DNA." *Nature* 356, 126–131.

D. Sen and W. Gilbert. 1990. "A Sodium-Potassium Switch in the Formation of Four-Stranded G4-DNA." *Nature* 344, 410–414.

F. W. Smith and J. Feigon. 1992. "Quadruplex Structure of *Oxytricha* Telomeric DNA Oligonucleotides." *Nature* 356, 164–168.

W. Sundquist and A. Klug. 1989 "Telomeric DNA Dimerizes by Formation of Guanine Tetrads Between Hairpin Loops." *Nature* 342, 825–829.

J. R. Williamson, M. K. Raghuraman, and T. R. Cech. 1989. "Monovalent Cation-Induced Structure of Telomeric DNA: The G-Quartet Model." *Cell* 59, 871–880.

Unwinding and Bending of DNA Modified by Platinum Anticancer Drugs

S. F. Bellon and S. J. Lippard. 1990. "Bending Studies of DNA Site-Specifically Modified by Cisplatin, *trans*-Diamminedichloroplatinum (II), and *cis*-[Pt(NH$_3$)$_2$(N3-cytosine)Cl]$^+$." *Biophys. Chem.* 35, 179–188.

S. F. Bellon, J. H. Coleman, and S. J. Lippard. 1991. "DNA Unwinding Produced by Site-Specific Intrastrand Cross-Links of the Antitumor Drug *cis*-Diamminedichloroplatinum (II)." *Biochemistry* 30, 8026–8035.

S. L. Bruhn, P. M. Pil, J. M. Essigmann, D. E. Housman, and S. J. Lippard. 1992. "Isolation and Characterization of Human cDNA Clones Encoding a High Mobility Group Box Protein that Recognizes Structural Distortions to DNA Caused by Binding of the Anticancer Agent Cisplatin." *Proc. Natl. Acad. Sci. USA* 89, 2307–2311.

P. M. Pil and S. J. Lippard. 1992. "Specific Binding of Chromosomal Protein HMG1 to DNA Damaged by the Anticancer Drug Cisplatin." *Science* 256, 234–237.

Metal-Organized Structures as Probes for Conformation

J. K. Barton. 1986. "Metals and DNA: Molecular Left-Handed Complements." *Science* 233, 727–734.

A. M. Pyle and J. K. Barton. 1990. "Probing Nucleic Acids with Transition Metal Complexes," in S. J. Lippard, ed., *Progress in Inorganic Chemistry*, Volume 38. Wiley-Interscience, New York, 413–475.

D. S. Sigman. 1986. "Nuclease Activity of 1,10 Phenanthroline–Copper Ion." *Acc. Chem. Res.* 19, 180–186.

T. D. Tullius, ed. 1989. *Metal-DNA Chemistry*, Number 402. ACS Symposium Series, American Chemical Society, Washington, DC.

CHAPTER 8

Binding of Metal Ions and Complexes to Biomolecule-Active Centers

---

Principles: *Biomolecules can select metal ions in terms of their charge, radius, lability, hard-soft properties, position in the Irving-Williams series, ligand and other coordination preferences, and bioavailability. Some systems are clearly under thermodynamic control; for others, kinetic factors are important. Shape selectivity is used to discriminate between related but stereochemically different metal complexes when binding to biopolymers such as proteins and nucleic acids. Kinetic barriers to the insertion of metals into their binding sites can be overcome by specialized enzymes that distort the receptor to facilitate coordination. Since most metal-binding sites are buried within proteins, near overall charge neutrality is usually maintained. Ligand protonation/deprotonation equilibria play an important role in maintaining charge balance.*

8.1. Selection and Insertion of Metal Ions for Protein Sites

We have seen in Chapters 5 and 6 how specific metal ions are selected and accumulated by cells for incorporation into metalloproteins or nucleic-acid structures. In subsequent chapters we shall discuss how particular metal ions can execute distinct functions with fine-tuning by the biological host. The relationship between the two themes of accumulation and function requires the insertion of "correct" metal ions into their biological binding sites. Incorporation of incorrect metal ions can have detrimental consequences. For example, binding of zinc into a copper site in an electron-carrier protein (see Section 9.1.b) would produce a molecule incapable of undergoing redox chemistry, since, under normal conditions, zinc has only one accessible oxidation state. Similarly, misincorporation of a redox-active metal ion into a zinc-finger protein could, in principle, induce oxidative cleavage reactions close to the key DNA recognition sites within the genome. Thus proper functioning of metalloproteins within biological systems requires specific binding of appropriate metal ions and exclusion of inappropriate ones. In this chapter we examine the strategies adopted by nature to achieve this specificity.

8.1.a. Thermodynamic Control. For some systems, thermodynamic factors determine which metal ion(s) are incorporated into metal-binding sites. We shall discuss three examples that illustrate the chemical features employed to achieve the appropriate thermodynamic preferences. The first involves the zinc-finger proteins discussed in Section 7.1.b. Early experiments demonstrating the role of metal ions in proper functioning of the TFIIIA transcription factor revealed that added Zn^{2+} could reactivate protein treated with EDTA. Other metal ions such as Co^{2+}, Ni^{2+}, Fe^{2+}, and Mn^{2+} were not able to restore activity at similar concentrations. The question of whether differences in metal-ion affinity for the cysteine- and histidine-binding sites in the protein can explain these observations has been probed by direct spectroscopic determination of the dissociation constants for the complexes of a single zinc-finger peptide with a variety of metal ions. The dissociation constants were measured to be 2 pM, 50 nM, 2 μM, 3 μM, and > 10 μM for Zn^{2+}, Co^{2+}, Ni^{2+}, Fe^{2+}, and Mn^{2+}, respectively. Thus the unique role of zinc ion in reactivating TFIIIA can be explained by the fact that it binds 1,000 to 100,000 times more tightly to the proteins than do the other metals that were examined.

This zinc-ion specificity has been rationalized in terms of changes in ligand-field stabilization energy (LFSE) accompanying metal binding to the tetrahedral zinc-finger sites (Figure 8.1). LFSE is the energy associated with unequal populations of d-orbitals arising from d-orbital splitting in a site of nonspherical symmetry (Section 2.3). For a d^{10} ion such as Zn^{2+}, the LFSE is zero for any geometry. For ions with an incompletely filled d shell, however, the LFSE is different from zero and depends on the geometry and the nature of the ligands. As an example, let us consider Co^{2+}, a d^7 ion. In an octahedral site such as that in the $[Co(OH_2)_6]^{2+}$ ion, the LFSE has the value $-4/5\Delta_0$, where Δ_0 is the splitting between the t_{2g} (d_{xy}, d_{xz}, and d_{yz}) and e_g (d_{z^2} and $d_{x^2-y^2}$) orbitals. For the aqua ion, Δ_0 has the value 9300 cm^{-1}, so that the LFSE is -7440 cm^{-1}, or -21.3 kcal/mol. For Co^{2+} in a tetrahedral site, the LFSE is $-6/5\Delta_t$, where Δ_t is the d orbital splitting for a tetrahedral site, which is approximately half that for an octahedral site. For a zinc-finger site containing two cysteine thiolate and two histidine imidazole ligands, Δ_t is 4900 cm^{-1}, so that the LFSE is 5880 cm^{-1} or -16.8 kcal/mol. These values show that a Co^{2+} ion loses $21.3 - 16.8 = 4.5$ kcal/mol in LFSE in going from an octahedral site in aqueous solution to the tetrahedral site in a zinc-finger domain. The Zn^{2+} ion, in contrast, has no such price to pay, since it has no LFSE in any site. This difference in LFSE change semiquantitatively accounts for the observed difference in the dissociation constants for the zinc-finger peptide complexes of Zn^{2+} or Co^{2+}.

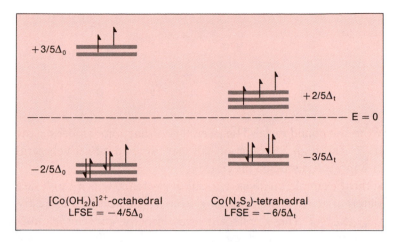

Figure 8.1
Ligand field stabilization energy (LFSE) for cobalt(II) (d^7) in an octahedral environment in aqueous solution and in a tetrahedral Cys_2His_2 site within a zinc-finger protein.

LFSE effects do not alone dictate metal-binding preferences, however. Other factors, including hard-soft acid-base effects and the position of the metal ions within the Irving-Williams series, can be important determinants as well. As an illustration of this point, consider the binding of Zn^{2+} and Cd^{2+} to a Cys_2His_2 site such as that discussed earlier and to Cys_4 sites such as those found in the steroid receptors. Both of these ions have d^{10} electronic configurations, so that LFSE effects are immaterial. For the Cys_2His_2 site, Zn^{2+} binds approximately two orders of magnitude more tightly than does Cd^{2+}, whereas, for the Cys_4 sites, Cd^{2+} binds two to three orders of magnitude more tightly. This large difference can be explained in terms of hard-soft acid-base effects. Since cadmium is below zinc in the periodic table, it is softer and therefore prefers the softer sulfur ligands over the harder nitrogen donors (Section 2.1).

The second example of the role of thermodynamics in metal-site selectivity is the binding of calcium to EF-hand proteins. Since these proteins respond to changes in intracellular calcium-ion concentration, they must undergo relatively rapid metal-exchange reactions. For this reason, metal binding must be under thermodynamic rather than kinetic control. Furthermore, as was discussed in Section 7.1.c, the proteins must be reasonably specific for Ca^{2+} over other ions, particularly Mg^{2+} and monovalent ions such as Na^+ and K^+. Much of the metal-binding specificity can be accounted for in terms of hard-soft acid-base and coordination-number effects. Calcium, a relatively hard metal, has a strong preference for oxygen ligands. The EF-hand sites are made up entirely of such ligands from carboxylate and carboxamide side chains, from peptide carbonyl groups, and from water molecules that are not displaced from the bound metal. The overall coordination number for these sites is usually 7. This high coordination number favors the relatively larger Ca^{2+} ion over Mg^{2+}, as discussed previously.

The third example involves iron-sulfur proteins, which can house a variety of different active centers, ranging from single-iron ions to Fe_2S_2, Fe_3S_4, and Fe_4S_4 clusters. These units are bound to the proteins almost always through the thiolate groups of cysteinate side chains. Because these different centers have distinct electronic properties, it is important that the "correct" cluster be inserted into each protein site. The mechanisms of iron-sulfur protein biosynthesis are not yet well understood. A variety of studies has revealed that the clusters will spontaneously self-assemble, even in aqueous environments; so it seems plausible that the clusters are built up from mononuclear iron precursors and sulfide ligands during the process of protein folding. An alternative possibility is that cluster transfer reactions occur, whereby a preassembled cluster is transferred from solution to a site in a protein by exchanging its simpler thiolate ligands for cysteinate groups. Such reactions have formed the basis for the core-extrusion method of iron-sulfur cluster identification discussed in Section 5.3.a.vi. Core-extrusion methods depend on the reaction shown in Equation 5.9, in which, under appropriate conditions, an excess of thiol can remove a cluster intact from its protein environment. Not only can these extruded clusters be characterized in their free form, in which they are bound to simple thiolate ligands, but they can alternatively be captured by addition of the appropriate apoprotein. For example, apoadrenal ferredoxin, an Fe_2S_2-cluster-binding protein, and apo B. polymyxa ferredoxin, an Fe_4S_4-cluster-binding protein, have been reconstituted in this manner and refolded by dilution of the organic solvent. The overall reaction is shown in Equation 8.1. The inclusion of the strong

$$[Fe_2S_2(SR)_4]^{2-} + \text{apo adrenal Fd} \xrightarrow[\text{dithionite}]{\text{refolding}} \text{adrenal Fd}_{red}$$

$$[Fe_4S_4(SR)_4]^{2-} + \text{apo B. polymyxa Fd} \xrightarrow[\text{dithionite}]{\text{refolding}} \text{B. polymyxa Fd}_{red}$$

$$(8.1)$$

reductant dithionite produces the reduced forms of the reconstituted proteins, which can be identified by EPR spectroscopy. Experiments used in the development of these methods have demonstrated that the ferredoxin apoproteins show strong preferences for their cognate clusters. Since the disposition of the four thiolate ligands is rather different for the Fe_2S_2 versus the Fe_4S_4 clusters, it seems likely that shape differences in the folded structures of these proteins provide the basis for this recognition.

8.1.b. Kinetic Control. Passive metal binding, in which insertion of a metal is not facilitated by any external agent, appears to account for the incorporation of metal ions into many systems. Some other demetallated proteins or prosthetic groups do not bind metal ions at rates that would be biologically viable. For these systems, metal incorporation must be facilitated by some additional mechanism or factor and can be under kinetic rather than thermodynamic control. Although no system of this type is yet thoroughly understood in detail, we will discuss here one of the better-characterized examples.

The copper-containing nitrous oxide reductase from *Pseudomonas stutzeri* is an enzyme that functions in the process of denitrification, or removal of nitrate from the soil. Genetics experiments revealed that certain mutants produced the copper-free, apo form of this enzyme. One of these mutants lacks the product of the *nosA* gene, which acts as an "insertase," taking up copper from the external medium and transferring it to the apo enzyme. The gene product, NosA, which has been purified and partially characterized, is found in the outer membrane of the cell, whereas the nitrous-oxide reductase enzyme is present in the periplasm. NosA has a subunit molecular mass of 65 kilodaltons, with 1 copper bound per subunit. Expression of NosA is highly sensitive to exogenous copper concentration, being significantly reduced by levels of copper in the micromolar range. Since the ability of this protein to transfer copper to the apoenzyme *in vitro* has not been demonstrated, other factors may be required. Recently, three additional genes, *nosD*, *nosF*, and *nosY*, have been identified, all of which are required for copper acquisition or processing. Their gene products are NosF, a 33.8-kDa protein with a nucleotide (for example, ATP or GTP) binding site; NosY, 29.4 kDa, which may be a transmembrane spanning protein; and NosD, a 48.2-kDa periplasmic component. Although specific functions for these components have not been delineated, all are required for insertion of copper into the N_2O reductase. The family of Nos proteins represents just one of many potential insertase systems that have been identified by such genetic techniques. As these systems are more fully characterized, through expression and studies of the proteins, much new information can be expected. It should be noted that metal-insertion mutants may be related to a variety of different defects in mechanisms, including metal uptake, modification of protein side chains, protein folding and unfolding, and biosynthesis of cofactors or effector molecules. The systems used to transport and insert metal ions into protein-active sites probably have much in common with other bacterial systems that have evolved for transporting small molecules.

Another example where kinetic rather than thermodynamic factors dictate metal-ion binding to a biomolecule-active center is the ferrochelatase system, to be discussed in Section 8.4.a.

8.1.c. Bioavailability. Whether a system is under thermodynamic or kinetic control, enough of the cognate metal must be present to bind and possibly to compete against any other metals that might fit into the receptor site. In Chapters 5 and 6 we discussed some of the mechanisms by which organisms can accumulate and regulate metal-ion concentrations. Such systems are clearly crucial for metallobiomolecule biosynthesis. Direct evidence for the importance of the bioavailability of metal ions is provided by observations about different proteins with highly similar metal-binding sites that come from different organisms and, hence, experience different metal-ion concentration pools. Two such proteins are rubredoxin (Section 5.3.a.i) from bacteria such as *Clostridium* and *Desulfovibrio*, and aspartate transcarbamoylase (Section 7.1.a) from *E. coli*. Each of these proteins has a metal-binding site comprising four cysteinate thiolate groups with very similar overall structures. Only the iron complex of rubredoxin and the zinc complex of ATCase, however, have been reported as being isolated from natural sources. Thermodynamic studies of metal binding to four cysteinate sites reveal that zinc(II) binds very much more tightly than does iron(II). Thus the observation that rubredoxin contains only iron as isolated strongly suggests that zinc concentrations are limited relative to iron concentrations in the species of bacteria that use this protein. This point has received further support in studies of heterologous expression of rubredoxin in *E. coli.*, where it was found that much of the protein produced contained zinc rather than iron. The alternative explanation, that one or both bacterial systems contain auxiliary factors involved in metal incorporation into these sites, has not been excluded.

8.2. Preservation of Electroneutrality

Many metal-binding sites are buried deep within proteins. Because metal ions are positively charged, such placement can be quite costly energetically. Thus, if much of the metal-ion charge were not neutralized by negatively charged ligands, binding to proteins would be unfavorable. Examination of well-defined metal-binding sites reveals that most of them have overall charges that are very close to zero. Exceptions to this observation occur when the metal ions are located close to or on the protein surface.

8.2.a. Mononuclear Sites. Many sites in metallobiomolecules contain single metal ions with formal oxidation states that range from $+1$ through $+6$, charges that can be neutralized in several different ways. We first consider sites that consist only of ligands contributed from protein side chains. Examples include the zinc-finger protein sites (Section 7.1.b), the blue-copper protein sites (Section 9.1.b), and the iron site in rubredoxin (Section 5.3.a.i). In the first example, the metal is Zn(II), and the ligands are derived from two cysteine and two histidine residues. The cysteines are deprotonated upon metal binding, whereas the histidines are neutral, so that the overall charge on the ligands is -2, and the charge on the complex is 0. For the blue-copper sites, the metal ion bound is copper, which alternates between the $+1$ and $+2$ oxidation levels. The ligands are one cysteinate, two histidines, one methionine, and, perhaps, a peptide carbonyl, so that the overall charge contributed by the ligands is -1. The charge on the complex is therefore 0 in the reduced, Cu(I), form and $+1$ in the oxidized, Cu(II), form. For rubredoxin, the metal is Fe(II, III), and the ligands are four cysteinates. The overall charge on the complex is -2 for the reduced form and -1 for the oxidized form. Here residual charge persists in either oxidation state. This site is close to the protein surface, however, and some of the charge is neutralized by hydrogen bonding between the peptide amide and the cysteinate sulfur atoms.

Other mononuclear sites have some open coordination positions that can be occupied by substrate or solvent molecules. Of necessity, these sites are near the protein surface, but the tendency to achieve charge neutrality is preserved to some extent. In the active site of carbonic anhydrase (Section 10.1), however, none of the $+2$ charge of the zinc is neutralized by ligands contributed by the protein, since the amino-acid side chains utilized are three neutral histidine groups. The remaining site is occupied by a water molecule that has quite a low pK_a value, however, so that it is easily deprotonated to form OH^-, neutralizing some of the charge. In general, deprotonation equilibria of bound water molecules provide an additional mechanism for the adjustment of the charge at a metal site.

Many mononuclear sites have ligands that are contributed by prosthetic groups such as porphyrins. Here again, electroneutrality is maintained. Free-base porphyrins are neutral, discounting charges on peripheral propionic-acid groups. Upon binding a metal, two protons are released (Equation 8.2, in Section 8.4.a). Thus, for a metal with a $+2$ charge, the metalloporphyrin complex is neutral; for a $+3$ charged metal, the overall charge is $+1$. This remaining charge can be neutralized by cysteinate or tyrosinate axial ligands. A final example is provided by the molybdenum-cofactor-containing enzymes (Section 5.4.d). In their oxidized forms, these enzymes usually have molybdenum in $+6$ formal oxidation state. This charge is neutralized by two negative charges from the molybdopterin dithiolate group and by two terminal oxo or one oxo and one sulfido groups. Here the prosthetic group and deprotonated water act in concert to produce an uncharged species.

8.2.b. Polynuclear Sites.   Polymetallic centers present a potentially even greater problem of maintaining charge neutralization, because of the presence of multiple units of positive charge. The ability of ligands to bridge two or more metal ions, however, facilitates deprotonation and the formation of negatively charged residues, providing an additional means of charge neutralization. Moreover, the higher overall charge on a polynuclear center is better tolerated, because it can be delocalized over several atoms. Let us consider three examples.

The first is Cu-Zn superoxide dismutase (SOD). The active site of this enzyme consists of a zinc(II) ion coordinated by three histidines and one aspartate and a copper (I, II) ion coordinated by four histidines. Under normal circumstances, this arrangement of ligands would be expected to lead to an overall charge of $+2$ or $+3$ on the dimetallic center, with only one negative charge provided by the ligands. In SOD, however, one of the histidine imidazole ligands is deprotonated and shared by the two metal ions, which contributes an additional negative charge. As a consequence, the overall charge on the complex is $+1$ for the Cu(I) form. It should be noted that this residual positive charge is functionally important. Together with additional positive charges from nearby amino-acid side chains, it helps to generate an electrostatic potential that attracts the negatively charged substrate O_2^- into the active site (Section 11.4.a).

The second example involves iron-sulfur clusters. The iron centers in a four-iron cluster carry a total positive charge ranging from $+9$ to $+11$, depending on their oxidation levels. This charge is neutralized by a combination of bridging sulfido groups, providing two negative charges each, and anionic terminal ligands such as cysteinate. Terminal sulfido groups are very rare in metallobiomolecules, the active site of xanthine oxidase being one example. Doubly or triply bridging sulfido groups are much more common. For a four-iron cluster, four triply bridging sulfido groups account for a -8 charge contribution. With four terminal cysteinate ligands, the overall charge on an $Fe_4S_4(SR)_4$ ranges from -3 to -1, or less than -1 per metal ion. Much like rubredoxin, this negative charge is compensated for by peptide amide hydrogen bonding to sulfido or terminal thiolate groups.

The third example involves the two-iron center of hemerythrin. In the deoxy form, this core consists of two Fe(II) ions with five imidazole and two carboxylates as protein-derived ligands (see Figure 1.2). A solvent-derived hydroxide ion bridges the two irons for an overall charge of $+1$. Binding of O_2 results in significant electronic rearrangement. The irons are oxidized to Fe(III), the O_2 is reduced to peroxide and protonated, and the hydroxide is deprotonated to form oxide. There is now a total of five negative charges (two from carboxylates, one from hydroperoxide, and two from oxide) and six positive charges from the two ferric ions, so that the overall charge is maintained. This example again illustrates the role that ligand protonation and deprotonation equilibria can play in maintaining approximate charge balance.

8.3. Metal-Ion and Metal-Complex Binding to Nucleic Acids

Many of the principles described in Section 8.1 for metal ion-protein interactions appear also to apply to metal-nucleic acid binding and will not be further elaborated. As was discussed in Section 1.2, however, nucleic acids are the biological target molecules for metal-based pharmaceuticals, and we might inquire whether these unnatural metal complexes follow the same principles in binding to their cellular receptors. As a paradigm for this group of drugs, we will consider the reaction chemistry of cisplatin with its major biological target, DNA. This example draws attention to another aspect of selectivity of the binding of metal ions to biopolymers. Once inside the cell and, if necessary, metabolized to an active form, metal ions or complexes must find their way to the target sites necessary to afford a chemotherapeutic response, but with minimal binding to alternative targets, which could lead to undesired toxicity. Although this subject is in its infancy, some information about the strategies adopted is available through studies of cisplatin.

We have already seen in Chapters 5 and 7 that it is the hydrolyzed, cationic forms of cisplatin that interact with DNA to form intrastrand d(GpG) and d(ApG) cross-links, together comprising 90 percent of all the adducts. This high degree of regiospecificity, which far exceeds the frequency of nearest neighbor pairs of purine nucleosides, appears to be dictated by both thermodynamic and kinetic factors. Theoretical studies of the electrostatic potential of double-stranded DNA reveal that the purine N7 positions are the most nucleophilic. As the positively charged cis-$[Pt(NH_3)_2Cl(H_2O)]^+$ complex approaches the duplex, it is presumed to migrate along the sugar-phosphate backbone until it finds the most negative electrostatic potential, a pair of adjacent purine residues. Thus the regiospecificity of the drug is dictated in part by thermodynamic preferences for the most negative patches on the DNA. The preference for purine nitrogen rather than phosphate oxygen donors reflects the strong preference of Pt(II) to coordinate to the softer donor ligand.

Kinetic factors are also important, however. The time-dependent formation of monofunctional adducts, in which only one site in the platinum coordination sphere is occupied by a DNA donor atom, followed by closure to form intrastrand cross-links, has been monitored by using ^{195}Pt NMR spectroscopy (Figure 8.2). The chemical shift of the platinum-195 nucleus ($I = 1/2$) is quite sensitive to the nature of the coordination environment and can be used to monitor the stepwise substitution of nonammine ligands by nitrogen donors on the DNA. As Figure 8.2 shows, well-separated ^{195}Pt resonances appear over the time course of the reaction corresponding to the two different kinds of adducts. Kinetic analysis revealed the half-life for closure of the monofunctional to bifunctional adducts to be two hours. This value is characteristic of the time required for hydrolysis of chloride ligands from Pt(II) complexes and strongly implies that the rate-determining step is replacement of Cl$^-$ with an aqua ligand, which is then more rapidly (~ 8 min) displaced by a purine N7 donor ligand. Since X-ray structural studies have shown that hydrogen bonding occurs between the phosphate oxygen atom on the 5' side of the d(pGpG) unit and the ammine ligand on the platinum atom (Figure 7.18), a reasonable hypothesis is that this interaction helps guide the cis-diammineplatinum(II) unit to its site on the DNA.

This postulate has received recent support through an investigation of the binding of cis-[Pt(NH$_3$)(CyNH$_2$)(H$_2$O)$_2$]$^{2+}$, where CyNH$_2$ is cyclohexylamine, to DNA. This complex is the presumed active metabolite of a new, orally active, anticancer platinum(IV) compound, cis,trans,cis-[Pt(NH$_3$)(CyNH$_2$)(CH$_3$CO$_2$)$_2$Cl$_2$], that has recently entered clinical trials. When cis-[Pt(NH$_3$)(CyNH$_2$)(H$_2$O)$_2$]$^{2+}$ binds to DNA, two orientational isomers of the intrastrand d(GpG) cross-link are formed in a 2:1 ratio as the major adducts, depending on whether the cyclohexylamine ligand is directed toward the 5' or the 3' nucleobase. From NMR measurements of the $J_{N,N}$ coupling constants for ^{15}N-enriched ammine and guanine ligands in cis-[Pt(NH$_3$)(CyNH$_2$)d(GpG)]$^{2+}$, it was determined that the major isomer has its ammine ligand oriented to the 5' side, which maximizes the hydrogen-bonding interaction with the phosphate oxygen atom (compare Figure 7.18).

These results reveal that many of the same principles that dictate how metal ions are inserted into their natural binding sites in proteins also apply to the interaction of cisplatin with its target sites in DNA.

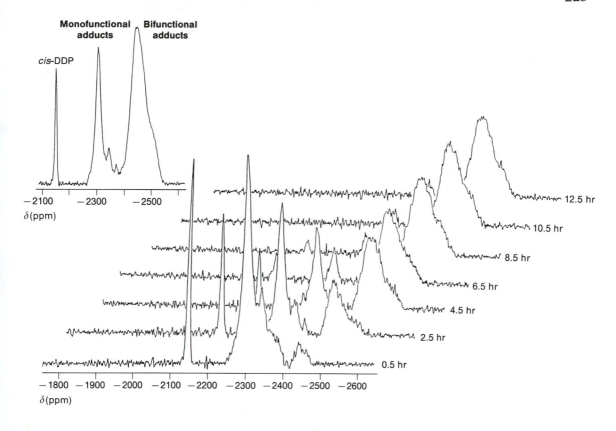

Figure 8.2

Time-dependent ^{195}Pt NMR spectral changes of cisplatin in the presence of sonicated calf thymus DNA. The signal at $\sim -2300\ \delta$ corresponds to monofunctional adducts, while that at $\sim -2450\ \delta$ corresponds to bifunctional adducts on the nitrogen donor atoms of the DNA bases adenine and guanine. (Adapted from D. P. Bancroft et al., *Journal of the American Chemical Society* **112**, 6860–6871 (1990)).

8.4. Biopolymer-Promoted Metal-Ligand Interactions

Proteins and nucleic acids can themselves catalyze or promote specific reactions of metal ions with ligands, a principle illustrated here by two examples. The first such system involves proteins that insert metal ions into porphyrins; the second employs double-stranded DNA as a template for promoting the substitution of a ligand on platinum by the amino substituent of an organic intercalator.

8.4.a. Ferrochelatase. The metallation reaction exhibited in Equation 8.2 is actually quite difficult to perform *in vitro*, often requiring high

$$\text{(porphyrin)} + Fe^{2+} \longrightarrow \text{(Fe-porphyrin)} + 2H^+ \qquad (8.2)$$

temperatures to achieve appreciable rates. An enzyme, ferrochelatase, has been characterized that catalyzes such reactions. The bovine enzyme has a molecular mass of 40 kilodaltons and will turn over approximately 250 to 2,500 times per hour in the presence of saturating amounts of substrate. This velocity represents a rate enhancement of about four orders of magnitude over the uncatalyzed rate at millimolar metal-ion concentrations. The enzyme is not, however, highly specific for a particular metal ion, for it will insert Zn^{2+} at essentially the same rate as Fe^{2+}. Although the mechanism of action of this enzyme has not been elucidated in detail, identification of a potent inhibitor has led to some exciting results. N-methylprotoporphyrin IX is highly effective in inhibiting ferrochelatase, with a K_i of 7 nM. Structural characterization of N-alkylporphyrins has revealed that they are significantly distorted from planarity, or "domed," because of steric interactions with the N-alkyl group, as shown for a zinc(II) complex in Figure 8.3. Based on this observation, N-methylmesoporphyrin IX was used as a hapten to produce monoclonal antibodies as shown in Figure 8.4. It was hypothesized that such

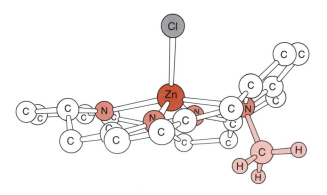

Figure 8.3
Structure of the zinc complex of an N-methyl porphyrin showing the distortion of the porphyrin ring.

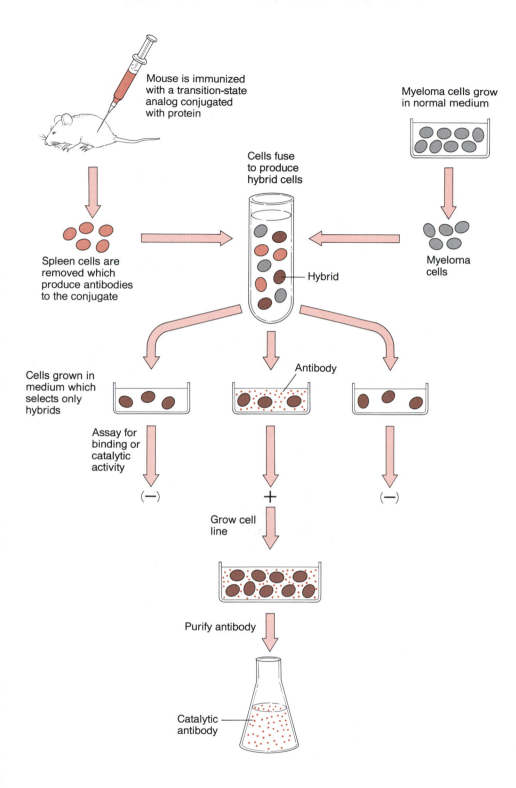

Mouse is immunized with a transition-state analog conjugated with protein

Myeloma cells grow in normal medium

Cells fuse to produce hybrid cells

Spleen cells are removed which produce antibodies to the conjugate

Hybrid

Myeloma cells

Cells grown in medium which selects only hybrids

Antibody

Assay for binding or catalytic activity

(−)

+

(−)

Grow cell line

Purify antibody

Catalytic antibody

Figure 8.4
Scheme depicting the route used to prepare catalytic monoclonal antibodies.

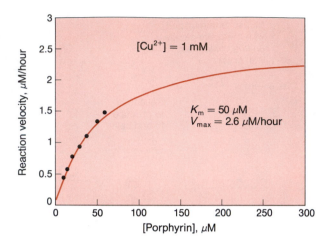

Figure 8.5

The kinetics of the insertion of Cu^{2+} into mesoporphyrin promoted by a catalytic antibody. The points represent the experimental data. The curve corresponds to a fit to Michaelis-Menten kinetics with the K_m and V_{max} values indicated. (Adapted from A. G. Cochran and P. G. Schultz, *Science* **249**, 781–783 (1990)).

antibodies might catalyze the metallation of related nonalkylated porphyrins by binding them in domed conformations with the pyrrole nitrogen atoms presented for metal binding in the manner displayed in Figure 8.3. This approach did, indeed, produce such an antibody. The kinetics of the reaction catalyzed by this antibody could be fit by Equation 8.3, where Ig is the antibody, P is the unmetallated porphyrin, and PM is the metallated

$$\text{Ig} + \text{P} \xrightarrow{K_m} \text{Ig.P} \xrightarrow[M^{2+}]{k_{cat,app}} \text{Ig.PM} \xrightarrow{K_P} \text{Ig} + \text{PM} \qquad (8.3)$$

porphyrin (Figure 8.5). With $M^{2+} = Zn^{2+}$ at 1 mM concentration, the kinetic parameters were found to be $K_M = 49$ μM and $K_P = 2.9$ μM, with the turnover number being 80 per hour. The catalytic antibody thus shares many properties with ferrochelatase. As expected, it is strongly inhibited by the hapten N-methylmesoporphyrin. In addition, porphyrin substrate and product binding are similarly in the 10 μM range in all cases. The biggest differences involve kinetic parameters relating to the metal ion. Whereas both ferrochelatase and the antibody will insert Zn^{2+}, the enzyme will not insert Cu^{2+}, which is a good substrate for the antibody. This difference in metal specificity may relate to the fact that the enzyme must have a metal-binding site in addition to the one for the porphyrin. Maximal turnover of the enzyme occurs at low metal-ion concentration ($K_M{}^{Zn} = 32\ \mu M$), whereas the turnover of the antibody shows no sign of saturation at the 1 mM metal-ion concentrations used in the assay. This example shows how the catalytic antibody approach can shed light on the functioning of a naturally occurring enzyme and demonstrates one mechanism by which an insertase can work.

8.4.b. DNA-Promoted Reaction Chemistry. Many compounds used in chemotherapy are administered as part of a regimen of drugs, which together provide better results than expected from the sum of their individual potencies. This synergism led investigators to evaluate the relative effects of cisplatin and intercalators in binding to DNA in order to find out whether cooperative effects might occur during reaction with the presumed target molecule in the cell. In particular, the binding of ethidium (Etd, Figure 4.3) and cisplatin to DNA was studied. In the absence of DNA and at low concentrations, 10 μM, there is very little covalent binding of Etd to cisplatin. When the two are mixed in the presence of DNA, however, a fraction of the platinum forms covalent cross-links between purine nucleobases on DNA and the exocyclic amino groups on the Etd ring. Analysis of the geometry of cisplatin bound monofunctionally to DNA and ethidium bound intercalatively at the adjacent position by molecular mechanics methods led to a reasonable hypothesis to explain this DNA-promoted reaction chemistry. As shown in Figure 8.6, the exocyclic amino group in such an arrangement is perfectly positioned to carry out an S_N2 displacement of chloride ion bound to the platinum atom. Moreover, since the Etd can migrate along the DNA, the frequency of encounter with bound platinum could be significantly greater than the collision frequency for the two freely diffusing molecules in solution. This phenomenon could also increase the rate of reaction. More recently, it has been found that a monofunctional cisplatin adduct with DNA reacts with the intercalating drug N-methyl-2,7-diazapyrenium and that the platinum complex of this intercalating ligand can dissociate from the biopolymer. Thus DNA might even catalyze the formation of a metal complex. Although much more work needs to be done to quantitate and extend these discoveries, they illustrate that biopolymers can serve as templates for the facilitated diffusion of two potentially reactive small molecules and can even orient them in a way that promotes a catalytic reaction between them. These principles could be used to design new drug molecules by using the target biopolymer to facilitate the formation of a reagent that could, ultimately, lead to its very destruction.

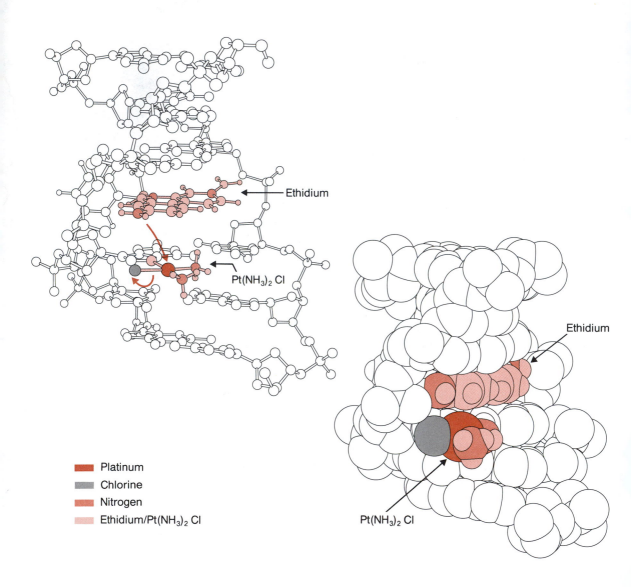

Figure 8.6
Ball-and-stick (left) and space-filling (right) views of the ternary complex between duplex DNA, ethidium, and *cis*-diamminechloroplatinum(II) leading to covalent bonding of ethidium to the platinum complex. The DNA holds the exocyclic amino group of the ethidium in nearly perfect position to initiate an S_N2 substitution of the chloride ligand on platinum.

Study Problems

1. Suppose that you wanted to build a metalloprotein-active site that contained the inert metal ion cobalt(III). Design a strategy by which you would incorporate this metal ion into the site.

2. If the site designed in the previous problem were to function as a reversible carrier of dioxygen, what amino-acid residues might you employ? In answering this question, bear in mind especially the charge-neutrality principle discussed in this chapter and the fact that the formal oxidation state of the host molecule might change upon binding to the metal.

3. It has recently been speculated that ceruloplasmin, a redox-active copper protein thought to be required for copper transport and storage, might play a critical function in the loading of iron into ferritin (see Chapter 6). Suggest how ceruplasmin might accomplish this function.

4. The reaction between electron-donor and electron-acceptor complexes such as ML_3^{n+}, where L is bipyridine or a related ligand and M is photoactivated Ru(III) (donor) or Co(III) (acceptor), has been studied in the presence of double-stranded DNA. Would you expect the rate constant for photoinduced electron transfer to differ from that found in solution in the absence of the nucleic acid? Discuss possible experiments that you might carry out to investigate the role that DNA might be playing.

5. The metal-binding selectivity of the copper protein azurin (see Section 9.1.b) has been probed by using scanning calorimetry. In this method, the temperature of a sample containing the protein is raised, and the amount of heat absorbed is monitored. At a certain temperature, corresponding to that at which the protein "melts" (becomes unfolded), a large amount of heat is absorbed. For native, Cu(II) azurin, a total of 21 cal/gram of heat is absorbed. For Ni(II)-substituted azurin, only 10 cal/gram are absorbed. Which metal ion is bound more tightly? Explain the observations in terms of ligand-field stabilization energies for an approximately tetrahedral site.

Bibliography

Selection and Insertion of Metal Ions for Protein Sites

D. Chatterji and F. Y.-H. Wu. 1982. "Selective Substitution In Vitro of an Intrinsic Zinc of *Escherichia coli* RNA Polymerase with Various Divalent Metals." *Biochemistry* 21, 4651–4656.

R. P. Hausinger. 1990. "Mechanisms of Metal Ion Incorporation into Metalloproteins. *Biofactors* 2, 179–184.

D. K. Lavallee. 1987. *The Chemistry and Biochemistry of N-Substituted Porphyrins.* VCH, Weinheim, 181–209.

K. Mokhele, Y. J. Tang, M. A. Clark, and J. L. Ingraham. 1987. "A *Pseudomonas stutzeri* Outer Membrane Protein Inserts Copper into N_2O Reductase." *J. Bact.* 169, 5721–5726.

W. G. Zumft, A. Viebrock-Sambale, and C. Braun. 1990. "Nitrous Oxide Reductase from Denitrifying *Pseudomonas stutzeri*: Genes for Copper-Processing and Properties of the Deduced Products, Including a New Member of the Family of ATP/GTP-Binding Proteins." *Eur. J. Biochem.* 192, 591–599.

Metal-Ion and Metal-Complex Binding to Nucleic Acids

S. L. Bruhn, J. H. Toney, and S. J. Lippard. 1990. "Biological Processing of DNA Modified by Platinum Compounds," in S. J. Lippard, ed., *Progress in Inorganic Chemistry*, Volume 38. Wiley-Interscience, New York, 477–516.

J. F. Hartwig and S. J. Lippard. 1992. "DNA Binding Properties of *cis*-[Pt(NH$_3$)(C$_6$H$_{11}$NH$_2$)Cl$_2$], a Metabolite of an Orally Active Platinum Anticancer Drug." *J. Am. Chem. Soc.* 114, 5646–5654.

Biopolymer-Promoted Metal-Ligand Interactions

A. G. Cochran and P. G. Schultz. 1990. "Antibody-Catalyzed Porphyrin Metallation." *Science* 249, 781–783.

H. A. Dailey and J. E. Fleming. 1983. "Bovine Ferrochelatase-Kinetic Analysis of Inhibition by N-Methylprotoporphrin, Manganese, and Heme." *J. Biol. Chem.* 258, 11453–11459.

D. K. Lavallee. 1988. "Porphyrin Metalation Reactions in Biochemistry." *Molecular Structures and Energetics.* 9, 279–314.

J.-M. Malinge, M. Sip, A. J. Blacker, J.-M. Lehn, and M. Leng. 1990. "Formation of a DNA Monofunctional *cis*-Platinum Adduct Cross-Linking the Intercalating Drug N-methyl–2,7–diazapyrenium." *Nucleic Acids Res.* 18, 3887–3891.

W. I. Sundquist, D. P. Bancroft, L. Chassot, and S. J. Lippard. 1988. "DNA Promotes the Reaction of *cis*-Diamminedichloroplatinum(II) with the Exocyclic Amino Groups of Ethidium Bromide." *J. Am. Chem. Soc.* 110, 8559–8560.

N. J. Turro, J. K. Barton, and D. A. Tomalia. 1991. "Molecular Recognition and Chemistry in Restricted Reaction Spaces: Photophysics and Photoinduced Electron Transfer on the Surfaces of Micelles, Dendrimers, and DNA." *Acc. Chem. Res.* 24, 332–340.

CHAPTER 9

Electron-Transfer Proteins

Principles: *The metal-binding sites in electron-carrying proteins are tailored to minimize structural reorganization accompanying changes in oxidation level. One-electron transfer processes are generally preferred. Coupling of proton and electron transfer affords a measure of redox potential control. Iron and copper are the most commonly utilized metals. Electrons can be transferred over long distances, > 10 Å. Reorganization energy, distance, and driving force are the most important parameters in determining protein electron-transfer rates, with medium effects playing a more subtle but potentially crucial role in some systems.*

At this point in our discussion, we have seen how metal ions are taken up by cells and inserted into their appropriate sites in proteins, nucleic acids, and cofactors. We turn in this and the following three chapters to a discussion of specific functions performed by metalloproteins, namely, electron transfer, the binding and activation of substrates, and atom- and group-transfer chemistry. The ability of the biopolymer to tune the properties of the same or similar metal centers in a variety of proteins is an important property that is also discussed in detail.

9.1. Electron Carriers

In a formal sense, all chemical transfer reactions can be classified broadly as acid-base and/or oxidation-reduction in character. In Chapter 10 we shall consider reactions catalyzed by metalloenzymes that are purely acid-base with no changes occurring in the oxidation-reduction state of the substrate, whereas in Chapter 11, we shall discuss atom- and group-transfer reactions. These latter transformations often have both acid-base and redox components. In this chapter we examine reactions that are purely oxidation-reduction in nature and involve the transfer of electrons from one protein to

231

another. Several classes of such electron-transfer proteins have been well characterized, and these will be discussed and compared.

9.1.a. Iron-Sulfur Proteins.

The metal-containing units in iron-sulfur proteins have been discussed extensively in Section 5.3.a. Table 5.2 lists the known categories of iron-sulfur proteins and some of their physical properties. In most of these proteins, the major function of the iron-sulfur cluster is to facilitate electron transfer, but Fe_nS_m units that carry out well-defined catalytic functions such as dinitrogen or nitrite reduction, interconversion of citrate and isocitrate, and even the site-specific hydrolysis of DNA have now been found in several enzymes. These enzymes usually, but not always, have additional metal centers at which atom-transfer reactions take place. The best-characterized example of an enzyme that has only an iron-sulfur center at its active site is aconitase. Interestingly, the reaction catalyzed by this enzyme does not appear to involve electron-transfer processes at any stage; yet a unit nearly identical to one found in pure redox iron-sulfur proteins is utilized. The functioning of a similar unit in the IRE-binding protein was discussed in Chapter 6, and aconitase itself will be addressed in Chapter 12.

Here we focus exclusively on iron-sulfur proteins that function as electron-transfer agents. The methods used to isolate and purify these proteins are of historical interest. Classically, in biochemistry, proteins have been identified by measuring an enzymatic activity that increases with purification by methodologies such as column chromatography, gel electrophoresis, and fractional precipitation. With the tools of modern molecular biology, however, pure proteins are often prepared from expression of a cloned gene or cDNA fragment. For iron-sulfur proteins having no enzymatic activity, isolation and purification relied heavily on physical rather than chemical methods, especially electron-spin resonance spectroscopy (Table 5.2). Fortunately, as already mentioned (Section 4.4.a.), the high sensitivity of the EPR method enabled impurities to be detected and removed, leading to some crystalline samples of purified iron-sulfur proteins.

The oxidation-reduction chemistry of iron-sulfur proteins and the corresponding synthetic models for their nFe-mS cores have been studied extensively. The four known classes of iron-sulfur proteins will be discussed in order of increasing number of iron atoms in the cluster.

(i) [1Fe-0S]. As noted in Section 5.3.a., the rubredoxins have active sites that consist of a high-spin Fe(II, III) ion coordinated to four cysteinate sulfur atoms in an approximately tetrahedral array. The crystal structures of several rubredoxins from different bacterial sources have now been solved and refined with the use of high-resolution (1.0 — 1.5 Å) X-ray diffraction data. The

mean iron-sulfur distances are all nearly identical within experimental error, having values near 2.28 Å. This distance is in good agreement with the value for the synthetic analog *bis*(*o*-xylyldithiolato)iron(III), in which the mean Fe−S distance is 2.267 Å. The structure of the reduced, Fe(II) forms of rubredoxin has not been as extensively studied. EXAFS studies of oxidized and reduced *P. aerogenes* rubredoxin indicated an increase in Fe−S distance from 2.26 to 2.32 Å. The latter distance is also consistent with data from model complexes such as *bis*(*o*-xylyldithiolato)iron(II) and tetrakis-(benzenethiolato)iron(II), which have mean Fe−S distances near 2.36 Å. Taken together, the structural data on the rubredoxins and model complexes indicate that the iron center does not change much upon reduction. The difference between the ferric and ferrous forms amounts to a 2−3 percent increase in Fe−S distance with no change in coordination number or spin state. The reduction potentials for rubredoxins are generally between −50 and +50 mV.

   (ii) [2Fe-2S]. The crystal structure of *S. platensis* ferredoxin revealed the presence of an $\{Fe_2S_2(S\text{-cys})_4\}$ unit. As discussed in Section 5.3.a., the structure of this dinuclear cluster was deduced from physicochemical experiments prior to its determination by X-ray crystallography. Moreover, synthetic model complexes had been prepared and structurally and spectroscopically characterized prior to the report of the protein structure. The oxidized form of this site contains two Fe(III) ions, each of which is tetrahedrally coordinated by two bridging sulfido groups and by two terminal cysteine thiolates. The iron-iron distance is 2.70 Å with Fe−S−Fe angles near 75°, reflecting the presence of significant metal-metal bonding. All physiological processes involving this type of center appear to undergo one electron transfer to form a mixed-valence Fe(III)-Fe(II) state. No detailed structural information on this state is available for a protein or for models, although EXAFS studies suggest an increase in the iron-iron distance to 2.76 Å. The two iron centers are antiferromagnetically coupled, and the odd electron is delocalized over both iron sites. Mössbauer spectroscopic studies, however, indicated that the electron exchange is slow on this time scale (see Table 4.1). Similar studies for a reduced model complex revealed the same behavior, indicating that the asymmetry, that is, the existence of discrete but interconverting Fe(III) and Fe(II) centers rather than a completely delocalized system, is an inherent property of the cluster and is not imposed by the protein. The reduction potentials for proteins containing these two iron sites are more negative than the mononuclear centers and range from −280 to −490 mV.

   (iii) [3Fe-4S]. The events leading to the characterization of three-iron centers in iron-sulfur proteins were discussed in Section 5.3.a. Well-refined crys-

tal structures are now available for several proteins that contain the $Fe_3S_4(S-cys)_3$ unit. This species is most easily visualized as an Fe_4S_4 cube missing one corner (see Equation 5.2). The oxidized form of the three-iron center contains three Fe(III) ions. EPR and Mössbauer spectra are consistent with three equivalent iron atoms and an overall spin of one-half. Thus the three high-spin ferric ions are antiferromagnetically coupled to yield a system with one unpaired electron in the ground state. The structure of this unit is consistent with a significant amount of iron-iron bonding, as found in the dinuclear sites discussed above. The reduced form of this center contains two Fe(III) ions and one Fe(II) ion. Mössbauer spectra indicate the presence of one high-spin Fe(III) center and a pair of equivalent sites with a net oxidation state near $Fe^{+2.5}$. The overall spin state of the trinuclear cluster is $S = 2$. These results can be interpreted in terms of antiferromagnetic coupling between the iron centers.

The potential of *D. gigas* ferredoxin II is -130 mV. Recently, the redox potentials in *A. vinelandii* ferredoxin I have been measured by direct electrochemical experiments (see Section 4.8). Originally it had been proposed that the [4Fe–4S] and [3Fe–4S] clusters in this protein underwent one-electron transfer reactions at potentials of $+0.320$ and -0.424 V, respectively. These results raised the question as to what possible function could require such disparate potentials. The initial supposition proved to be erroneous, however, when it was discovered that $K_3[Fe(CN)_6]$, the reagent used in the redox titrations, destroys the [4Fe–4S] cluster, producing a thiol radical. This example illustrates the importance of assuring that a metalloprotein retains its native composition and structure following any physical measurement. Figure 9.1 displays the cyclic voltammogram of *D. africanus* Fd III at pH 7.0 at a pyrolytic graphite electrode. Couple A' corresponds to one-electron reduction of the [3Fe–4S] cluster, $E_{1/2} = -110 \pm 10$ mV, which is pH-dependent with $dE_{1/2}/d(pH) = -55$ mV. The latter result indicates that one proton is taken up per electron. Couple B' reveals the reduction of the $[4Fe–4S]^{2+}$ to the $[4Fe–4S]^+$ cluster at a potential of -400 ± 10 mV (see next section). The redox process corresponding to couple C' at -695 mV has tentatively been assigned as the $[3Fe-4S]^{0/2-}$ reduction.

(iv) [4Fe–4S]. The final type of well-characterized iron-sulfur center has the Fe_4S_4 cubane-type structure. These clusters can be visualized as a dimer of two Fe_2S_2 units, of the kind found in plant-type ferredoxins, joined so that the sulfido groups now bridge three rather than two iron centers. This structure has been observed in several different proteins, either alone or in the presence of other prosthetic groups. Once again, the iron-iron separations are approximately 2.7 Å. One of the most striking properties of this unit that

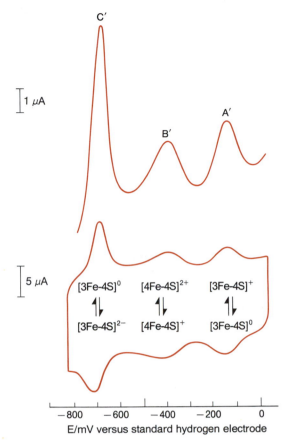

Figure 9.1
Differential pulse polarogram (top) and cyclic voltammogram (bottom) of
ferredoxin from *D. africanus* on a pyrolytic graphite electrode. The assignments for
the three electron transfer processes are shown. (Adapted from A. J. Thomson et al.
Biochemical Society Transactions **19**, 594–599 (1991))

distinguishes it from the other centers we have discussed is that it can exist
in three rather than two oxidation levels. Moreover, it does so in different
proteins. Thus, for *P. aerogenes* ferredoxin, which contains two $Fe_4S_4(S\text{-cys})_4$
clusters, the oxidized state of each cluster formally contains two Fe(III) ions
and two Fe(II) ions, corresponding to an overall charge of $+2$ on the Fe_4S_4
unit. The system is actually electronically delocalized, with four equivalent
iron centers of average oxidation state $+2.5$. These clusters can be reduced
to $Fe_4S_4^{1+}$ units that now formally contain one Fe(III) and three Fe(II) ions,
although again electron delocalization occurs. In contrast, the reduced form
of *Chromatium* ferredoxin (formerly called the high-potential iron protein or
HiPIP) contains an $Fe_4S_4^{2+}$ unit, whereas the oxidized form now contains an
$Fe_4S_4^{3+}$ unit formally comprised of three Fe(III) and one Fe(II) ions.

The realization that the $Fe_4S_4(S\text{-cys})_4$ unit could exist in more than two oxidation levels rationalized a large amount of spectroscopic and chemical data that distinguished the two classes of bacterial ferredoxins. The preparation and characterization of synthetic models for these sites allowed determination of their overall charge and oxidation levels, which could not be determined from the protein crystal structures alone. These model complexes have also revealed some important properties of the various oxidation levels, including:

(i) $Fe_4S_4^{2+}$ cores are relatively uniform in properties with structures slightly compressed along one axis, lowering the symmetry from T_d to D_{2d};

(ii) $Fe_4S_4^{+}$ cores are much more susceptible to environmental influences, with the structure and even the spin of the ground state changing with crystalline environment;

(iii) the $Fe_4S_4^{3+}$ form is difficult to stabilize, but, with very bulky thiolate groups, models can be prepared and characterized, even by X-ray crystallography.

The structural changes at each iron atom are smaller for the $Fe_4S_4(SR)_4^{n-}$ clusters than they are for the single-iron-type centers. The mean iron-sulfur distances are 2.25 Å, 2.28 Å, and 2.31 Å for typical $[Fe_4S_4(SR)_4]^-$, $[Fe_4S_4(SR)_4]^{2-}$, and $[Fe_4S_4(SR)_4]^{3-}$ clusters, respectively, corresponding to 1.3 percent changes per added electron compared with the 2−3 percent change noted previously for the mononuclear species. The structural changes associated with oxidation and reduction observed for R = Ph (or substituted Ph) are shown in Figure 9.2.

The paramagnetism arising from the odd-electron character of the $Fe_4S_4^{+}$ and $Fe_4S_4^{3+}$ units facilitates their identification by EPR spectroscopy (Table 5.2). This paramagnetism also produces greater paramagnetic shifts of the $\beta\text{-CH}_2$ resonances of the cysteinyl residues than for the $[4Fe-4S]^{2+}$ state. Extensive resonance Raman spectral studies have also been carried out for these proteins, with several Fe−S, Fe−SR, and cluster breathing modes having been assigned in the 245−400 cm^{-1} region of the spectrum. These comprehensive studies of the physical properties of [4Fe−4S] proteins now make it relatively easy to identify them from their characteristic magnetic and spectroscopic signatures.

The redox potentials of proteins containing [4Fe−4S] units span a very wide range because of the accessibility of three oxidation levels. For proteins that operate between the $[Fe_4S_4]^+$ and $[Fe_4S_4]^{2+}$ levels, the potentials range from −650 to −280 mV, whereas those that use $[Fe_4S_4]^{2+}$ and $[Fe_4S_4]^{3+}$ fall near +350 mV. Studies with synthetic models illustrate some of the effects that can modulate these potentials. For example, the potentials for $[Fe_4S_4(SCH_2CH_2OH)_4]^{2-/3-}$ and $[Fe_4S_4(S\text{-cys-}(Ac)NHMe)_4]^{2-/3-}$ transitions

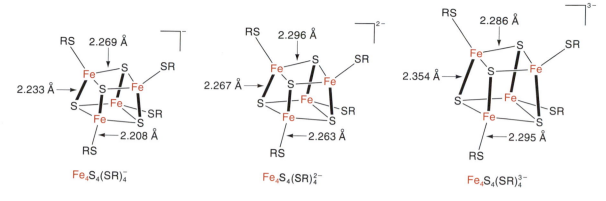

Figure 9.2
Structures of $[Fe_4S_4(SR)_4]$ clusters, R = aryl, as a function of overall charge as deduced from crystallographic studies of synthetic models. The two more highly oxidized clusters are slightly compressed along one direction, whereas the most reduced one is elongated along one direction.

are somewhat more negative than the midpoint of the corresponding protein transitions, indicating that the protein matrix shifts the potentials by $\approx 60-80$ mV to more positive values.

Another important quantity that has been measured on the model compounds is their electron-transfer rate constant. As will soon be developed more fully, long-range electron transfer from the surface to a redox-active metal center in a metalloprotein is a subject of much current interest in bioinorganic chemistry. For the iron-sulfur proteins, the major functions of which involve electron transfer, it is essential to understand the factors that determine the rates of such reactions. Studies of the bimolecular, outer-sphere rate constants of $[Fe_4S_4(SR)_4]^{n-}$ have been conveniently carried out by observing the ^1H NMR line-shape changes that result from electron transfer between the $n = 2$ and $n = 3$ states of the cluster. The rate constants fall in the 10^6 to 10^7 M^{-1} s^{-1} range, making electron transfer for the $[Fe_4S_4]$ cubes among the faster known self-exchange processes in inorganic chemistry. These values are $> 10^3$ larger than the fastest known electron-transfer rate constants for $[4Fe-4S]^{n-}$ proteins undergoing the $n = 2$ to $n = 3$ transition. The results of these studies with model compounds thus lead to the following picture for the [4Fe−4S] clusters. The rates of electron transfer are fast, consistent with minimal cluster reorganizational energy. Actual rates of electron transfer from the proteins are therefore controlled by factors extrinsic to their iron-sulfur cores.

9.1.b. Blue Copper Proteins. A second important class of proteins that carry out electron-transfer reactions are the so-called blue copper pro-

238

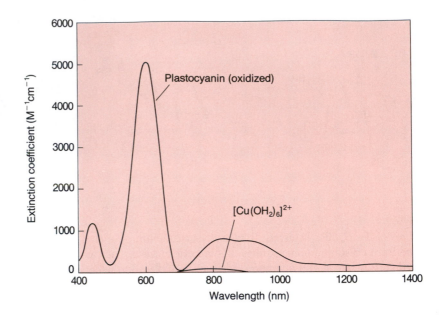

Figure 9.3

Comparison of the electronic spectra of oxidized plastocyanin and hexaaquacopper(II) ion.

teins. This term may appear to be superfluous, since many copper(II) complexes, such as $[Cu(OH_2)_6]^{2+}$, are blue. The absorption bands responsible for the color of simple cupric compounds, however, typically have extinction coefficients of $5-10$ M^{-1} cm^{-1}. In contrast, the blue copper proteins are characterized by bands appearing around 600 nm that are much more intense, with extinction coefficients of >3000 M^{-1} cm^{-1}. An example is displayed in Figure 9.3. Interest in the origin of this remarkably intense color has sparked much research into the nature of metal-binding sites in these proteins. Further studies revealed other noteworthy spectroscopic features. Most notably, the EPR spectrum of the blue copper proteins (see Section 4.4 and Figure 4.5) exhibit unusual features, including high g values and low A values, suggesting that the unpaired electron is more delocalized than in other copper(II) complexes. Blue copper centers occur in relatively small proteins that contain single copper ions such as plastocyanin (from plants) and azurin (from bacteria), as well as in complex enzymes (the multicopper oxidases) that contain four or more copper ions in a variety of distinguishable sites. The copper sites from the blue copper proteins are often referred to as type 1, as previously discussed in Chapter 4.

Before the structures of several blue copper proteins were determined by X-ray crystallography, many spectroscopic investigations were carried out in order to elucidate their copper-site geometries. Optical, photoelectron, nuclear magnetic resonance, and X-ray absorption spectroscopy all played significant roles. These studies have generally depended on the availability of more than one redox state of the protein for comparison of spectroscopic properties. Optical studies of native blue copper proteins as well as of derivatives in which the copper ion had been replaced with other metals led to important conclusions regarding the origin of the intense visible absorption band, implicating thiolate as a ligand to the copper(II) ion. As noted above, the position of the absorption band that characterizes these proteins occurs at an energy that is similar to that of d-d bands seen in many five- and six-coordinate copper(II) complexes, but its high intensity made the assignment as a d-d transition implausible. Examination of the cobalt(II) derivatives of several blue copper proteins revealed the presence of bands of intensity similar to those in the copper(II) analogs, but appearing at approximately 300–350 nm. This shift to higher energies on going from Cu(II) to Co(II) is consistent with assignment of the transitions as ligand-to-metal charge-transfer bands. Thiolate was identified as the probable donor on the basis of the position of this charge-transfer transition. In addition to providing an assignment of the unusual visible band in the copper(II) complex, analysis of the ligand field transitions in the near-infrared region for the Cu(II) form, and in the visible and near infrared for the Co(II) form, suggested a distorted tetrahedral geometry for these metal ions. Other evidence supporting thiolate coordination came from analysis of the sulfur region of the photoelectron spectra of the metalated and apo proteins. In these studies, one sulfur atom was found to undergo significant shifts in the different forms. These results suggested that one cysteine sulfur atom is in quite different environments in the different forms, being protonated and free in the apo form, deprotonated and bound to metal in the metal-containing forms.

NMR spectral studies of reduced spinach plastocyanin in D_2O revealed resonances that could be assigned to two histidine residues. When the protein was oxidized to the Cu(II) state, these resonances disappeared. Their loss could be attributed to shifting and broadening due to the proximity of the protons to the paramagnetic metal center. Furthermore, studies of the spectra of the reduced protein as a function of pH revealed unusually low pK_a values for these two histidines. These observations provided strong evidence that two of the donor atoms to the copper were imidazole nitrogens from histidine residues. Taken together, these data led to the proposal that the metal environment in the blue copper proteins was distorted tetrahedral, with one cysteine thiolate, two histidine nitrogens, and a fourth ligand proposed to be a deprotonated amide nitrogen based on optical and infrared spectroscopic studies. X-ray absorption studies of oxidized azurin provided support for this overall structure. EXAFS data could be fit with one sulfur atom at the strikingly short distance of 2.10 Å and two to three nitrogen atoms at 2.00 Å. These results were clearly consistent with the main features of the proposed structure, but they could not conclusively address the nature of the fourth ligand.

Definitive structural characterization of a blue copper site came from the report of the crystal structure of oxidized poplar plastocyanin at 2.7 Å resolution. The copper ion was found to be coordinated in a distorted tetrahedral site with one cysteine, two histidines, and a methionine providing the ligands. Subsequent refinement of this structure at higher resolution has confirmed this initial interpretation. The structure of this protein and its copper site are shown in Figure 9.4. The protein structure is of the beta-barrel type first observed in immunoglobulin domains. Examination of the copper site reveals several interesting features. First, the conclusions drawn from spectroscopic studies were remarkably close to the actual structure. Second, the copper-cysteinate sulfur bond is, indeed, quite short, as anticipated from the EXAFS results. In contrast, the copper-methionine bond is very long (2.9 Å) and is not detectable by EXAFS studies despite considerable effort. The significance of this weak bonding interaction is not clear. Third, the site is near the surface of the protein, with the edge of one of the coordinated histidine residues exposed to solvent. The surface of the protein in the vicinity of the copper site includes a hydrophobic patch. It has been proposed that this patch is the likely site of interaction between plastocyanin and its redox-partner proteins, an aspect to which we shall return later.

The structures of several other blue copper proteins have now been determined by crystallographic and NMR methods. For example, the structure of *Pseudomonas aeruginosa* azurin reveals a quite similar overall protein fold. The copper site is similar to that in plastocyanin, although the metal is now five-coordinate, with the cysteinate and two histidines forming a trigonal plane, and a methionine sulfur and a backbone amide oxygen forming long (~ 3 Å) axial bonds. The structure of French bean plastocyanin has also been determined by NMR methods, revealing a structure essentially identical with that for poplar plastocyanin. Although this observation is not surprising, it is of some importance, since most of the spectroscopic and chemical studies have been performed on the French bean protein. Poplar plastocyanin was chosen for crystallographic studies because it was the only one of fourteen different plastocyanins that could be coaxed to crystallize! Finally, the structure of ascorbate oxidase, an enzyme that contains four copper atoms, has recently been determined. Three of the coppers are part of

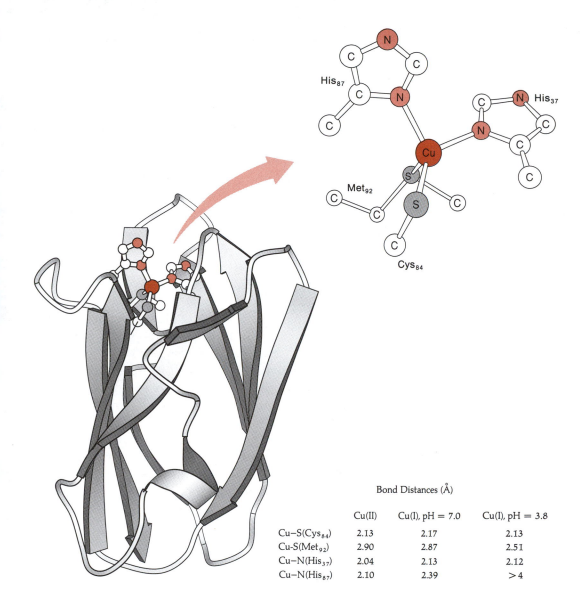

Bond Distances (Å)			
	Cu(II)	Cu(I), pH = 7.0	Cu(I), pH = 3.8
Cu–S(Cys$_{84}$)	2.13	2.17	2.13
Cu-S(Met$_{92}$)	2.90	2.87	2.51
Cu–N(His$_{37}$)	2.04	2.13	2.12
Cu–N(His$_{87}$)	2.10	2.39	>4

Figure 9.4
Structure of oxidized poplar plastocyanin and its copper site and metrical
parameters for the copper sites of various states of plastocyanin.

a trinuclear unit; the remaining copper is in a type 1 site that is quite similar
to that in plastocyanin.

Since these proteins participate in electron-transfer reactions, it was of
great interest to examine the structure of a blue copper protein in its reduced
form. This goal has been achieved for poplar plastocyanin. At higher pH
values (pH = 7.8), the copper site is very similar to that in the oxidized
protein with only slight elongation of the Cu–N bonds (Figure 9.4). This

241

result again reflects the relatively low reorganizational energy associated with metal centers involved in biological electron-transfer reactions. Interestingly, when the structure was reexamined at lower pH (to pH 3.8) values, one of the histidine ligands was found to be protonated and dissociated from the copper center. Concomitantly, the Cu-methionine sulfur bond shortened to 2.5 Å. This result explains earlier observations that the reduction potential of the protein is pH dependent. The coupling of proton and electron transfer illustrates an important principle, namely, that tuning of the redox potential of a metal center in a protein can be affected by the acid-base properties of coordinated ligands.

The copper ion can often be removed from blue copper proteins by treatment with cyanide. The apo protein of poplar plastocyanin obtained in this manner was crystallized and found to be isomorphous to the metal-containing forms, indicating that the structure undergoes very little change upon metal removal. Detailed analysis of the crystal structure confirmed this expectation. The only significant change observed was reorientation of one of the histidine side chains. These results clearly indicate that the protein, including its metal-binding site, is preorganized prior to metal binding. Thus, blue copper sites may be good examples of entatic states (see Section 12.1), wherein the protein structure is sufficiently rigid to impose constraints on the metal center that would be difficult to achieve with less-constrained molecules. In this example, the constraints may afford a distorted tetrahedral (or trigonal bipyramidal for azurin) ligand arrangement and stabilization of the Cu(II)-thiolate unit. For many simpler ligands, the Cu(II) and Cu(I) complexes are quite different in structure, since Cu(II) often prefers higher coordination numbers (usually 5 or 6) and square-planar geometries for four-coordinate species, whereas Cu(I) prefers lower coordination numbers and tetrahedral four-coordinate stereochemistry.

Many attempts have been made to produce synthetic analogs for the blue copper sites, with only limited success. Undoubtedly, the greatest impediment to progress has been the tendency for potential thiolate ligands to reduce Cu(II) to Cu(I) with concomitant coupling of thiolate to disulfide. The protein obviates this reaction by burying the cysteinate ligand deeply enough in the polypeptide chain that coupling does not occur. The best synthetic analogs of the blue copper sites involve sterically demanding substituted tris(pyrazoyl)borate ligands, which provide three donor atoms, and an aromatic thiolate. Under appropriate conditions, the Cu(II) complexes of these ligands do mimic the optical and EPR properties of the type 1 sites, but they generally decompose at temperatures above -40 °C. Recently, however, it has been possible to crystallize cupric complexes with coordinated thiolates $C_6F_5S^-$ or Ph_3CS^-. Their structural properties are as expected, with short Cu^{II}-SR distances.

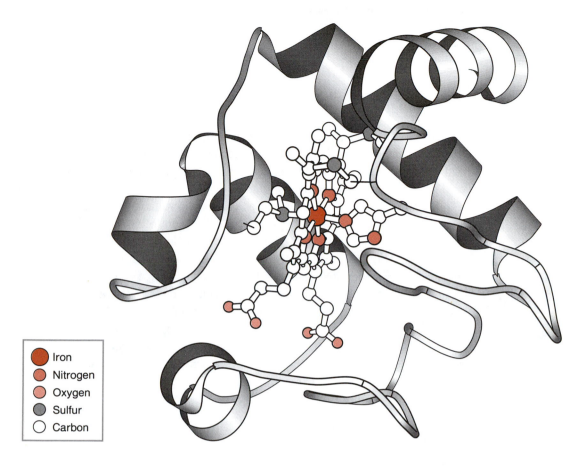

●	Iron
●	Nitrogen
●	Oxygen
●	Sulfur
○	Carbon

Figure 9.5
Tuna cytochrome *c* structure showing coordination of the iron porphyrin group by protein residues.

9.1.c. Cytochromes. The final class of electron-transfer proteins to be considered are the cytochromes. Like hemoglobin and the cytochromes P-450, these proteins contain iron porphyrins as the redox-active centers. Whereas hemoglobin and cytochrome P-450 have only one protein-derived ligand coordinated to the iron, leaving an open coordination site for oxygen binding, most cytochromes have two, affording coordinatively saturated octahedral complexes. For eukaryotic cytochrome *c*, which we shall discuss in detail, these axial ligands are a histidine nitrogen and a thioether sulfur from methionine. Other cytochromes have two histidine ligands.

The cytochrome *c* from tuna was the first member of this family to have its structure determined. The protein, shown in Figure 9.5, consists of both alpha helices and extended chains that wrap around the porphyrin. Unlike hemoglobin and cytochrome P-450, where the prosthetic group is held in place only by interactions with the axial ligands and by noncovalent forces, the porphyrins in cytochromes c are bound covalently via two thioether linkages formed by addition of two cysteinate side chains to the vinyl groups of the protoporphyrin IX. The edge of the porphyrin bearing the propionic-acid substituents is exposed to solvent.

High-resolution crystal structures are now available for cytochromes c from several sources. Comparison of the structures of oxidized, Fe(III), and reduced, Fe(II), cytochromes reveals that very little structural change accompanies the redox reaction. Both forms are low-spin complexes, so that the electronic difference between the two involves a t_{2g}-type electron that is nonbonding in a sigma-only molecular orbital sense. This result illustrates another strategy that nature has utilized to generate an oxidation-reduction system that undergoes very little structural rearrangement upon electron transfer. Various cytochromes c have been used extensively for studies of electron-transfer reactions between redox partners at known distances from one another. Such studies bear on the biological function of all the electron-transfer proteins discussed here.

9.2. Long-Distance Electron Transfer

Like many biomolecules, electron carriers must satisfy two sets of constraints that work in opposite directions. First, they must be reasonably *fast*. This constraint is based on the fact that electrons must move through pathways at rates that are biologically useful. Second, the reactions must be *specific*. Electron-transfer pathways are coupled to other biological processes such as proton translocation. Utilization of the reducing equivalents generated at one end of a pathway requires sequential transfer of electrons through a series of molecules. If these reactions were not specific, transfer of electrons from the most reducing member to the most easily reduced member would short-circuit the intervening steps and their coupled processes. Thus electron transfer generally occurs through specific protein-protein complexes. These complexes must be sufficiently stable that recognition of the cognate partner is possible, but not so stable that the lifetime of the complex is so long that it significantly limits overall electron flux. Because of the structural constraints inherent in such a process, biological electron-transfer reactions often must occur with relatively large distances (> 10 Å) between donor and acceptor.

Such long-distance electron-transfer processes have been probed in several ways. The first involves studies of the self-exchange rates for electron-transfer proteins. This approach has the advantage that the driving force for a self-exchange reaction is identically zero, thereby removing any dependence of the electron-transfer rate on this quantity. Rapid self-exchange rates can be measured by NMR line-broadening effects on mixtures of oxidized and reduced protein. For *Pseudomonas aeruginosa* azurin, the self-exchange reaction is quite rapid, with a rate constant of 1.3×10^6 M^{-1}s^{-1}. The self-exchange process is quite sensitive to the structure at the surface of the protein near the copper site. Replacement of a methionine residue in a hydrophobic patch with a positively charged lysine residue by site-directed mutagenesis resulted in a large decrease in the self-exchange rate, to less than 1×10^3 M^{-1}s^{-1}. Thus, while observation of a fast self-exchange process clearly demonstrates the facility of a protein to execute electron-transfer reactions, slow self-exchange may reflect protein self-aggregation modes rather than the properties of the redox center itself.

The second method involves preparation of artificial donor-acceptor pairs. Heme proteins such as cytochrome *c* have been most extensively studied by this approach. The first example involved attachment of a $\{Ru(NH_3)_5\}^{3+}$ group to histidine 33 on the surface of horse-heart cytochrome *c* (Figure 9.6). The distance between redox centers is approximately 12 Å. Such systems are studied via the scheme shown in Equation 9.1. Initial fast reduction of the surface-bound ruthenium complex is accomplished via

$$\text{Cyt } c(\text{Fe(III)})\text{-Ru(III)} \underset{}{\overset{\text{fast}}{\rightleftharpoons}} \text{Cyt } c(\text{Fe(III)})\text{-Ru(II)} \underset{\text{e}^-\text{transfer}}{\overset{\text{intramolecular}}{\rightleftharpoons}} \text{Cyt } c(\text{Fe(II)})\text{-Ru(III)}$$

$$(9.1)$$

photochemical or pulse radiolysis means. The intramolecular electron transfer, presumed to occur over the known long distance, is then monitored as reduction of the Fe(III) center by the reduced ruthenium complex. For the system noted above, the rate for the intramolecular electron transfer is approximately 30 s^{-1} and is essentially independent of temperature. This result demonstrates that such long-distance electron-transfer processes do, indeed, occur. In addition, investigations with other systems are allowing such factors as distance, driving force, and the nature of the intervening medium to be studied more or less systematically. A key question involves whether such electron transfers occur through bonds, wherein the protein acts as a large bridging ligand, through space, or through a combination of these two types of pathway.

Another method involves studies with actual protein-redox pairs. One system that has been investigated is the yeast cytochrome *c*/cytochrome *c* peroxidase pair. Replacement of the iron porphyrin in cytochrome *c* peroxidase by the corresponding zinc porphyrin results in a species that can

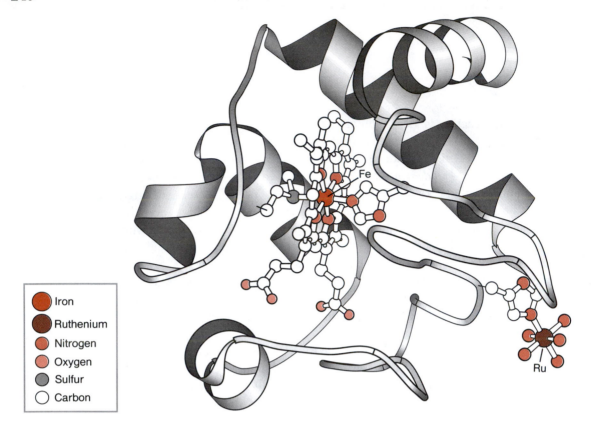

Figure 9.6
Structure showing attachment of the $[Ru(NH_3)_5]^{3+}$ moiety to His 33 on the surface of horse-heart cytochrome *c*.

be photoexcited to a triplet state that is capable of reducing other species at long distances. The availability of this light-triggered system allows the electron-transfer step itself to be studied, as shown in Equation 9.2. Here,

$$[ZnP, Fe^{III}P] \xrightarrow{h\nu} [^3ZnP, Fe^{III}P] \rightarrow [ZnP \cdot^+, Fe^{II}P] \rightarrow [ZnP, Fe^{III}P] \quad (9.2)$$

light produces the triplet excited state of the zinc porphyrin in the cytochrome *c* peroxidase. This species is a strong reductant that transfers an electron to the ferric porphyrin in the cytochrome *c*. Thermal electron transfer then occurs between the zinc porphyrin radical cation and the ferrous porphyrin in the cytochrome *c* to return to the initial state. The last step corresponds to the physiological reaction, that is, the oxidation of reduced cytochrome *c* by an oxidized form of the cytochrome *c* peroxidase (see Section 11.4.b.). For the yeast cytochrome *c*/cytochrome *c* peroxidase pair, the rate of this electron transfer is approximately 1.1×10^4 s^{-1} over a distance estimated to be 17 Å. If the yeast cytochrome *c* is replaced by horse-heart cytochrome c, the rate is reduced to 12 s^{-1}, again illustrating the importance of protein-protein recognition in facilitating intermolecular electron transfer.

An exciting recent advance in this area is the determination of the crystal structures of complexes between yeast cytochrome c peroxidase and cytochrome c molecules from yeast and from horse heart. With yeast cytochrome c, a complex is formed with an iron-iron separation of 26.5 Å and several van der Waals interactions between a short stretch of polypeptide (residues 191–194 of cytochrome c peroxidase) that is close to the heme group and portions of the yeast cytochrome c, including one of the heme methyl groups. In contrast, for the horse-heart cytochrome c structure, the iron-iron separation is 3.5 Å longer, and there are no close contacts suggestive of an electron-transfer path. The availability of such detailed structural information for protein-protein complexes involved in biological electron transfer should be of great utility in designing experiments to test long-distance electron-transfer mechanisms.

Another system that has been extensively investigated is the photosynthetic reaction center from bacteria. The reaction center functions naturally as a light-induced electron-transport system. A major advantage of the reaction center is that it exists as a tightly bound assembly of proteins that has had its crystal structure determined to reasonably high resolution. This structural information allows features such as the distances and relative orientations of electron donors and acceptors to be known with certainty. The rates of electron-transfer reactions between many of the prosthetic groups in this complex have been measured. Intriguingly, the reaction center has approximate twofold symmetry with two potential electron-transfer networks, yet only one of these is actually utilized. In addition, this system executes unidirectional electron transfer with quite high efficiency. We shall return to this system in the next section.

9.3. Distance and Driving-Force Dependence of Electron Transfer

Theories of electron transfer make predictions about the dependence of electron-transfer rates on the difference in structure between the oxidized and reduced species, the distance between donor and acceptor, and on the driving force of an electron-transfer reaction. These theories are based on simple quantum-mechanical expressions for the wave functions involved in donating and accepting the electron. The medium in between serves to couple weakly these two wave functions. The rate of an electron-transfer reaction is given by Equation 9.3. Here, T_{DA} is the tunneling matrix element, a measure of

$$k_{et} = (4\pi^2/h)\, T_{DA}^2 (FC) \tag{9.3}$$

the electronic coupling between the reductant (donor, D) and the oxidant (acceptor, A), and FC is the so-called Franck-Condon factor, a measure of the nuclear motion involved in the electron-transfer process. Each of the terms can be approximated in various ways. The simplest representation of T_{DA}^2 shows an exponential decay with distance, as described by Equation 9.4,

$$T_{DA}^2 = T_{DA}^{0\ 2} \exp(-\beta(R - R_0)) \tag{9.4}$$

where T_{DA}^0 is the tunneling matrix element at $R = R_0$, R is the distance between the closest atoms in the donor and acceptor, R_0 is the van der Waals contact distance (generally 3.6 Å), and β is a parameter that describes the effects of the intervening medium. More complicated forms of this expression that explicitly involve the structure of the intervening medium have been developed. These expressions will be discussed shortly. The simplest form of the FC term constitutes the basis of Marcus theory. The classical expression is given by Equation 9.5, where λ is the so-called reorganization energy, k is

$$FC = (4\pi\lambda kT)^{-1/2} \exp[-(-\Delta G^0 - \lambda)^2/4\lambda kT] \tag{9.5}$$

the Boltzmann constant, T is the absolute temperature, and ΔG^0 is the standard free energy for the reaction. This expression has several consequences. Perhaps the most interesting, and initially counterintuitive, result is that the rate of electron transfer should reach a maximum value at some driving force and then decrease as the driving force is further increased.

The distance dependence of the electron-transfer rate has been examined via compilation of optimal rate data from several systems. In some, such as the photosynthetic reaction center, the distance is well known from X-ray crystallographic data. For other systems, such as ruthenium-modified cytochrome c, distances are estimated by molecular modeling methods. Some of the data available from such investigations are shown in Figure 9.7. An approximately linear decrease in the log of the rate with distance is observed, as predicted by Equations 9.3 and 9.4, although some of the rates differ from the calculated values by one or two orders of magnitude. The exponent β, determined from the slope of the plot, is 1.4 Å^{-1}, which corresponds to a tenfold decrease in rate for every 1.7 Å. This value should be compared to

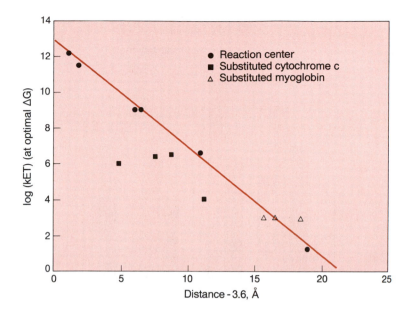

Figure 9.7

Plot depicting the electron transfer rate constant as a function of distance for several natural and ruthenium-modified proteins. The structures of the reaction center, cytochrome *c*, and myoglobin are depicted in Figures 9.10, 9.5 and 11.1, respectively. An approximately linear dependence of the log of the rate constant on the distance is observed over a wide range of rates and distances although some significant deviations do occur. (Adapted from C. C. Moser et al., *Nature* **355**, 796–802 (1992)).

those of 2.8 $Å^{-1}$ expected for electron transfer through a vacuum, and 0.7 $Å^{-1}$ obtained from studies of electron-transfer rates for systems linked by covalent bonds. The maximum rate, at van der Waals contact, is approximately 10^{13} per second.

The intermediate value for β obtained from the protein data suggests that electron transfer takes place by a combination of through-bond and through-space steps. A model that explicitly incorporates these effects has been developed. Three types of steps are included, through covalent bonds, through hydrogen bonds, and through space. The lengths of the steps through the latter two types of transitions are included as parameters. The method is implemented by considering, with the aid of a computer program, all potential paths from the donor to the acceptor and calculating the expected overall rate. The approach has two appealing features. First, it provides a rationalization for the intermediate value of β for proteins noted above. Second, it can potentially account for some of the deviations from the simple exponential distance dependence for some systems. Some pairs of sites within proteins are coupled via quite efficient paths that include mostly covalent bonds and only

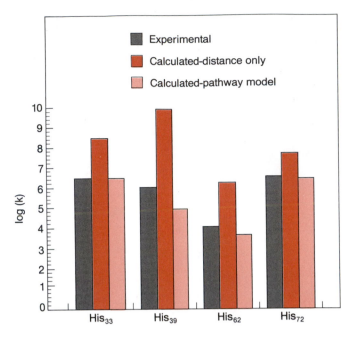

Figure 9.8

A comparison of the experimental rate constants for electron transfer for a series of pentaammineruthenium-modified cytochrome c derivatives with those calculated based on the distance only and on a model that explicitly includes possible electron-transfer pathways. (Adapted from D. S. Wuttke et al., *Science* **256**, 1007−1009 (1992)).

a small number of short through-space jumps, whereas others, even though they are separated by comparable overall distances, are expected to be much less efficiently coupled. The detailed nature of this model allows the design of experiments in which the electron-transfer rate predicted from simple distance dependence and that predicted from the pathway model are quite distinct. A comparison of the experimental rates for a series of ruthenium-modified cytochrome c derivatives and the predictions of the pathway model, as well as those based on simple distance dependence, is shown in Figure 9.8. Note that, in many instances, the pathway model predicts rates similar to those expected from simple distance dependence. This result arises not because the rate of electron transfer is independent of the nature of the path, but rather because the protein structures are such that paths with similar mixtures of through-bond and through-space steps are often available.

The driving-force dependence predicted by Equation 9.5 has also been experimentally probed (Figure 9.9). For example, for ruthenium-modified cytochrome c derivatives, including those in which zinc has been substituted for iron in order to generate a relatively long-lived photoexcited state, rates were determined for driving forces spanning a range of 0.4 eV, corresponding to approximately 10 kcal/mol. The rate data can be fit to Equation 9.5 to

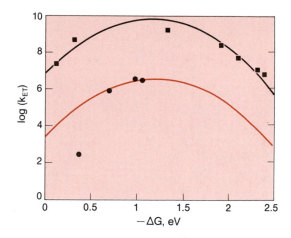

Figure 9.9
Plot of effect of driving force on rate constant for electron transfer reactions illustrating the inverted Marcus region. Data are shown for a series of ruthenium-modified cytochrome c derivatives (lower) and for a series of covalently linked organic donor/acceptor compounds (upper). (Adapted from T. J. Meade et al., *Journal of the American Chemical Society* **111**, 4353–4356 (1989) and G. L. Closs and J. R. Miller, *Science* **240**, 440–447 (1988)).

yield a value of 1.1 to 1.2 eV for the reorganizational energy λ, which corresponds to the energy required to distort the starting complex into the geometry of the product without electron transfer occurring. This parameter embodies several components, including the difference in geometry between the donor and acceptor complexes themselves, the reorganization of the protein, including, for example, peptide dipole reorientation, and solvent reorganization. As noted earlier in this chapter, the first term seems to be quite small for natural groups involved in electron transfer. For ruthenium-modified cytochrome c, the first term is estimated to be 0.2 eV. The protein and solvent terms are estimated to be 0.2 eV and 0.6 eV, respectively. Thus, the overall value of λ is approximately 1.0 eV, in reasonable agreement with the experiment.

As indicated above, one of the best-characterized electron-transfer systems is the photosynthetic reaction center. This protein complex transports electrons across a membrane in response to light absorption. The light is initially absorbed by the so-called special pair, a set of two closely spaced bacteriochlorophylls, reduced porphyrins with bound Mg(II). The reaction center also contains two additional bacteriochlorophylls, two bacteriopheophytins [bacteriochlorophyll without the Mg(II)], a nonheme iron that does not appear to have any redox function, and one or two bound quinones. These prosthetic groups are arranged into two quite similar networks that extend from one side of the membrane to the other. The structure is shown schematically in Figure 9.10 (see also Figure 12.11). Fast kinetic studies have

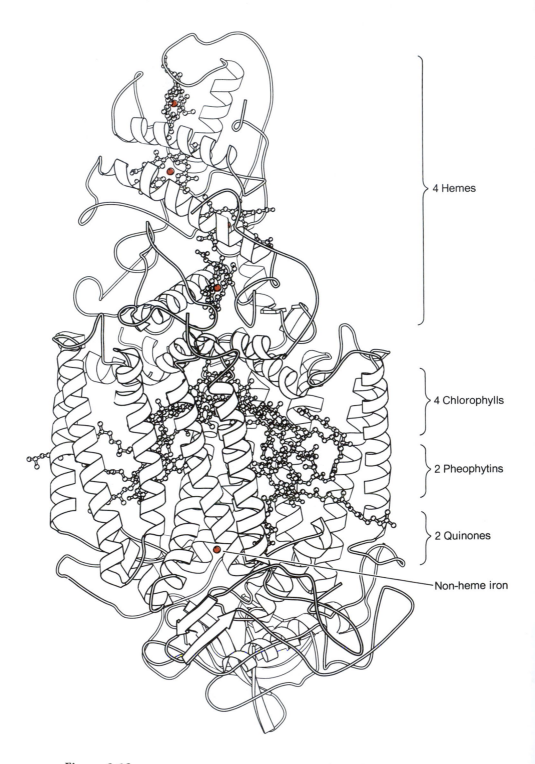

4 Hemes

4 Chlorophylls

2 Pheophytins

2 Quinones

Non-heme iron

Figure 9.10
View of the photosynthetic reaction center from *Rhodopseudomonas viridis* showing the overall structure and positions of the prosthetic groups.

revealed that the initial electron transfer occurs in less than five picoseconds. This and other observations indicate that this step is optimized, with the driving force and reorganizational energy well matched. The driving force for this forward reaction includes the energy of the absorbed photon. The presence of the inverted region is important for avoiding the back-reaction of electron transfer from the reduced bacteriopheophytin to the oxidized special pair. This reaction has a large driving force, significantly greater than the reorganizational energy, diminishing the rate so that transfer along the pathway across the membrane can compete successfully. Thus the inverted region is not just a theoretical curiosity, but plays a central role in making photosynthetic charge separation possible.

Study Problems

1. Through the use of site-directed mutagenesis, it has been possible to replace the cysteine bound to copper in azurin with aspartic acid. What differences in the spectroscopic properties would you predict for the copper(II) complex of the mutant versus the wild-type protein?

2. Electron-transfer rates have been studied in ruthenium-modified myoglobin. Reduced myoglobin is a high-spin five-coordinate complex, whereas the oxidized form binds a water molecule to form a six-coordinate complex. Would you expect the rate of electron transfer from this center to a bound surface ruthenium to be similar to that for the corresponding cytochrome c derivative given the same distance and driving force? Explain.

3. Estimate the optimal rate for an electron transfer occurring over a distance of 30 Å through a system linked by covalent bonds. Repeat the calculation assuming the electron transfer occurs through a protein.

4. One of the striking features of the photosynthetic reaction-center structure is the presence of a parallel pair of electron-transfer pathways, one of which appears not to be used. Given that the data indicate the rate down the used path to be at least 100 times faster than that for the unused one, calculate the differences in Marcus theory parameters, assuming that the difference is due (a) entirely to a distance difference, (b) entirely to a driving force difference, or (c) entirely to a reorganizational energy difference. For the used path, assume the distance is 10 Å, the driving force is 0.2 eV, and the reorganizational energy is 0.2 eV. Note: kT is equal to 0.026 eV at room temperature.

5. Suppose that the electron-transfer rate constant for two proteins in which the electron donor and acceptor are 20 Å apart is 200 s^{-1}. Calculate the rate constant for the same two proteins if the distance increases to 25 Å, assuming β, the medium parameter, to be 1.4 Å^{-1}.

Bibliography

General References

H. Sigel, ed. 1991. *Electron Transfer Reactions in Metalloproteins. Metal Ions in Biological Systems*, Volume 27. Marcel Dekker, New York.

Electron Carriers

E. T. Adman. 1991. "Copper Protein Structures," in C. B. Anfinsen, J. T. Edsall, D. S. Eisenberg, and F. M. Richards, eds., *Advances in Protein Chemistry*, Volume 42. Academic Press, San Diego, 145–198.

H. Beinert. 1990. "Recent Developments in the Field of Iron-Sulfur Proteins." *FASEB J.* 4, 2483–2491.

S. K. Chapman. 1991. "Blue Copper Proteins," in R. W. Hay, J. R. Dilworth, and K. B. Nolan, eds., *Perspectives in Bioinorganic Chemistry*, Volume 1. JAI Press, London, 95–140.

W. Cramer and D. B. Knaff. 1991. *Energy Transduction in Biological Membranes.* Springer-Verlag, New York.

J. B. Howard and D. C. Rees. 1991. "Perspectives on Non-Heme Iron Protein Chemistry," in C. B. Anfinsen, J. T. Edsall, D. S. Eisenberg, and F. M. Richards, eds., *Advances in Protein Chemistry*, Volume 42. Academic Press, San Diego, 199–280.

W. L. Lovenberg, ed. 1973–1977. *Iron-Sulfur Proteins*, Volumes 1–3. Academic Press, New York.

G. R. Moore and G. W. Pettigrew. 1990. *Cytochromes C.* Springer-Verlag, New York.

E. I. Solomon and M. D. Lowery. 1993. "Electronic Structure Contributions to Function in Bioinorganic Chemistry." *Science* 259, 1575–1581.

A. G. Sykes. 1991. "Plastocyanin and the Blue Copper Proteins." *Struct. and Bond.* 75, 175–224.

Long-Distance Electron Transfer

P. Betrand. 1991. "Applications of Electron Transfer Theories to Biological Systems." *Struct. and Bond.* 75, 1–48.

B. E. Bowler, A. L. Raphael, and H. B. Gray. 1990. "Long-Range Electron Transfer in Donor(Spacer)Acceptor Molecules and Proteins." *Prog. Inorg. Chem.* 38, 259–322.

S. G. Boxer. 1990. "Mechanisms of Long-Distance Electron Transfer in Proteins: Lessons from Photosynthetic Reaction Centers." *Ann. Rev. Biophys. Biophys. Chem.* 19, 267–299.

G. H. Closs and J. R. Miller. 1988. "Intramolecular Long-Distance Electron Transfer in Organic Molecules." *Science* 240, 440–447.

B. M. Hoffman and M. A. Ratner. 1987. "Gated Electron Transfer: When Are Observed Rates Controlled by Conformational Interconversion?" *J. Am. Chem. Soc.* 109, 6237–6243.

R. J. P. Williams. 1989. "Electron Transfer in Biology." *Molec. Phys.* 68, 1–23.

Distance and Driving-Force Dependence of Electron Transfer
D. Beratan, J. N. Onuchic, J. R. Winkler, and H. B. Gray. 1992. "Electron Tunneling Pathways in Proteins." *Science* 258, 1740–1741.
C. C. Moser, J. M. Keske, K. Warncke, R. S. Farid, and P. L. Dutton. 1992. "Nature of Biological Electron Transfer." *Nature* 355, 796–802.
J. N. Onuchic, D. N. Beratan, J. R. Winkler, and H. B. Gray. 1992. "Pathway Analysis of Protein Electron-Transfer Reactions." *Annu. Rev. Biophys. Biomol. Struct.* 21, 349–377.
H. Pelletier and J. Kraut. 1992. "Crystal Structure of a Complex Between Electron Transfer Partners, Cytochrome c Peroxidase and Cytochrome c." *Science* 258, 1748–1755.

CHAPTER 10

Substrate Binding and Activation by Nonredox Mechanisms

Principles: *The positive charge of a metal ion is a key feature in metalloenzymes that lowers the pK_a of coordinated water and provides a locally high concentration of the otherwise unavailable reagent OH^-. In a similar manner, the positively charged metal center can also serve as a general Lewis acid for activation of a substrate molecule, modulating its reactivity following coordination. Rate acceleration can occur by internal attack within the coordination sphere of the metal ion or by positioning of a substrate ligand near an essential group at the active site. Protein side chains greatly assist the assembly of the activated complex for catalysis, working in concert with the metal center. Electrostatic interactions are of fundamental importance in all these processes.*

In this chapter we explore the manner in which metal ions bind and activate substrates without themselves undergoing redox reactions during any step of their catalytic mechanism (Table 1.3). Many metalloenzymes that function in this manner employ redox-inactive metal ions such as Mg(II) and Zn(II), either as intrinsic components of their active sites or as cofactors for binding and activating a substrate molecule. Zinc(II) ion, in particular, is uniquely suited among biological transition-metal ions to serve as a Lewis acid without participating in electron-transfer reactions. The first recognition of a specific function for zinc in a biochemical pathway was the discovery in 1939 that zinc is an essential component of carbonic anhydrase, a lyase that catalyzes the reaction shown in Equation 10.1. In the absence of enzyme, the reaction rate is slow, $\sim 10^{-2}$ s^{-1} at physiological pH, but it rises to a turnover number of as high as 10^6 s^{-1} in the presence of the enzyme. In 1955, a second

$$CO_2 + H_2O \rightleftharpoons HCO_3^- + H^+ \qquad (10.1)$$

zinc-requiring metalloenzyme was discovered, carboxypeptidase (CPD) A, which cleaves peptide bonds according to Equation 10.2. In this reaction, R' is

$$R-CO-NH-CHR'-CO_2^- + H_2O \rightleftharpoons R-CO_2^- + H_3N^+-CHR'-CO_2^- \quad (10.2)$$

the C-terminal aromatic residue of a protein or peptide; a related protein known as carboxypeptidase B prefers R' groups that are basic. Both enzymes are synthesized in the pancreas and often isolated from this source. Other hydrolase enzymes in which zinc plays a catalytic role include alkaline phosphatase, which promotes the reaction in Equation 10.3, snake-venom

$$R-O-PO_3^{2-} + H_2O \rightleftharpoons ROH + HPO_4^{2-} \quad (10.3)$$

phosphodiesterase, and various nucleases. In addition, zinc is a functionally significant component of transferases, such as DNA and RNA polymerases and reverse transcriptase, and of both isomerases and lyases, including some aminoacyl tRNA synthetases, although for some of these we have too little information to distinguish catalytic from structural roles. Clearly zinc(II) plays an essential role in the many hydrolysis and group-transfer enzymes in biology and is therefore one of the most studied elements in bioinorganic chemistry. Three examples of zinc enzymes to be discussed in detail are carboxypeptidase, a hydrolytic enzyme; carbonic anhydrase, a lyase; and alcohol dehydrogenase, an oxidoreductase in which zinc activates the substrate for redox chemistry that involves the cofactor NAD^+.

10.1. Hydrolytic Enzymes

We begin our discussion with carboxypeptidase A and thermolysin, two important hydrolytic enzymes the structures and mechanisms of which are now reasonably well understood. Peptide hydrolysis is a rather difficult reaction, despite being quite thermodynamically favorable in aqueous solution. In order to catalyze this reaction, an enzyme must accomplish several things. First, it must facilitate the nucleophilic attack on the peptide carbonyl group by a nucleophile. This function can be accomplished by producing a highly reactive nucleophile or by activating the carbonyl for attack by polarization. Second, it must stabilize the tetrahedral intermediate or transition state that is generated following nucleophilic attack at the carbonyl carbon. Finally, it must stabilize the amide nitrogen atom to make it a suitable leaving group, so that the tetrahedral intermediate can collapse upon C–N bond cleavage. In principle, a metal ion such as zinc might play a role in any or all of these

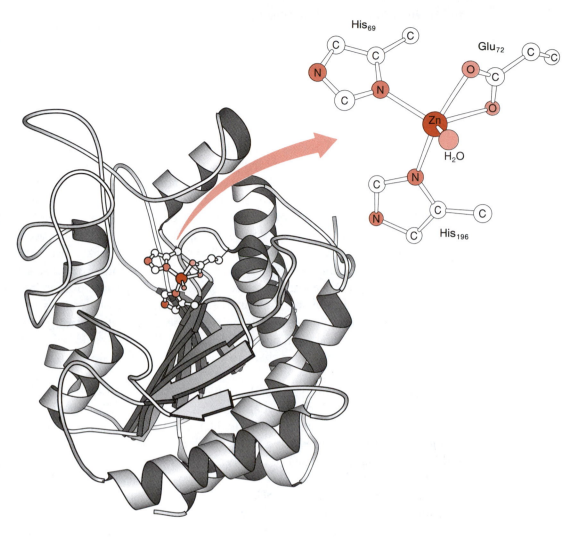

Figure 10.1
Structure of carboxypeptidase A from bovine pancreas and its zinc-based active site.

processes. Over the years, many studies have been directed toward deducing the mechanisms of peptidases in order to elucidate the functions of the active-site zinc ion.

10.1.a. Carboxypeptidase A and Thermolysin: Structural Studies.
Three forms of bovine pancreatic carboxypeptidase A, designated A_α, A_β, and A_γ, with 307, 305, and 300 amino-acid residues, respectively, have been obtained. The latter two forms are shortened at the N-terminus. The structure of carboxypeptidase A_α has been determined in a high-resolution X-ray crystal structure study that forms the basis for much of the mechanistic discussion about the enzyme. Depicted in Figure 10.1 is the structure of the protein as well as a close-up view of the active site. The zinc atom is situated well inside

259

the protein surface and is coordinated by two imidazole side chains, from His-69 and His-196, and to the bidentate carboxylate group of Glu-72. The coordination geometry is completed by a water molecule, resulting in penta-coordinate Zn(II). In the presence of substrates or inhibitors, the Glu-72 carboxylate group becomes coordinated in a more nearly unidentate fashion. The ability of Glu or Asp residues to rearrange in such a manner, a process that has been termed the *carboxylate shift*, allows maintenance of a constant or nearly constant coordination number even when various forms of substrate are bound to the metal. Near the active site is a hydrophobic pocket into which protrudes the aromatic ring on the C-terminal end of the substrate molecule. There are hydrogen bonds between the coordinated water molecule and Glu-270. Several arginine and tyrosine residues are positioned in the active site in a way that allows them to participate in substrate binding and activation.

High-resolution X-ray structural studies of a number of substrate and inhibitor complexes of carboxypeptidase A, including Gly-Tyr, $(-)$-2-benzyl-3-p-methoxybenzoylpropionic acid (an ester analog), and a 39-amino-acid inhibitor from potatoes, have been reported. Studies at 1.6 Å resolution and low temperature (-9 °C) of the glycyl-L-tyrosine adduct reveal several important features (Figure 10.2). The tyrosine ring displaces water molecules and resides in the hydrophobic pocket mentioned above. The glycine residue forms a bidentate chelate with zinc using both its amide carbonyl oxygen atom and the terminal amino nitrogen atom. The latter replaces the water molecule ordinarily coordinated to zinc, rendering this enzyme-substrate complex incapable of undergoing peptide hydrolysis; otherwise, the Gly-Tyr complex would be hydrolyzed in the crystal during the time required to collect the X-ray data. These results provide indirect evidence that activation of the water molecule by coordination to the zinc ion is a key feature of the enzyme mechanism. Arg-145 is hydrogen-bonded to the terminal R-CO$_2^-$ residue. The γ-carboxylate group of Glu-270 is within hydrogen-bonding distance of the peptide amide nitrogen atom. Finally, Arg-127 is located in the active site on the side opposite Glu-270.

The structure of the endopeptidase thermolysin has also been determined to high resolution. The overall structure of the enzyme, shown in Figure 10.3, bears little resemblance to that of carboxypeptidase. Closer examination of the active site reveals many striking similarities, however. In the unligated form, the zinc ion is coordinated by three side chains, from His-142, His-146, and Glu-166, and a water molecule. The noncoordinating amino-acid side chains in the two proteins are also similar. Glu-143 is in a position analogous to that of Glu-270 in carboxypeptidase, whereas that of His-231 matches

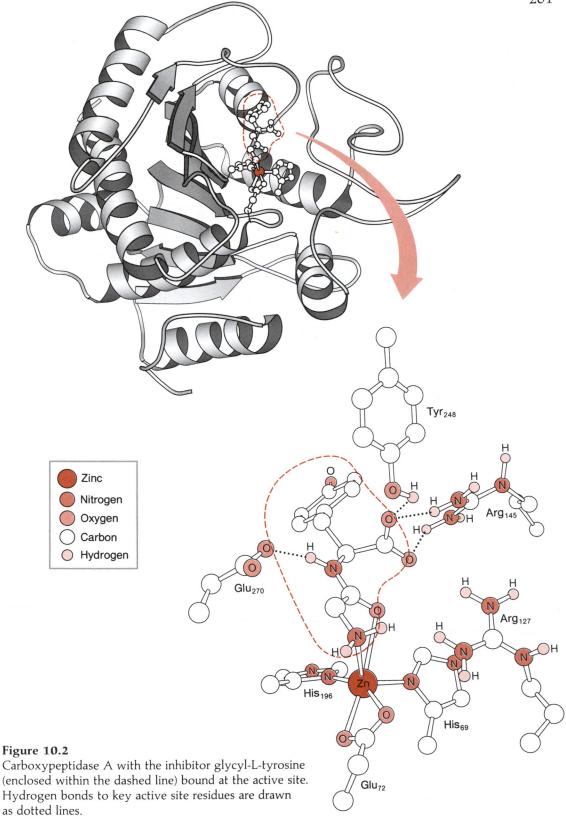

Figure 10.2
Carboxypeptidase A with the inhibitor glycyl-L-tyrosine
(enclosed within the dashed line) bound at the active site.
Hydrogen bonds to key active site residues are drawn
as dotted lines.

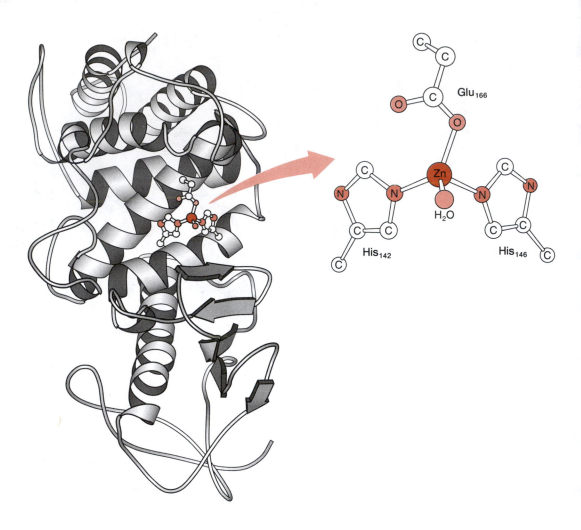

Figure 10.3
Structure of thermolysin and its active site.

Arg-127. These similarities and the corresponding mechanistic consequences reveal that this pair of enzymes constitutes an example of convergent evolution, whereby a nearly identical structural and chemical solution has evolved to facilitate the catalysis of two closely related reactions in otherwise unrelated protein frameworks.

10.1.b. Carboxypeptidase A: Metal Substitution and Spectroscopy. Carboxypeptidase represents a classic case in bioinorganic chemistry of the application of metal-substitution probes. Because Zn(II) is magnetically and spectroscopically silent, extensive studies of various metal-substituted derivatives have been carried out. The cobalt(II) derivative, in particular, can be prepared from the apoprotein and is more active than the native enzyme in

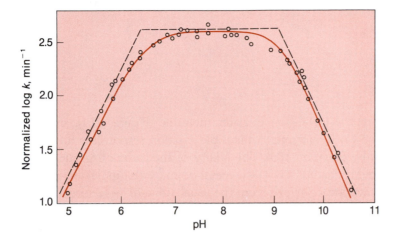

Figure 10.4
pH dependence of log k for hydrolysis of a set of four substrates by
carboxypeptidase A. The rate decrease at low pH is ascribed to protonation of the
zinc-bound hydroxide. (Adapted from D. S. Auld and B. L. Vallee, *Biochemistry* **9**,
4352–4359 (1970)).

hydrolyzing peptides and equally good at ester hydrolysis. Derivatives containing Mn(II) and Cd(II) show less activity, those of Ni(II), Hg(II), and Pb(II)
restore only a little activity, and the Cu(II) and Co(III) enzymes, like the
apoprotein, are totally inactive. The optical absorption, MCD, and EPR
spectra of Co(II)-CPD resemble those of a 4- or 5-coordinate cobaltous
complex, with $\lambda_{max} \sim 550$ and 572 nm ($\varepsilon \sim 150$ M^{-1} cm^{-1}), a shoulder at
500 nm, and peaks at 940 and 1570 nm. The g-values occur at 5.53, 2.94, and
2.01. These spectral features are uncommon for Co(II), however, and have
been used to formulate the entatic state hypothesis (see Chapter 12). They do
not change upon addition of anions (Cl$^-$, Br$^-$, F$^-$, SO$_4^{2-}$, NO$_2^-$) or even
very much when Gly-Tyr is added. Major alterations do occur when Glu-270
is modified by a carbodiimide reagent. In hindsight, and with the benefit of
X-ray crystallography, the spectra can be judged to arise from the distorted
pentacoordinate structure depicted in Figure 10.1. Although these studies did
not produce definitive results concerning the mechanism, they did provide a
number of constraints that need to be satisfied by any mechanistic proposal.

10.1.c. Kinetic Studies and the Postulated Enzyme Mechanism.
Carboxypeptidase A hydrolyzes tri- and tetrapeptides according to Michaelis-
Menten kinetics with k_{cat} values of ~ 20 to 500 s^{-1} and K_m values of 0.01 to
1 mM, depending on the substrate. Here, $k_{cat} = V_{max}/E_0$, the so-called turnover number, where V_{max} is defined in Section 3.1.f. and E_0 is the total enzyme
concentration. The kinetic profiles are bell-shaped (Figure 10.4), with apparent pK_a values of 6.1 and 9.0. The former value is almost certainly a conse-

quence of the metal-bound water molecule (Figure 10.1), since it shifts to 5.3 in the Co(II) enzyme. A plausible reaction responsible for this ionization is given by Equation 10.4. The ability of the carboxyl group of Glu-270 to form a hydrogen bond with the metal-bound water molecules may contribute to the lowering of its pK_a from the usual value of ~ 10 (see Table 2.2).

$$(10.4)$$

From this kinetic information and the X-ray structural results for carboxypeptidase A and its inhibitor complexes, the following detailed picture of the enzyme mechanism has emerged (Figure 10.5). The terminal amino-acid residue of a polypeptide chain binds to the active site, orienting itself through interaction of the terminal carboxyl group with Arg-145. Additional stabilizing hydrogen bonding interactions with Asn-144 and Tyr-248 can also occur. If the terminal group is aromatic, it resides in the hydrophobic pocket. The carbonyl group of the scissile peptide bond, that is, the one being hydrolyzed, interacts with Arg-127, increasing its Lewis acidity. Next there is an attack of the coordinated water molecule on the peptide carbonyl group, with concomitant transfer of a proton to Glu-270. This key step in the enzyme mechanism is effected by the increased nucleophilicity of the water molecule as a result of deprotonation upon coordination to Zn(II) and the general base character of the properly positioned residue Glu-270. This nucleophilic attack produces the tetrahedral intermediate. Crystallographic studies of certain inhibitor complexes strongly suggest that this intermediate now becomes bound to the zinc in a bidentate fashion. Consider, for example, N-(tert-butoxycarbonyl)-5-amino-2-benzyl-4-oxo-6-phenylhexanoic acid, shown in Figure 10.6, a substrate analog in which the amide NH has been replaced by CH_2. The carbonyl group of this simple ketone is less than 0.2 percent hydrated in aqueous solution. Crystallographic studies of the complex of this compound with carboxypeptidase, however, revealed that it is fully hydrated and bound as a bidentate gem-diol(ate). This result suggests that either the enzyme selected the low concentration of hydrated ketone from solution or the enzyme binds the unhydrated substrate and executes the first step of the amide hydrolysis mechanism, addition of hydroxide ion to the carbonyl group.

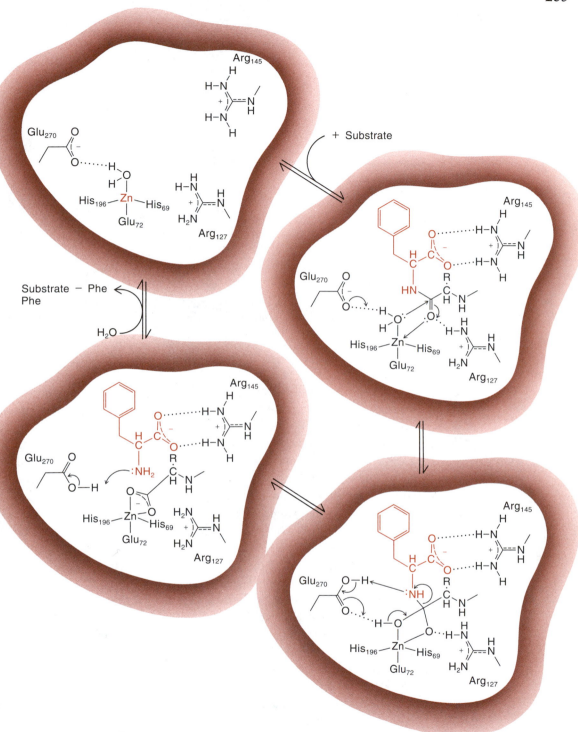

Figure 10.5
Mechanism for peptide bond hydrolysis by carboxypeptidase A involving carboxylate-assisted generation of a zinc-bound hydroxide ion which attacks the peptide carbonyl group.

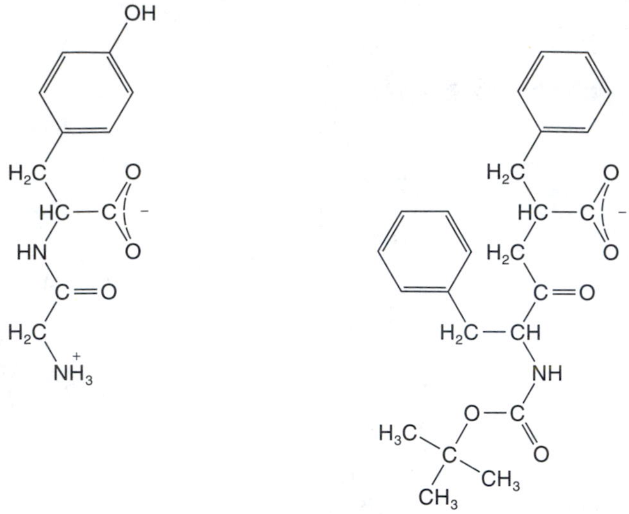

(a) Gly-Tyr

(b) N-(*tert*-butoxycarbonyl)-5-amino-2-benzyl-4-oxo-6-phenylhexanoic acid

Figure 10.6
Structures of two inhibitors of carboxypeptidase A.

The reaction cannot proceed because there is no appropriate leaving group. For the normal substrate, collapse of the tetrahedral intermediate requires protonation of the amide nitrogen. Initially, from the X-ray structural analysis, it appeared that Tyr-248 might be involved in carrying a proton to the leaving group. Site-directed mutagenesis studies, however, have revealed that replacing Tyr-248 with Phe changed the activity of the enzyme only slightly. An attractive alternative is Glu-270, now protonated, having extracted a proton from the zinc-bound water. This hypothesis is supported by comparison with thermolysin, where Glu-143 is well positioned for this role and for which there are no other candidate proton donors. Carbon-nitrogen bond cleavage results in production of the free carboxyl terminal amino acid and the shortened peptide chain with its terminal carboxylate group bound to the zinc. Displacement of these by water or by additional substrate completes the cycle. This mechanism is summarized in Figure 10.5.

The above mechanism illustrates several aspects of how a metal ion functions to promote a hydrolytic reaction. The zinc(II) ion serves as a template to assemble the reactive groups involved in peptide-bond cleavage, especially the critical carboxyl functionality of Glu-270. The Lewis acidity of the metal ion and the general base character of the Glu-270 carboxyl group lower the pK_a of the bound water molecule, which attacks the scissile peptide bond. It must be stressed, however, that although the zinc ion plays a critical role in effecting peptide hydrolysis, amino-acid residues supplied by the surrounding polypeptide chain are also of great importance. Glu-270 and Arg-127 perform the critical functions leading to peptide-bond cleavage. Thus nature has assembled an efficient chemical system to promote the catalytic hydrolysis of peptide bonds, an essential metabolic reaction, at rates far exceeding the uncatalyzed hydrolysis at neutral pH.

10.1.d. Alkaline Phosphatase. A remarkable number and variety of zinc hydrolases have been characterized. In one such enzyme, alkaline phosphatase, a pair of zinc(II) ions binds the terminal phosphate group of a substrate, typically a monoester such as *p*-nitrophenyl phosphate. A serine hydroxyl group at the active site then attacks the phosphoryl group, cleaving the ester functionality; in the process, the phosphate is transferred to the enzyme, forming a phosphorylserine residue. Hydrolysis of this phosphate ester by coordinated hydroxide ion completes the catalytic cycle. These steps are drawn out in detail in Figure 10.7. In alkaline phosphatase, the pair of zinc ions serves as a general Lewis acid, polarizing the substrate and rendering it a better electrophile. Positioning of the phosphate ester at the active site by the arginine residue is analogous to Arg-145-facilitated terminal carboxyl orientation of substrate in carboxypeptidase. Finally, cleavage of the phosphorylserine ester by a zinc-bound hydroxide is directly analogous to steps in Figure 10.5. Interestingly, in alkaline phosphatase, there is a third metal, a magnesium ion, within ~5 Å of one zinc and 7 Å of the other. Although this metal ion does not appear to participate directly in the catalytic mechanism, it may contribute by shaping the structure and the electrostatic potential of the active site (see Chapter 12).

10.2. A Lyase, Carbonic Anhydrase

As noted earlier, carbonic anhydrase was the first biomolecule discovered to contain zinc. Like carboxypeptidase, this enzyme has been extensively investigated by a variety of techniques, and a consensus has now been reached on the mechanism and the role of the zinc ion. Several different isozymes have been characterized crystallographically, and the overall structure of one of these is depicted in Figure 10.8. The zinc(II) ion is coordinated by three histidine imidazole side chains, namely, those of His-94 and His-96 (through their N_ε atoms) and His-119 (through N_δ). In addition, a water/hydroxide is bound to the metal to complete a distorted tetrahedral coordination sphere.

The kinetics of the carbonic anhydrase reaction (Equation 10.1) are pH-dependent, being faster at high pH and under the control of a group having an apparent pK_a value near 7.0. The simplest and most widely accepted interpretation of these observations is that this reaction, like that for car-

268

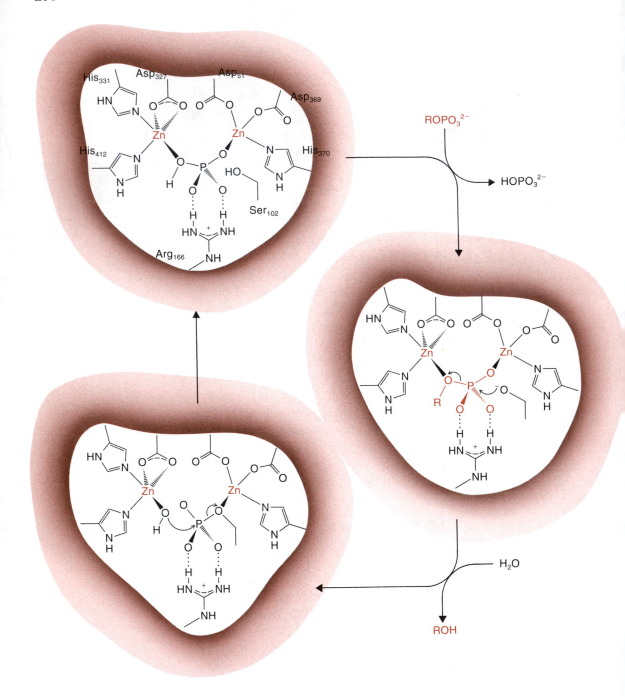

Figure 10.7
Phosphate monoester hydrolysis by alkaline phosphatase.

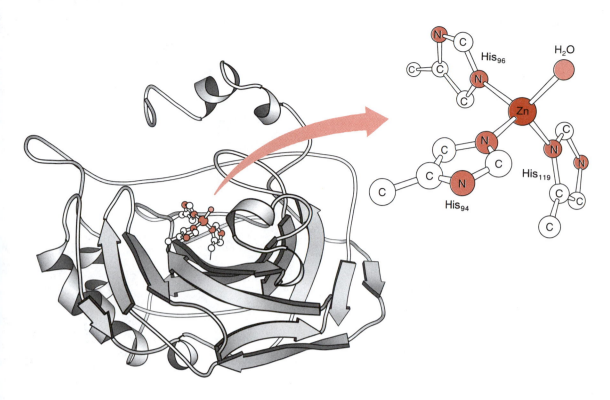

Figure 10.8
Structure of carbonic anhydrase and the geometry of its zinc site.

boxypeptidase discussed above (Equation 10.2), involves formation of a zinc-bound hydroxide ion. This proposal is also supported by the observation that many anions inhibit the reaction, presumably by competing with water ligands for the open coordination site on zinc. A simple mechanism based on the activity of the zinc hydroxide is shown in Figure 10.9.

Although this mechanism can explain many aspects of the carbonic anhydrase reaction, it cannot account for the extremely rapid rate of the reaction. As noted, the enzyme can turn over at rates of up to a million times per second, a value that is remarkable for the following reason. As was discussed earlier, the rate depends on a group, probably the zinc-bound water, that titrates with a pK_a value near 7. Consider the equilibrium shown in Equation 10.6. If the overall equilibrium constant for this reaction is 10^{-7}, and the

$$PZn\text{-}(OH_2) \rightleftharpoons PZn(OH^-) + H^+ \qquad (10.6)$$

reverse reaction is diffusion-limited, then the forward reaction cannot proceed at a rate faster than, at most, 10^4 sec^{-1}. In order for the reaction to proceed more rapidly than this rate, components of the buffer other than water or hydroxide must participate in deprotonating the zinc-bound water. This con-

269

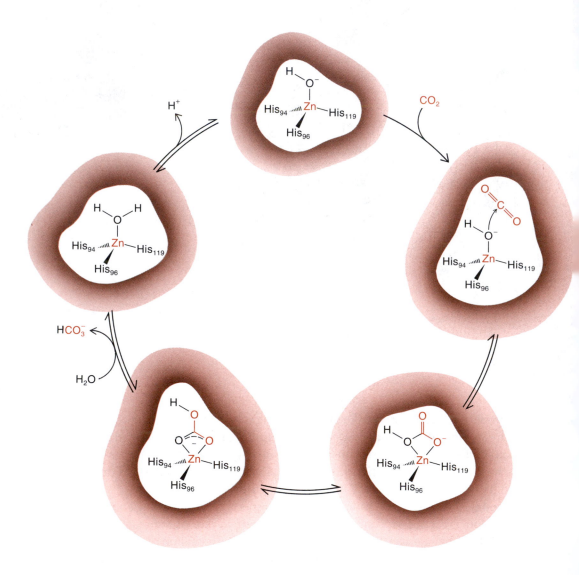

Figure 10.9
A mechanistic scheme for the hydration of carbon dioxide by carbonic anhydrase.

clusion has been verified experimentally; at low buffer concentrations, the reaction rate does depend on the buffer concentration. Additional studies suggest that the deprotonation may proceed in several steps, with other protein side chains, such as histidine, shuttling protons for equilibration with the medium.

Once the metal-hydroxide unit is formed, binding of carbon dioxide (probably not directly at the metal site) can occur, followed by nucleophilic attack of the hydroxide to produce a metal-bound bicarbonate. Displacement of this product by water completes the cycle, except for the proton shuttling noted above.

The reduced pK_a of the zinc-bound water is crucial for this reaction. As discussed in Chapter 8, the low-pH carbonic anhydrase site is one of the most highly charged mononuclear sites, since all the ligands are neutral. This feature favors release of a proton, which reduces the positive charge. As a consequence, the pK_a value for this coordinated water molecule is near 7, which is very low compared with the values of 14 for free water and 10 for $[Zn(OH_2)_6]^{2+}$ in aqueous solution.

10.3. An Oxidoreductase, Alcohol Dehydrogenase

This enzyme catalyzes the NAD^+-dependent conversion of primary alcohols to aldehydes (Equation 10.7). Although the reaction involves redox

$$RCH_2OH + NAD^+ \rightleftharpoons RCHO + NADH + H^+ \tag{10.7}$$

chemistry, the enzyme contains zinc, which serves to bind and activate the substrate molecule prior to a hydride-transfer step; it does so much as it does in carboxypeptidase and carbonic anhydrase, and is thus of interest here.

Liver alcohol dehydrogenase (LADH) comprises two 40-kilodalton single polypeptide subunits, each of which contains two zinc(II) ions in separate domains. One of the Zn(II) ions resides in the catalytic domain, which also binds the essential coenzyme NAD^+, and is coordinated to two cysteinate sulfur atoms, one histidine nitrogen atom, and a water molecule. The other zinc is bonded to four cysteine thiolate ligands and is inaccessible to solvent. The latter metal ion is believed to play a structural role, the nature of which will not be further discussed here. The structure of horse-liver alcohol dehydrogenase, determined by X-ray crystallography, is displayed in Figure 10.10.

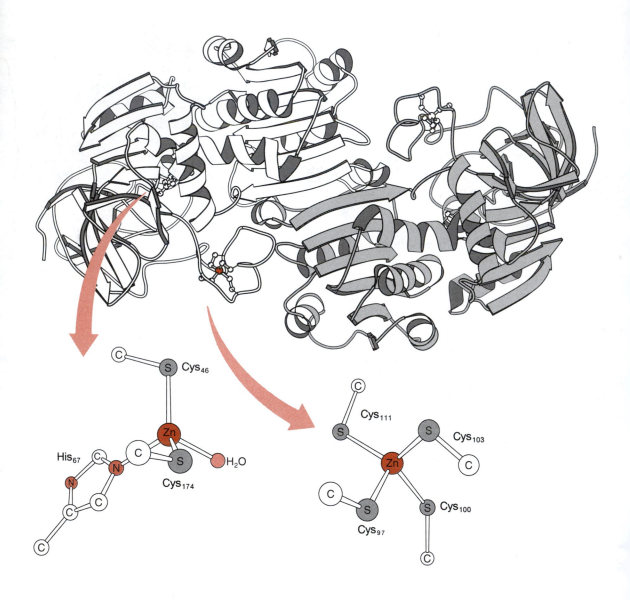

Figure 10.10
Structure of the dimer of liver alcohol dehydrogenase and view of the catalytic
(left) and structural (right) zinc sites.

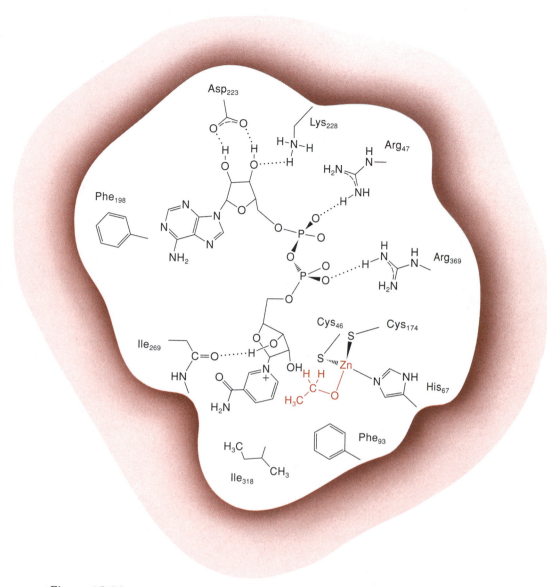

Figure 10.11
Schematic view of the NAD$^+$-binding site in liver alcohol dehydrogenase.

The catalytic zinc is located in a pocket buried deep within the protein about 20 Å from the surface near the junction of two domains, one of which performs the catalysis and the other of which houses the coenzyme. This pocket binds substrate as well as the nicotinamide portion of the coenzyme. A schematic drawing of the substrate-binding pocket is presented in Figure 10.11. Considerable information is now available about the role of the zinc ion in LADH. Zinc is not required for coenzyme binding, since NADH occupies its normal site even in an enzyme from which the catalytic zinc has been removed. Alcohol substrates, on the other hand, bind through their oxygen atom directly to the zinc, a process accompanied by loss of water and deprotonation of the ligand to form coordinated RO$^-$. The X-ray structure of the p-bromobenzylalcohol complex of LADH has been determined, from which a Zn—O bond distance of 2.1 Å was measured. The remaining ligands at zinc complete a distorted tetrahedral structure with Zn—N and Zn—S bond lengths of 2.1 to 2.3 Å.

The binding of coenzyme NAD$^+$ to LADH is accompanied by a change in conformation of the protein from an open to a closed form. These conformations differ by a small, 10° rotation of the catalytic domains with respect to the core of the dimer. The conformational change serves to position the substrate binding pocket at the appropriate distance and orientation with respect to the coenzyme in order to facilitate the critical step in the enzyme mechanism, namely, the transfer of the hydride ion from the α-carbon of the alcohol to the nicotinamide ring. A schematic drawing of the hydride transfer step of the enzyme mechanism is given in Figure 10.12. Following hydride transfer, the product and NADH dissociate from the enzyme, which can then bind additional molecules of substrate and coenzyme. The role of the zinc is thus seen to be twofold, namely, to facilitate deprotonation of the alcohol, rendering it a better hydride-ion donor because of the accumulated negative charge, and to position the substrate in the active site in a way that maximizes the stereoelectronic factors leading to direct hydride transfer to the coenzyme. Extrusion of water from the active site upon substrate binding both retains the coordination number and geometry of the zinc ion and removes a potentially reactive molecule from the vicinity of the activated substrate molecule. The dual role of zinc as a general Lewis acid and in positioning the alcohol substrate mimic its functions in the enzymes discussed previously in this chapter. The fact that one histidine nitrogen and two cysteine sulfur ligands occur in most known zinc alcohol dehydrogenases suggests that they optimize the electronic properties of the metal ion for its required functions in these enzymes.

10.4. Nucleotide Activation

Several zinc enzymes are involved in nucleic-acid metabolism, especially phosphodiesterases and nucleases. Belonging to the class of zinc enzymes known as transferases are DNA and RNA polymerase and reverse transcriptase, proteins that catalyze the formation of polymers from mononucleoside triphosphate building blocks. Thus zinc-catalyzed chemistry is fundamental to the key reactions of molecular biology. Detailed mechanisms for most of these important systems remain to be established.

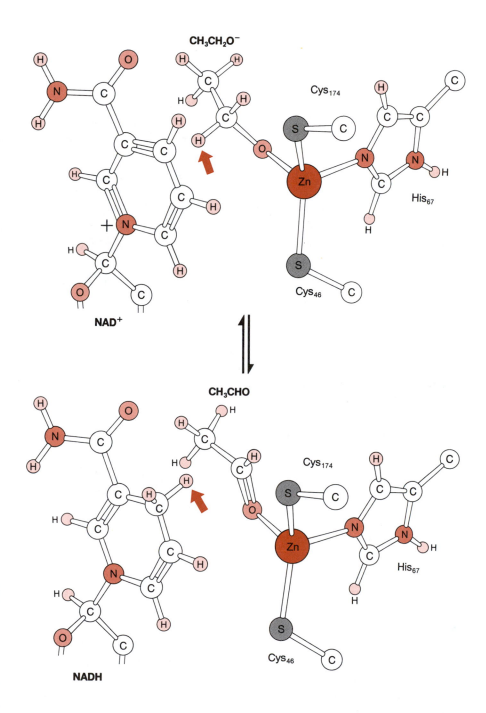

Figure 10.12
Hydride transfer from the zinc-bound ethanolate ion to NAD$^+$ forming acetaldehyde and NADH. The hydride that is transferred is indicated by an arrow.

Essentially all enzymes that act on nucleotides such as ATP require Mg^{2+} (or, as a substitute, Mn^{2+}) for activity, because of the unique roles that these metal ions can play in binding and stabilizing polyphosphate moieties. By forming relatively weak complexes ($K_a \approx 10^4 \, M^{-1}$) with these groups, these metals can perform one or more of the following roles. They can stabilize the substrate in a conformation suitable for binding to the enzyme, a function analogous to the role such metals play in stabilizing nucleic-acid structures, as discussed in Chapter 7. They can orient the substrate in a productive conformation for reaction to occur. And, finally, they can stabilize the leaving group in the reaction by binding to polyphosphate groups.

The complexes formed in these cases are often chelates, as shown in Figure 5.18. Unfortunately, partly because of their kinetic lability, it has been difficult to obtain detailed structural information about these magnesium complexes. Several kinetically inert Cr(III) and Co(III) nucleotide complexes have been prepared and structurally characterized, however. These complexes often serve as inhibitors of enzymes that require Mg(II) nucleotide substrates or cofactors and are thus considered good models for the types of structures that will form with magnesium. The metal is always coordinated to the phosphate oxygen atoms rather than to the nucleobases, as expected from hard-soft acid-base considerations for the hard Mg^{2+} ion. The most common isomer is the β,γ-bidentate chelate, in which the metal ion is coordinated to a terminal phosphate oxygen atom from the end (γ) and the internal (β) phosphate residues of ATP. The resulting chelate ring renders the β-phosphate group chiral, and the resulting two isomers have been designated Λ and Δ, as depicted in Figure 10.13. This nomenclature should not be confused with the nomenclature for octahedral tris-chelate complexes (see Chapter 5).

There is evidence that specific isomers are used by enzymes that process Mg(ATP) complexes. The Λ and Δ isomers of $[Co(NH_3)_4(\beta,\gamma\text{-ATP})]$ were synthesized and allowed to react with the enzyme hexokinase, which normally uses Mg(ATP) to convert glucose to glucose-6-phosphate and produces Mg(ADP). When the products were analyzed, it was found that the Λ isomer of $[Co(NH_3)_4(\beta,\gamma\text{-ATP})]$ was converted to $[Co(NH_3)_4(\beta,\gamma\text{-ADP-}$ glucose-6-phosphate)], but the Δ isomer was unreactive. The absolute configuration of the cobalt-containing product was determined crystallographically after degradation to Λ-$[Co(NH_3)_4(\beta,\gamma\text{-PPP})]$. Thus magnesium, in much the same manner as zinc, can bind a substrate molecule, activating it for phosphodiester bond cleavage as a result of its general Lewis acid properties and orienting it in the proper conformation for binding to the catalytic unit

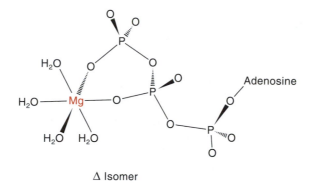

Figure 10.13
The Λ and Δ stereoisomers of Mg^{2+} bound to ATP
via its β and γ phosphates.

in the enzyme. It is likely that most magnesium-requiring enzymes utilize these principles of substrate binding and activation in a similar manner.

The ability of bioinorganic chemists to control the dependence of enzymes on substrate activation by Mg^{2+} ion can occasionally be used to advantage to study a particular feature of the mechanism. A striking example is the determination of the geometry of the complex formed between the restriction enzyme Eco RI and its substrate. In the presence of Mg^{2+}, the bound oligonucleotide would be cleaved, precluding crystallographic examination. By growing crystals in the absence of Mg^{2+}, however, a rare look at a protein-substrate complex was obtained; the structure is shown in Figure 10.14.

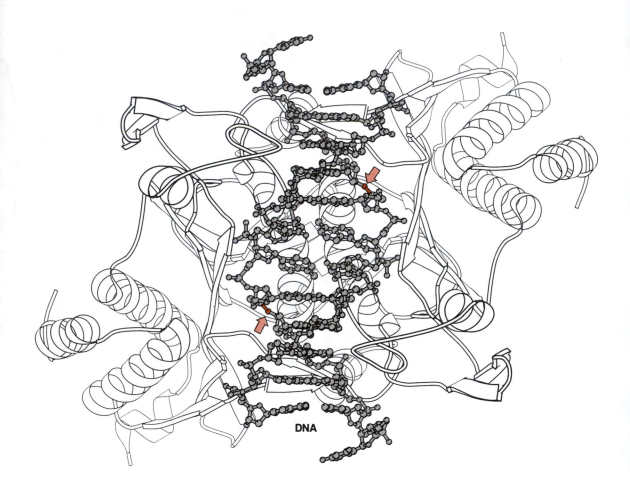

Figure 10.14
Crystal structure of the restriction enzyme Eco RI bound to duplex DNA determined in the absence of Mg^{2+}. The phosphodiester bonds that are cleaved upon Mg^{2+} addition are indicated by arrows.

Study Problems

1. Suggest possible reasons why trivalent ions, such as Fe(III) or Mn(III), have not been widely utilized in hydrolytic metalloenzymes for substrate binding and activation.

2. Figure 10.7 reveals how a pair of metal ions working together can activate a substrate for hydrolysis. Urease is an enzyme that catalyzes the hydrolysis of urea to form ammonium and carbamate ions, as shown in the equation. Urease has a pair of

nickel ions within 4 Å of each other in the active site. Propose a mechanism for the nickel-catalyzed hydrolysis of urea in this enzyme using the principles of this chapter.

3. Phosphorothioate groups (see Figure 3.21) have been incorporated into oligo-nucleotides. Discuss the potential ramifications of this substitution for the mechanism of phosphodiesterases, nucleases, and other enzymes that process phosphate esters.

4. In the examples discussed in this chapter, there were frequently amino-acid side chains that were not coordinated to the metal ion that participated in the enzyme mechanism. For any one of these examples, suggest specific changes in the residues that you might like to explore by site-specific mutagenesis, and discuss how these modifications could provide additional insight into the catalytic mechanism.

5. Some proteases such as chymotrypsin do not contain metal ions in their active sites. Instead, there are several other functionalities that facilitate peptide bond cleavage. The reaction proceeds through nucleophilic attack of a serine hydroxyl on the peptide carbonyl group. The carbonyl group of the substrate is positioned near a cluster of peptide NH moieties. Compare the roles that these groups might play in catalysis with those of zinc in carboxypeptidase.

Bibliography

General References

I. Bertini. 1983. "The Coordination Properties of the Active Site of Zinc Enzymes." In I. Bertini, R. S. Drago, and C. Luchinat, eds., *The Coordination Chemistry of Metalloenzymes: The Role of Metals in Reactions Involving Water, Dioxygen, and Related Species. Proc. NATO Adv. Stud. Inst.* D. Reidel, Dordrecht, Holland, 1–18.

D. W. Christianson. 1991. "Structural Biology of Zinc," in C. B. Anfinsen, J. T. Edsall, D. Eisenberg, and F. M. Richards, eds., *Advances in Protein Chemistry*, Volume 42. Academic Press, San Diego, 281–355.

J. E. Coleman. 1992. "Zinc Proteins: Enzymes, Storage Proteins, Transcription Factors, and Replication Proteins." *Annu. Rev. Biochem.* 61, 897–946.

T. G. Spiro, ed. 1983. *Zinc Enzymes*. Wiley-Interscience, New York.

J. Suh. 1992. "Model Studies of Metalloenzymes Involving Metal Ions as Lewis Acid Catalysts." *Acc. Chem. Res.* 25, 273–279.

B. L. Vallee and D. S. Auld. 1990. "Active-Site Zinc Ligands and Activated H_2O of Zinc Enzymes." *Proc. Natl. Acad. Sci. USA* 87, 220–224.

Hydrolytic Enzymes

D. S. Auld, J. F. Riordan, and B. L. Vallee. 1989. "Probing the Mechanism of Carboxypeptidase A by Inorganic, Organic, and Mutagenic Modifications," in H. Sigel, ed., *Action of Interrelations Among Metal Ions Enzymes, and Gene Expression. Metal Ions in Biological Systems*, Volume 25. Marcel Dekker, New York, 359–394.

D. W. Christianson and W. N. Lipscomb. 1986. "X-ray Crystallographic Investigation of Substrate Binding to Carboxypeptidase A at Subzero Temperature." *Proc. Natl. Acad. Sci. USA* 83, 7568–7572.

D. W. Christianson and W. N. Lipscomb. 1989. "Carboxypeptidase A." *Acc. Chem. Res.* 22, 62–69.

S. J. Gardell, C. S. Craik, D. Hilvert, M. S. Urdea, and W. J. Rutter. 1985. "Site-Directed Mutagenesis Shows That Tyrosine 248 of Carboxypeptidase A Does Not Play a Crucial Role in Catalysis." *Nature* 317, 551–555.

H. Kim and W. N. Lipscomb. 1990. "Crystal Structure of the Complex of Carboxypeptidase A with a Strongly Bound Phosphonate in a New Crystalline Form: Comparison with Structures of Other Complexes." *Biochemistry* 29, 5546–5555.

A Lyase, Carbonic Anhydrase

A. E. Eriksson, T. A. Jones, and A. Liljas. 1988. "Refined Structure of Human Carbonic Anhydrase II at 2.0 Å Resolution." *Proteins: Struc., Func. Genet.* 4, 274–282.

A. E. Eriksson, P. M. Kylsten, T. A. Jones, and A. Liljas. 1988. "Crystallographic Studies of Inhibitor Binding Sites in Human Carbonic Anhydrase II: A Pentacoordinate Binding of the SCN^- Ion to the Zinc at High pH." *Proteins: Struc., Func. Genet.* 4, 283–293.

O. Jacob, R. Cardensas, and O. Tapia. 1990. "An *ab initio* Study of Transition Structures and Associated Products in $[ZnOHCO_2]^+$, $[ZnHCO_3H_2O]^+$, and $[Zn(NH_3)_3HCO_3]^+$ Hypersurfaces: On the Role of Zinc in the Catalytic Mechanism of Carbonic Anhydrase." *J. Am. Chem. Soc.* 112, 8692–8705.

Y. Xue, J. Vidgren, L. A. Svensson, A. Liljas, B.-H. Jonsson, and S. Lindskog. 1993. "Crystallographic Analysis of Thr-200 → His Human Carbonic Anhydrase II and Its Complex with the Substrate, HCO_3^-." *Proteins: Struc., Func. Genet.*, 15, 80–87.

Nucleotide Activation

D. Dunaway-Mariano and W. W. Cleland. 1980. "Preparation and Properties of Chromium (III) Adenosine 5'-Triphosphate, Chromium (III) Adenosine 5' Diphosphate, and Related Chromium III Complexes." *Biochemistry* 19, 1496–1505.

R. B. Martin. 1990. "Bioinorganic Chemistry of Magnesium," in H. Sigel ed., *Compendium on Magnesium and Its Role in Biology, Nutrition, and Physiology. Metal Ions in Biological Systems*, Volume 26. Marcel Dekker, New York, 1–14.

An Oxidoreductase, Alcohol Dehydrogenase

H. Ekland, J.-P. Samama, L. Wallén, C.-I. Brändén, Å. Åkeson, and T. A. Jones. 1981. "Structure of a Triclinic Ternary Complex of Horse Liver Alcohol Dehydrogenase at 2.9 Å Resolution." *J. Mol. Biol.* 146, 561–587.

T. D. Hurley, W. F. Bosron, J. A. Hamilton, and L. M. Amzel. 1991. "Structure of Human $\beta_1\beta_1$ Alcohol Dehydrogenase: Catalytic Effects of Non-Active-Site Substitutions." *Proc. Natl. Acad. Sci. USA* 88, 8149–8153.

Y. Pocker. 1989. "Alcohol Dehydrogenase: Structure, Catalysis, and Site-Directed Mutagenesis," in H. Sigel ed., *Interrelations Among Metal Ions, Enzymes, and Gene Expression. Metal Ions in Biological Systems*, Volume 25. Marcel Dekker, New York, 335–358.

CHAPTER 11

Atom- and Group-Transfer Chemistry

Principles: *Both substrate binding and redox changes occur during atom- and group-transfer chemistry, even when the overall reaction does not involve a net redox change. Coupled proton-electron transfer steps place redox potentials in the correct biological windows. Different redox-active, Lewis acid or base, and atom-transfer units positioned near one another in the active site work in concert to achieve difficult overall transformations. Interaction with substrates and other proteins can gate electron transfer into active sites at the appropriate potential and time for enzyme catalysis. A variety of strategies has been developed for facilitating two-electron transfer processes, including the use of specialized metals, metal-porphyrin units, or dimetallic centers. Metal centers can be used to create or destroy radical species, and changes in the coordination geometry at a metal center that accompany redox reactions can facilitate allosteric functions of the protein. The bioinorganic chemistry of dioxygen is paramount in illustrating these principles.*

In the previous two chapters we examined systems in which the metal center has served either as a source of electrons or as a general Lewis acid. We now turn our attention to metalloproteins in which substrate binding is accompanied by redox changes, even when the overall reaction does not involve a net oxidation-state change. Into this category fall many of the most important metalloprotein functions, including the transfer of atoms, molecules, and groups. Reactions involving dioxygen and its metabolic products are the largest subset of such functions. Several key principles emerge from a consideration of these systems. Since removal of electrons from a metal ion, that is, oxidation, is accompanied by an increase in positive charge, simultaneous loss of a ligand proton will frequently occur. By coupled proton-electron transfer reactions of this kind, the system can be tuned to facilitate subsequent oxidation steps at the metal center without requiring redox potentials that lie outside the -0.8 to $+0.8$ V window available in most cells. Similarly, when a reduced, often negatively charged, substrate molecule remains bound to the metal center, it may become protonated, again to minimize charge separation.

Two-electron transfer processes can be achieved by employing specialty metals, such as molybdenum, which undergo facile multielectron redox level changes, for example, between $M^{(n+2)+}$ and M^{n+}. Alternatively, a dimetallic center involving redox metals such as iron or copper, or coupling of one of these metals with a redox-active ligand such as a porphyrin, provides additional strategies for accomplishing two-electron transformations. Another important principle that emerges from the examples discussed in this chapter is that geometric changes resulting from ligand-mediated redox reactions at a metal center can trigger coordinated movements of nearby protein chains, resulting in cooperative behavior.

Group-transfer reactions can also occur through the generation of radicals at a metal center. Included here are tyrosyl and carbon-centered radicals derived from protein or coenzyme constituents as well as hydroxyl and superoxide radicals generated from media constituents water or dioxygen, respectively. Again, the special property of bioinorganic systems that facilitates this chemistry is their ability both to bind and to transfer electrons through metal-mediated redox transformations.

11.1. Dioxygen Transport

We begin our discussion with the important but nonenzymatic role of reversible dioxygen binding, a function essential for the transport and storage of the O_2 molecule (Figure 1.2). Although cobalt complexes are well known to perform this reaction in inorganic chemistry, the process appears to be carried out exclusively by iron and copper centers in biology. There are three known O_2 transport and storage systems, and all employ metalloproteins. The dioxygen carriers are hemoglobin (Hb), hemocyanin (Hc), and hemerythrin (Hr). They function, respectively, in all vertebrates and many invertebrates, in arthropods and mollusks, and in four phyla of marine invertebrates. Only hemoglobin has an iron-porphyrin group, a unit that has become synonymous with the term *heme* despite the derivation of this word from the Greek root meaning "blood." Hemocyanin has two copper atoms and hemerythrin two nonheme iron atoms in the functional core. Some properties of these three dioxygen carriers are listed in Table 11.1. The O_2 molecule is stored in tissues such as muscle by the proteins myoglobin (Mb) and myohemerythrin (myo Hr), which have metal core structures analogous to those of the transport proteins. In their oxygenated states, the three protein systems are all intensely colored.

Table 11.1
Some properties of protein oxygen carriers

Property	Hemoglobin	Hemerythrin	Hemocyanin
Metal	Fe	Fe	Cu
Oxidation state of metal in deoxy protein	(II)	(II)	(I)
Metal:O_2	Fe:O_2	2Fe:O_2	2Cu:O_2
Color, oxygenated	Red	Violet-pink	Blue
Color, deoxygenated	Red-purple	Colorless	Colorless
Coordination of Fe	Porphyrin ring	Protein side chains	Protein side chains
Molecular weight	65,000	108,000	400,000 to 20,000,000
Number of subuits	4[a]	8	Many

[a] In some species (for example, *Glycera*), hemoglobins are monomeric; in others (for example, *Arenicola*), they are multisubunit oligomers with molecular weights in the millions.

In Hr and Hc, the dioxygen-binding reaction involves oxidative addition of O_2 to a dimetallic center to form a coordinated peroxide group. For Hb, there is less electron transfer from the metal to the bound dioxygen molecule, which is considered by some to be bound as the superoxide ion. The reduced, dioxygen-free forms of all three proteins are referred to as deoxy, the dioxygen-bound forms as oxy, and the functionally inactive, oxidized states as met. We begin our discussion with the metalloproteins from the most advanced organisms, Hb/Mb.

11.1.a. Hemoglobin and Myoglobin.

(i) Structure. Myoglobin (Mb) is the O_2-binding protein found principally in muscle tissues of vertebrates. It consists of a single polypeptide chain of 153 amino acids called globin, made up of seven α-helical and six non-helical segments (Figure 11.1). Attached to the chain by coordination to the imidazole ring of a histidine residue is the dioxygen-binding prosthetic group, iron(II) protoporphyrin IX. A sketch of this porphyrin ring is given in Figure 1.2. The structure of sperm-whale myoglobin has been determined in all three (deoxy, oxy, and met) forms. Deoxy Mb has a pentacoordinate iron(II) center in which the metal atom lies 0.42 Å out of the plane of the four pyrrole-ring donor nitrogen atoms. It is displaced toward the bonded imidazole group, which is called the *proximal* side of the heme. Upon binding of dioxygen, the iron atom moves toward the FeN_4 planes. In oxy Mb, the dioxygen molecule is bonded end-on to iron, forming a bent structure with a Fe—O—O bond angle of 115°.

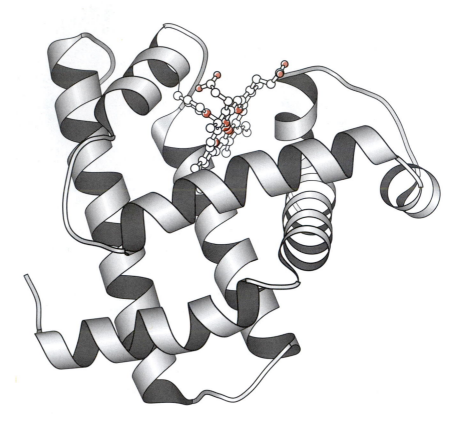

Figure 11.1
Structure of myoglobin.

Hemoglobin (Hb) transports O_2 in blood, taking dioxygen from air in the lungs and delivering it to Mb in tissues. Hemoglobin is a multisubunit protein, with two α and two β polypeptide chains. In each chain there is an iron protoporphyrin IX group held by a proximal histidine imidazole residue, as in Mb. In deoxy Hb, the iron lies 0.36 to 0.40 Å out of the porphyrin ring plane but moves within ± 0.12 Å of the plane upon binding of dioxygen.

In both Hb and Mb, there is a histidine situated in the O_2-binding pocket on the side away (distal side) from the coordinated imidazole base. X-ray and, for oxy myoglobin, neutron diffraction studies indicate the formation of a hydrogen bond between the coordinated dioxygen molecule and the N—H proton of the distal histidine residue. This Fe—O—O- - -H—N unit is nicely accommodated, owing to the angular bend at the coordinated dioxygen atom. In carbon-monoxide adducts of Mb and Hb, however, the X-ray data reveal a distorted structure in which the Fe—C≡O angle is either bent or, more likely, tilted off the proximal histidine N—Fe bond axis to avoid steric clash with the distal histidine. Thus nature has cleverly tailored the iron-

Table 11.2
Vibrational and geometrical properties of dioxygen species

Species	ν_{O-O} (cm^{-1})	d_{O-O} (Å)
O_2^+	1,905	1.12
O_2	1,580	1.21
O_2^-	1,097	1.33
O_2^{2-}	802	1.49

porphyrin centers in Hb and Mb to bind O_2 rather than its toxic surrogate $C\equiv O$, which often binds more tightly to metal complexes than does dioxygen.

(ii) The dioxygen-binding reaction. Hb and Mb, probably the most thoroughly investigated metalloproteins, have been the subjects of innumerable spectroscopic, thermodynamic, and kinetic measurements. Their optical spectra are dominated by intense porphyrin ring $\pi \to \pi^*$ transitions in the 400–600 nm range, known as the Soret, α, and β bands (see Table 4.6), that are sensitive to the state of oxygenation. From resonance Raman spectroscopic studies of the coordinated dioxygen molecule and its ^{18}O-substituted analogs, an O–O stretching band at ~ 1105 cm^{-1} has been identified. This value is characteristic of coordinated superoxide (O_2^-) ion (Table 11.2), suggesting that MbO$_2$ and HbO$_2$ adducts might best be assigned as iron(III) complexes of this ligand. Magnetic coupling between these ions leads to a diamagnetic, $S = 0$, ground state. Thus coordination of dioxygen to deoxy Hb or Mb is accompanied by electron transfer to form a superoxide ion, which in turn is stabilized by hydrogen bonding to the distal imidazole proton, as indicated above.

Of great importance to the physiological functions of Hb and Mb is the beautifully sophisticated bioinorganic system devised by nature in which dioxygen binds to Hb in the lungs and is transferred to Mb in tissues or is transferred to fetal Hb in the uterus of pregnant mammals. At the heart of this system is the motion of iron toward the plane of the porphyrin ring upon conversion of deoxy to oxy Hb, which serves as a trigger for cooperative binding of dioxygen by the multisubunit hemoglobin protein. The protein is presumed to have two different quaternary structures designated R, for relaxed, and T, for tense. The former has a high affinity for O_2, similar to that of isolated subunits, whereas the latter, tense state has a diminished O_2 affinity. These two conformational states are in equilibrium with one another. In the T state, prevalent when all four subunits are unligated, intersubunit interactions are believed to constrain the proximal histidine to resist movement into the porphyrin-ring plane and diminish the O_2 binding constant.

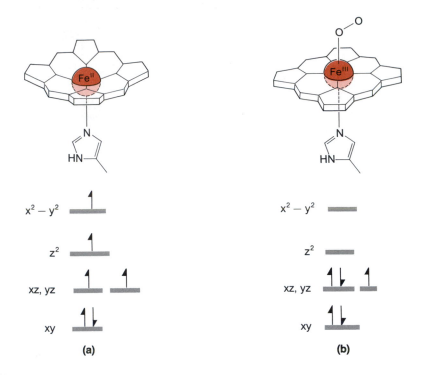

Figure 11.2
Diagram illustrating the structural and spin state changes that occur upon binding of dioxygen to an iron porphyrin. Shown are (a) the high-spin ferrous deoxy form and (b) the low-spin ferric oxy form.

After $\sim 2\ O_2$ molecules have bound a hemoglobin molecule, the quaternary structure of the protein switches to R, the state where such constraints are relaxed, and dioxygen binding to the remaining two subunits is facilitated.

The key to this mechanism is the requirement that, upon dioxygen binding, the iron atom moves into the porphyrin-ring plane, dragging the proximal histidine ligand along with it. The spin states for some iron porphyrins are given in Figure 11.2. In deoxy Hb, the iron(II) is high-spin ($S = 2$) and has a covalent radius too large to fit in the plane of the four nitrogen atoms. Addition of O_2 to form the Fe^{III}-O_2^- adduct diminishes the covalent radius of iron, which then moves into the porphyrin-ring plane. Although details of this model for the T state of Hb have been controversial, it seems likely that constraints at the iron-porphyrin site imposed by subunit interactions are responsible for the diminished binding affinity for dioxygen, a reaction that requires substantial structural reorganization of the iron coordination sphere. The result is a tailor-made fit of metal-coordination and protein-structural chemistry to provide a flexible system for O_2 binding and release under widely varying biological conditions. Low pH also favors the release of dioxygen from hemoglobin, further facilitating its transfer from blood to other tissues.

(iii) Model chemistry. It is relatively easy to synthesize iron(II)-porphyrin complexes with physical and structural properties similar to the metal centers in deoxy Mb and deoxy Hb, for example, [Fe(TPP)(N-MeIm)], where TPP = *meso*-tetraphenylporphyrin and N-MeIm = N-methylimidazole. Achieving reversible binding of dioxygen in such model complexes was much more of a challenge. Addition of O_2 to solutions of simple ferrous porphyrins results in immediate redox chemistry, leading, ultimately, to thermodynamically quite stable (μ-oxo)diiron(III) species. Kinetic studies suggested the mechanism shown in Equations 11.1 to 11.4, individual steps of which have been subsequently confirmed by various spectroscopic and

$$Fe^{II}P + O_2 \rightleftharpoons Fe^{III}P-O_2^- \tag{11.1}$$

$$Fe^{III}P-O_2^- + Fe^{II}P \rightleftharpoons PFe^{III}-O_2^{2-}-Fe^{III}P \tag{11.2}$$

$$PFe^{III}-O_2^{2-}-Fe^{III}P \rightarrow 2PFe^{IV}=O \tag{11.3}$$

$$PFe^{IV}=O + Fe^{II}P \rightarrow PFe^{III}-O-Fe^{III}P \tag{11.4}$$

structural studies. In these equations, P represents the porphyrin ring, and the Roman numerals indicate the iron oxidation state.

Reversible oxygenation of ferrous porphyrin Hb/Mb model complexes has been accomplished by several approaches. One involves the use of low temperatures to stabilize the dioxygen complex in dilute, nonaqueous solutions. Examples include $Fe^{II}TPP(L)$, where L is pyridine or 1-MeIm. This kinetic tactic avoids the thermodynamically favorable reactions 11.1 to 11.4. Above $-45\,°C$, however, the systems irreversibly oxidize. Since this decomposition reaction involves participation of two iron-porphyrin molecules in bimolecular steps (see Equations 11.2 and 11.4), steps that are prevented from occurring in Hb and Mb by the protein chains, tactics were devised to block approach of the reactive oxygenated iron centers. In one such approach, an iron-porphyrin impregnated in a polystyrene film reacted with O_2 to produce an oxy Hb-type optical spectrum; this adduct could be deoxygenated by flushing the films with N_2. The strategy of using steric factors to prohibit approach of two iron-porphyrin moieties was subsequently applied in the synthesis of sterically encumbered ligands to protect the O_2-binding site and thus stabilize the desired O_2 complex in solution at room temperature. Figure 11.3 displays the structure of two such porphyrins. One was named a *picket-fence* porphyrin because the four pivalamido substituents

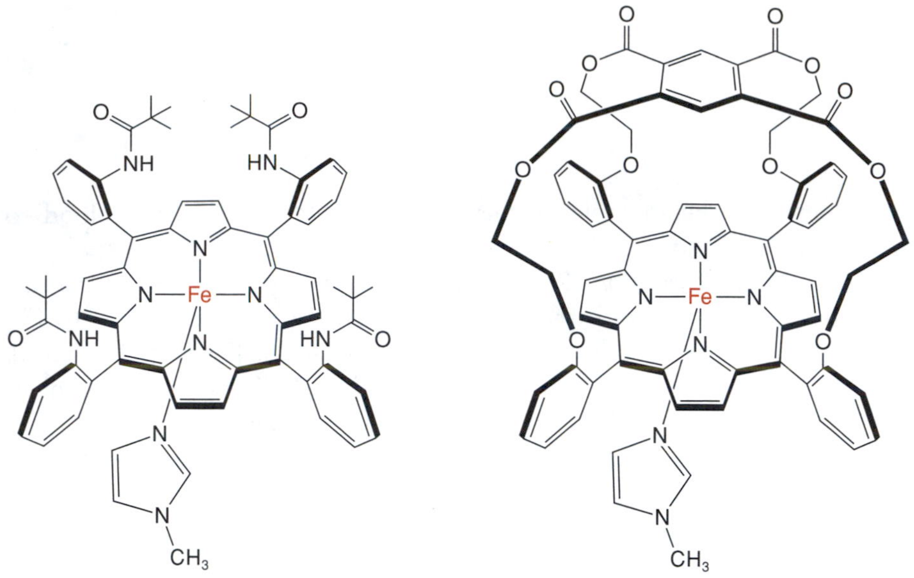

Picket fence porphyrin   Capped porphyrin

Figure 11.3
Structures of two globin model compounds capable of reversible dioxygen binding.

on the O-phenyl positions of TPP directed toward the same side of the ring resemble the pickets of a fence. In the presence of 1-MeIm or pyridine to serve as a base in the axial position on the other side of the ring from the pickets, the ferrous complex of this porphyrin binds dioxygen reversibly. The *capped* porphyrin, also shown in Figure 11.3, will similarly bind O_2 if a large excess of axial base is present. In these complexes, the phenyl-ring substituents provide a steric block to formation of the (μ-oxo)diiron(III) unit and also afford a pocket in which the O_2 molecule can reside.

Extensive spectroscopic, magnetic, and structural studies have been carried out on the O_2-binding Hb and for Mb model porphyrin complexes. Thermodynamic measurements reveal the dioxygen affinity of the models to be the same as for Mb and Hb in the relaxed state. Electronic, vibrational, and Mössbauer spectra as well as magnetic susceptibility results for the deoxy and oxy hemoproteins and their model complexes are also in good agreement. Moreover, the model complexes provide evidence to support the influence of a constrained axial base on the dioxygen-binding reaction in T-state hemoglobin. Coordination of a sterically hindered axial base such as 1,2-Me$_2$Im to the ferrous picket-fence porphyrin model complex restricts the amount of imidazole ring movement that accompanies oxygenation to ~ 0.3 Å, compared with an estimated motion of 0.3 to 0.5 Å for an unconstrained imidazole axial base such as 1-MeIm. The result is a decrease in dioxygen affinity. These findings clearly demonstrate the great value of synthetic models for providing insights into the intricate details of the chemistry of metalloprotein functional cores.

Remarkably, some of the picket-fence porphyrins with bulky axial ligands bind dioxygen in a cooperative manner in the solid state. Thus, with 2-methylimidazole as the axial ligand, a solid can be obtained that binds dioxygen with a low affinity at low O_2 partial pressures, higher affinity at high partial pressures, and cooperativity in the intermediate regime. Furthermore, in a subtly different form of the compound, which, in the solid state, has ethanol hydrogen bonded to the imidazole nitrogen, no cooperativity is observed, and the dioxygen affinity is very low. These observations illustrate the important influence of electronic changes at the iron center on the dioxygen-binding reaction and the ability of the accompanying internal electron-transfer reaction to effect structural changes that modify the behavior at nearby sites.

(iv) Dioxygen-activating metalloporphyrins. When suitably arranged in a protein matrix, dioxygen can be activated by iron-porphyrin prosthetic groups to carry out a variety of oxygenation reactions. In one celebrated example, that of cytochrome P-450, we now know a great deal about the details of the oxygenation reaction. Achieving an understanding of how the protein environments modulate the reactivity of iron-porphyrin cores to perform these different biological transformations is an exciting challenge for the bioinorganic chemist. As we shall see, a parallel situation occurs for the other dioxygen-transport centers in bioinorganic chemistry, the metal cores of which can also catalyze oxygen-atom-transfer reactions.

11.1.b. Hemerythrin.

(i) Structure. Hemerythrins are relatively easy to crystallize for protein molecules; the final purification step in the isolation of oxy Hr affords beautiful purple needles. High-resolution (1.66 to 2.0 Å) X-ray structural information is available for azidomet Hr and azidometmyo Hr, in which the azide anion (N_3^-) is bonded to iron at the site normally occupied by O_2 in oxy Hr. As shown in Figure 11.4, the structure consists of two octahedrally coordinated iron atoms joined by a μ-oxo and two bridging carboxylato groups, the latter being contributed by glutamate and aspartate amino-acid residues from the protein chain. Three terminal positions on one iron atom and two on the other are filled by imidazole groups of histidine ligands. The azide ion completes the octahedral coordination sphere of the second iron atom. Although not yet as well refined, the X-ray structures of deoxy and oxy Hr from *Themiste dyscritum* are consistent with the former, having a five-coordinate

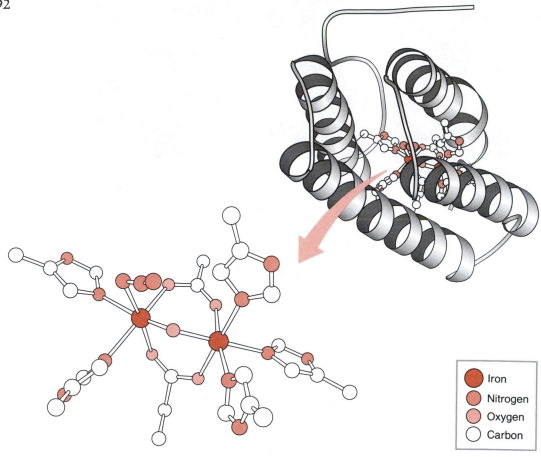

Figure 11.4
Structure of azidomethemerythrin and its dinuclear iron core. The N_3^- ion resides in the site occupied by dioxygen in oxyhemerythrin.

Iron
Nitrogen
Oxygen
Carbon

iron with a vacant site that is filled by a monodentate, end-bound dioxygen moiety in the latter, oxy form of the protein. The short (1.78 Å) Fe−O$_{oxo}$ distance and the 2.05−2.25 Å lengths for the remaining Fe−O and Fe−N bonds are consistent with the presence of a high-spin d^5 (μ-oxo)diiron(III) core center in the oxidized forms of the protein. EXAFS studies support this assignment and confirm that oxy and azidomet Hr have the same number and types of ligands bound to the $\{Fe_2O(O_2CR)_2\}^{2+}$ core.

(ii) Magnetic and spectroscopic properties. Table 11.3 summarizes magnetic and spectroscopic features of deoxy and oxy Hr. Deoxy Hr contains two high-spin ferrous ions, as established by Mössbauer and EPR spectroscopy. In met and oxy Hr, there are two high-spin ferric centers, antiferro-magnetically coupled owing to the presence of the μ-oxo group. The coupling is sufficiently strong ($J = -77$ or -134 cm^{-1} for oxy and met Hr,

Table 11.3
Selected structural, magnetic, and spectroscopic properties of dinuclear iron center-containing proteins and their models[a]

	Oxyhemerythrin	Deoxyhemerythrin	$[Fe_2O(O_2CCH_3)_2\text{-}(HBPz_3)_2]$	$[Fe_2O(O_2CCH_3)_2\text{-}(TACN)_2]$	$[Fe_2(OH)(O_2CCH_3)_2(Me_3TACN)_2]^+$
Structural properties					
Fe–O–Fe angle, deg	128	128	123.6	118.3	113.2
Fe–μO distance, Å	1.82	1.98	1.784	1.785	1.987
Fe$\cdots$Fe distance, Å	3.24	3.57	3.146	3.064	3.32
Mössbauer and magnetic properties					
Isomer shift, mm s^{-1}	0.52, 0.48	1.14	0.52	0.46	1.16
Quadrupole splitting, mm s^{-1}	1.92, 1.00	2.76	1.60	1.72	2.83
J, cm^{-1} ($\mathcal{H} = -2J|S_1 \cdot S_2|$)	−77	−13	−121		−13
Electronic spectra					
LMCT (O^{2-} to Fe), nm	330, 360 sh		262, 339, 358 sh	333, 368 sh	
d-d, nm	750, 990		528 sh, 695, 995		
Vibrational spectra					
Fe–O–Fe, symm. stretch, cm^{-1}	486		528		
Fe–O–Fe, asymm. stretch, cm^{-1}	757		751	749	

[a] HBPz$_3$ = hydrotris(pyrazoyl)borate, TACN = 1,4,7-triazacyclononane.

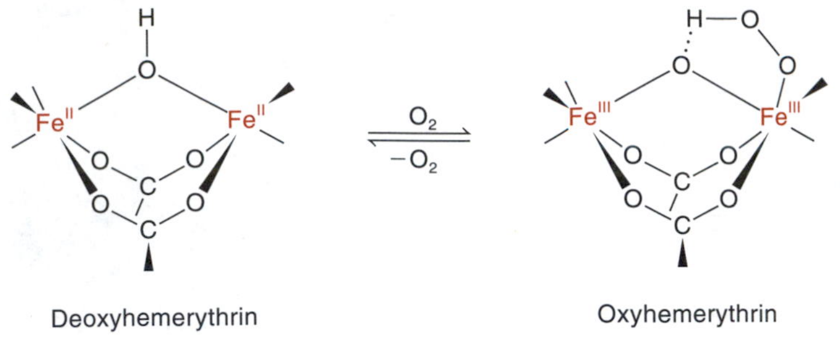

Deoxyhemerythrin   Oxyhemerythrin

Figure 11.5
Scheme depicting redox and structural changes that occur in the diiron core of hemerythrin upon reversible dioxygen binding.

respectively) that the ground state is diamagnetic; at liquid He temperature (4.2 K), there is no net magnetism. At room temperature, however, residual paramagnetism affects the solution properties of the protein. Resonance Raman spectroscopy has been especially valuable in characterizing the oxidized forms of Hr. Near-ultraviolet excitation of oxy or hydroxymet Hr produces a resonance-enhanced band at 436 or 492 cm^{-1}, assigned as the symmetric Fe$-$O$-$Fe stretching mode because it shifts to lower energy upon ^{18}O substitution. Raman studies also revealed that, in oxy Hr, O_2 is bound in an end-on configuration with an O$-$O stretching vibration at 844 cm^{-1}, characteristic of peroxide ion (Table 11.2). Exposure of oxy Hr to D_2O resulted in a 4 cm^{-1} shift of the Fe$-$O$-$Fe mode to higher energy, which was interpreted as indicative of hydrogen-bond formation to the bridging dioxygen atom.

(iii) The dioxygen-binding reaction. The structural, magnetic, and spectroscopic properties of deoxy and oxy Hr are consistent with the model shown in Figure 11.5 for the binding of dioxygen to the dinuclear iron core in Hr. In deoxy Hr, a (μ-hydroxo)bis(μ-carboxylato)diiron(II) center with a vacant terminal coordination position on one iron atom binds dioxygen and transfers two electrons and the proton from the bridging OH group. The resulting oxy Hr has a terminal hydroperoxy ligand bound to one iron in the now oxidized $\{Fe_2O(O_2CR)_2\}^{2+}$ core. A hydrogen bond is formed between the proton of the hydroperoxy group (the donor) and the μ-oxo group (the acceptor). This oxidative addition of dioxygen to the dimetallic center may occur by two sequential one-electron steps involving a mixed-valence $\{Fe^{II}Fe^{III}\}$ intermediate. As shown by studies of Hr model compounds (see below), the presence of the hydrogen bond to the μ-oxo ligand minimizes the degree to which the Fe$-$O bond is shortened in going from the deoxy to the oxy state. This phenomenon would limit the amount of redox-initiated inner-sphere reorganizational energy that accompanies the oxygenation/ deoxygenation reactions, facilitating the reversible binding of O_2. The hy-

drogen bond also accounts for the lesser magnetic exchange coupling in oxy Hr compared to met Hr.

(iv) Redox states of the protein. Deoxy and met or oxy Hr undergo one-electron oxidation and reduction reactions to give mixed-valence (Fe^{II}/Fe^{III}) forms designated as (semimet)$_O$ and (semimet)$_R$, respectively. Although these forms have no known role in the O_2-binding chemistry, they are useful analogs of related dinuclear iron-oxo proteins where mixed-valence forms are involved in the biological functions.

(v) Model compounds. Spontaneous self-assembly of the $\{Fe_2O(O_2CR)_2\}^{2+}$ core structure of Hr has been achieved in a variety of model complexes. One of the first such compounds, $[Fe_2O(O_2CCH_3)_2(HBpz_3)_2]$, where $HBpz_3^-$ is the hydrotris(pyrazolyl)borate anion, is depicted in Figure 11.6. As can be seen from Table 11.3, the structural and spectroscopic properties of this complex are in good agreement with those of the dinuclear iron core in oxy Hr. The bridging oxo group in the model compound can be reversibly protonated, as shown in Equation 11.5, to give the hydroxo-bridged

$$[Fe_2O(O_2CCH_3)_2(HBpz_3)_2] \underset{Et_3N}{\overset{HBF_4}{\rightleftharpoons}} [Fe_2(OH)(O_2CCH_3)_2(HBpz_3)_2]^+ \quad (11.5)$$

green-brown   orange

$J = -121 \text{ cm}^{-1}$ $J = -17 \text{ cm}^{-1}$

diiron(III) derivative. Protonation of the μ-oxo ligand lengthens the Fe—O bond from ~ 1.78 Å to ~ 1.96 Å with concomitant changes in the color and marked diminution of the antiferromagnetic exchange coupling constant.

When an attempt was made to reduce the (μ-oxo)bis(μ-carboxylato)-diiron(III) model complex, in an approach to deoxy Hr model chemistry, decomposition occurred to form mononuclear $[Fe(HBpz_3)_2]$. This problem was circumvented by using the N,N',N"-trimethyltriazacyclononane ligand (Figure 11.6) in place of hydrotris(pyrazolyl)borate. Because of unfavorable steric interactions, the $[Fe(Me_3TACN)_2]^{2+}$ bis-chelate cation cannot be formed. Thus, the target complex $[Fe(OH)(O_2CCH_3)_2(Me_3TACN)_2]^+$ was prepared by spontaneous self-assembly and its geometric and electronic structures determined. As indicated in Figure 11.6 and Table 11.3, the complex is an excellent spectroscopic and structural model for the reduced, or deoxy, form of hemerythrin. A major remaining objective in Hr model chemistry is to obtain a dinuclear complex containing the $\{Fe_2(OH)(O_2CR)_2\}^+$ or $\{Fe_2O(O_2CR)_2\}^{2+}$ core, in which one iron atom has three terminal N-donor ligands and the other iron atom has only two such ligands, and which can bind dioxygen or anions such as N_3^- reversibly in the available site.

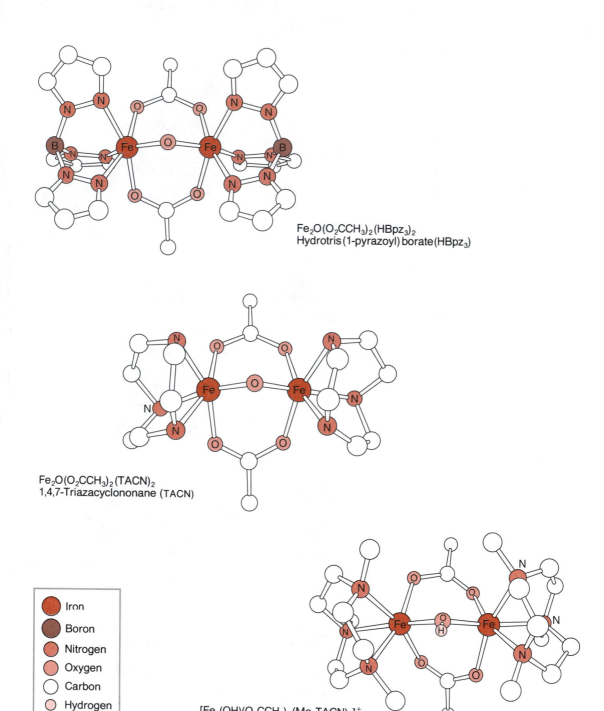

Fe$_2$O(O$_2$CCH$_3$)$_2$(HBpz$_3$)$_2$
Hydrotris(1-pyrazoyl) borate(HBpz$_3$)

Fe$_2$O(O$_2$CCH$_3$)$_2$(TACN)$_2$
1,4,7-Triazacyclononane (TACN)

●	Iron
●	Boron
●	Nitrogen
●	Oxygen
○	Carbon
○	Hydrogen

[Fe$_2$(OH)(O$_2$CCH$_3$)$_2$(Me$_3$TACN)$_2$]$^+$

Figure 11.6
Structures of two synthetic models (top, middle) for the oxidized form of
hemerythrin and one (bottom) for the reduced form.

(vi) Proteins with an Hr-type core. Following the structural elucidation of the bridged dinuclear core in Hr, several classes of nonheme iron proteins were identified as having spectroscopic properties analogous to those of met or semimet forms of the protein. Included in this category are mammalian and bacterial ribonucleotide reductase (RR) enzymes, purple acid phosphatases (PAP), including uteroferrin, and methane monooxygenase (MMO). Equations 11.6 to 11.8 illustrate the respective reactions catalyzed by these three

$$(11.6)$$

$$(RO)_2PO_2H + H_2O \xrightarrow{\text{PAP}} (RO)PO_3H_2 + ROH \qquad (11.7)$$

$$CH_4 + O_2 + 2H^+ + 2e^- \xrightarrow{\text{MMO}} CH_3OH + H_2O \qquad (11.8)$$

enzymes. The functional diversity of carboxylate-bridged diiron centers in these different proteins, including Hr, is fascinating.

11.1.c. Hemocyanin.

(i) Structure. Hemocyanins are large proteins containing many subunits. Recently, Hc crystals that were suitable for X-ray diffraction studies were grown from the spiny lobster *Panulirus interruptus*, an arthropod. The asymmetric unit in the crystals is comprised of six protein subunits and has a molecular mass of $\sim 460,000$ daltons, making a high-resolution structure difficult to obtain. The structure of the colorless, deoxy form of the protein is known, and the chemical contents of its dinuclear copper core are shown in Figure 11.7. In deoxy Hc, two cuprous ions are each coordinated by three histidine residues at an internuclear distance of 3.7 ± 0.3 Å. An empty cavity is available between the metals to accommodate the dioxygen molecule. Previous spectroscopic and magnetic studies of Hc from various sources had been interpreted in terms of a model structure for the oxy and met forms in which a tyrosine ligand bridged the two cupric ions. The closest such RO^- group is > 17 Å away from the dinuclear copper site in the *P. interruptus* crystal structure, however. It therefore appears that, in oxy Hc, no such bridge occurs for this and possibly all other Hc proteins.

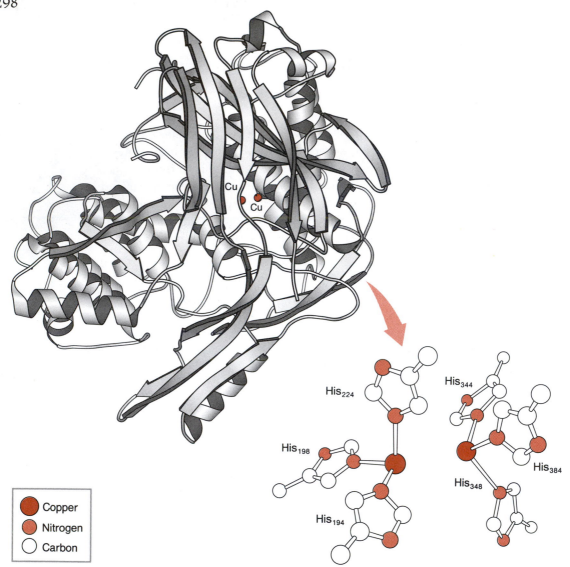

Figure 11.7
Structure of deoxyhemocyanin and its dinuclear copper core.

(ii) Spectroscopy and redox chemistry. The oxidized, oxy and met, forms of the protein have been the subject of intense scrutiny by a variety of physical methods. The two d^9 Cu(II) ions are so very strongly antiferromagnetically coupled that, at room temperature, the dinuclear center is essentially diamagnetic. Oxyhemocyanin has a characteristic, intense optical transition at 345 nm assigned as a ligand-to-metal charge transfer band. Peaks at $450 \leq \lambda \leq 700$ nm in the optical and circular dichroism spectrum of oxy

Hc, together with a 745–750 cm^{-1} band in the resonance Raman spectrum due to the O–O stretching vibration (Table 11.2), are consistent with the existence of a (μ-peroxo)dicopper(II) center. Isotopic labeling studies using $^{16}O-^{18}O$ dioxygen were used in an attempt to distinguish $Cu^{II}\cdots Cu^{II}-O^{-O}$ from $Cu^{II}-O\diagdown_O\diagup Cu^{II}$ binding modes for the O_2^{2-} ligand. The former would result in two O–O stretching bands, whereas the latter has only one, and only one has been observed. Recently, a model complex was synthesized in which two copper(II) ions were linked by an η^2,η^2-peroxide, as shown below.

$$Cu^{II}\diagup^{O}_{O}\diagdown Cu^{II}$$

The spectroscopic properties of this complex, including its Raman spectrum, were found to be strikingly similar to those of oxyhemocyanin. Subsequently, the crystal structure of horseshoe crab (limulus) hemocyanin in the oxy form has been determined, revealing the presence of just this type of η^2,η^2-peroxide moiety (Figure 11.8). Here the appropriate structural model compound preceded the determination of the protein structure.

(iii) *Model chemistry and the O$_2$-binding reaction.* Attempts to model the active site of hemocyanin have led to the discovery of some very interesting Cu(II)-dioxygen complexes. As with Hr, dioxygen adds to the dinuclear Cu(I) center of deoxy Hc in a formal oxidative addition reaction, but in oxy Hc, the O_2^{2-} ion bridges the two Cu(II) ions. One of the models for this chemistry, as indicated above, turned out to be a true replica of the active center in oxy Hc, and its structure is depicted above and in Figure 11.9. Another such model, also shown in the figure, was obtained by addition of O$_2$ to a copper(I) complex at low temperature in organic solvents to yield the $N_4Cu^{II}-O-O-Cu^{II}N_4$ unit. Crystals of this complex were grown, mounted, and studied at -98 °C by X-ray crystallography in experiments where the temperature was never allowed to rise above -30 °C! Although the resulting (μ-peroxo)dicopper(II) complex is not an accurate replica of oxy Hc, the latter having only three and not four nitrogen donor ligands at each Cu(II) site, its diamagnetism and spectroscopic properties made the $Cu^{II}-O-O-Cu^{II}$ unit a plausible model for oxy Hc at the time.

(iv) *Other proteins with Hc-type centers.* Dinuclear copper sites occur in several other metalloproteins besides hemocyanin, some of which fall into the spectroscopically defined type 3 copper category, and all of which are involved in some aspect of O$_2$ metabolism. Of special interest here is tyrosinase, a mammalian protein that has both catecholase (Equation 11.9) and

300

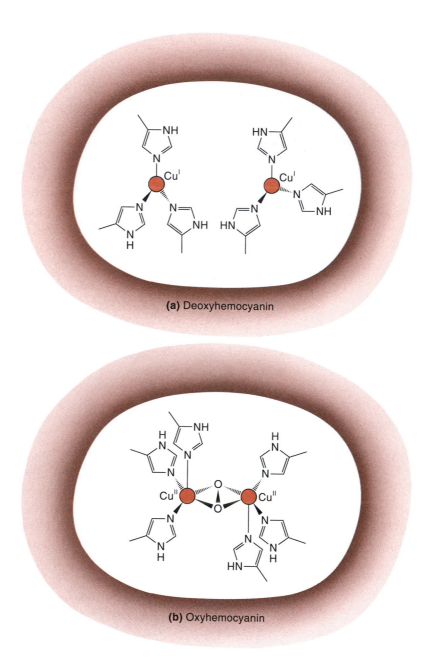

Figure 11.8
Schematic views of the structures of deoxyhemocyanin (top) and oxyhemocyanin (bottom) showing the binding of dioxygen as an η^2, η^2-peroxide.

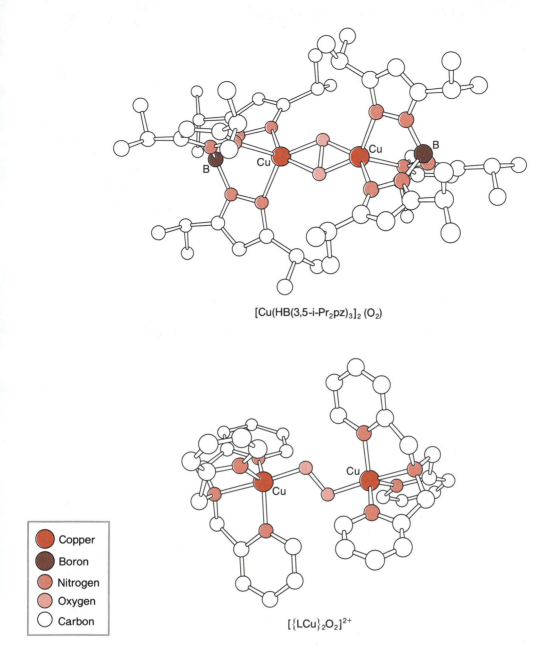

[Cu(HB(3,5-i-Pr$_2$pz)$_3$]$_2$ (O$_2$)

Copper
Boron
Nitrogen
Oxygen
Carbon

[{LCu}$_2$O$_2$]$^{2+}$

Figure 11.9
Two oxyhemocyanin models, one (top) containing the η^2, η^2-structure found in the protein and the other (bottom) having an alternative η^1, η^1-bridged attachment.

cresolase (Equation 11.10) activity. Oxidation of tyrosine leads to L-dopa, which in turn goes on to form the skin-pigment melanin. Equation 11.9 indicates tyrosinase to be a mixed-function oxidase. Thus, an interesting parallel emerges between the dinuclear copper centers in Hc and tyrosinase and the dinuclear iron centers in Hr and MMO. Both carry out reversible

$$2 \quad \text{[catechol]} \quad + \; O_2 \; \longrightarrow \; 2 \quad \text{[ortho-quinone]} \quad + \; 2H_2O \qquad (11.9)$$

$$CH_3-\text{[phenol]}-OH \; + \; O_2 \; + \; AH_2 \; \longrightarrow \; CH_3-\text{[catechol]}-OH \; + \; A \; + \; H_2O \qquad (11.10)$$

dioxygen binding and mixed-function oxidation catalysis, depending on the protein environment in which they are embedded.

Recently, mixed-function oxidase (or monooxygenase, as it is often called) chemistry has been observed in a dinuclear copper model complex. As shown in Figure 11.10, addition of O_2 to copper(I) in the binucleating ligand results in hydroxylation of the m-xylene group that forms part of the bridge linking the two copper-binding sites. The other atom of dioxygen becomes a hydroxide ion, resulting in a (μ-phenoxo)(μ-hydroxo)dicopper(II) complex. By use of $^{18}O_2$ and analysis of the products mass-spectrometrically, it was shown that both oxygen atoms were incorporated into the product. The relationship between this simple model chemistry for the O_2-binding and monooxygenase activities of Hc and tyrosinase is striking and further illustrates the value of small-molecule inorganic chemistry for providing important insights into the functions of metal centers in biology.

11.2. Oxygen-Atom-Transfer Reactions: Fe

As discussed in Chapter 1, there are many chemical reactions catalyzed by metalloenzymes in which an oxygen atom is transferred to a substrate molecule. Dioxygen is the most common reagent for achieving this chemistry. Those reactions in which oxygen atoms are inserted into C—H or C—C bonds fall into two categories, depending upon whether both (dioxygenases) or only one (monooxygenases) of the atoms of the dioxygen molecule ends up in the substrate. Interestingly, as noted above, each of the specific functional

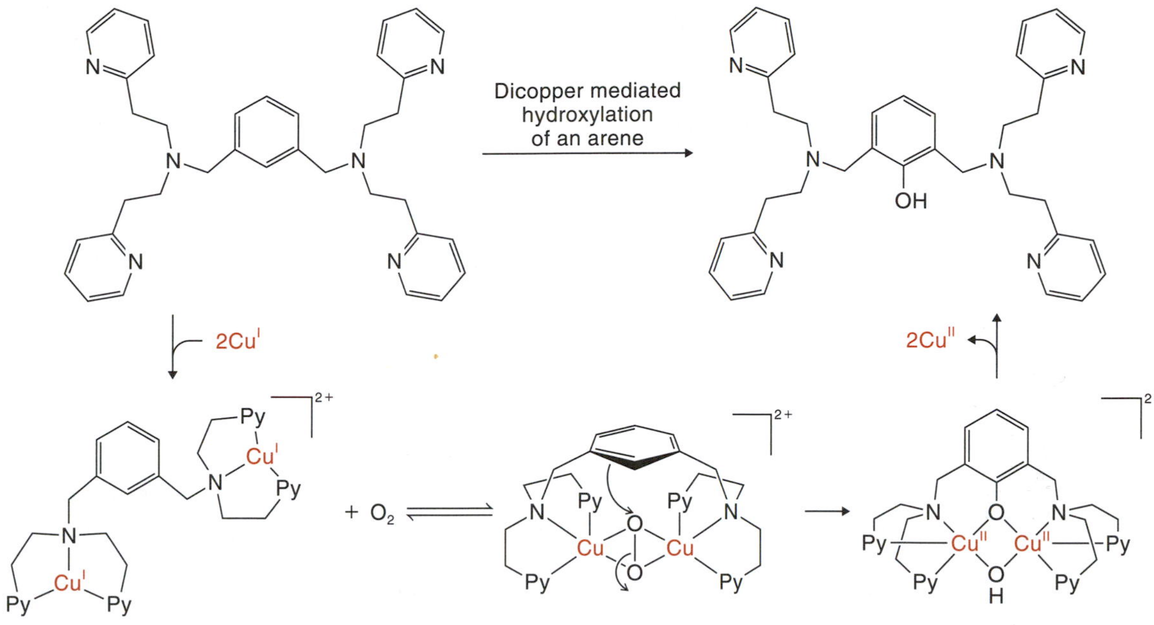

Figure 11.10
Monooxygenase activity observed in a synthetic model system. The dinuclear
copper(I) complex mediates oxygen insertion into an aromatic C-H bond of the
bound ligand with concomitant oxidation of the copper ions to Cu(II).

centers employed in the reversible binding of dioxygen also participates
in oxygen-atom-transfer reactions. Dinuclear centers analogous to the
dioxygen-binding sites of Hr, $\{Fe_2OH\}^+$, and Hc, $\{Cu_2\}^{2+}$, activate dioxygen in methane monooxygenase and tyrosinase, respectively, and the iron-
porphyrin centers in Hb/Mb bear a similar relationship to that in cytochrome
P-450. Here we explore in more detail iron metalloenzymes in which the
metal center serves as an activator, rather than reversible carrier, of dioxygen.

11.2.a. Cytochrome P-450.

(i) Discovery and biological function. This enzyme was first isolated from
rat-liver microsomes in 1958. At that time, the physical methods used to
characterize metalloproteins were far less sophisticated than now. From its
optical spectroscopic properties, the protein was known to be a cytochrome,
literally, a "cellular pigment," but this term has ultimately become synonymous with heme proteins. The Soret band at 450 nm of the ferrous-CO
derivative did not have its usual absorption maximum at ~420 nm; instead,
the band appeared at 450 nm, which led its discoverers to call the enzyme
cytochrome P-450 (P for "pigment"). Subsequent work revealed this mem-

Table 11.4
Major types of reactions catalyzed by cytochromes P-450

Reaction type	Simplified example	Typical substrate
Aliphatic hydroxylation	Cyclohexane → cyclohexanol	Pentobarbital
Aromatic hydroxylation	Benzene → phenol	Phenobarbital
Alkene epoxidation	Cyclohexene → cyclohexene oxide	Aldrin
N-dealkylation	$CH_3N(H)CH_3 \rightarrow CH_3NH_2 + H_2C=O$	Methadone
O-dealkylation	$C_6H_5OCH_3 \rightarrow C_6H_5OH + H_2C=O$	Codeine
Oxidative deamination	$(CH_3)_2CHNH_2 \rightarrow (CH_3)_2C=O + NH_3$	Amphetamine
S-oxidation	$CH_3SCH_3 \rightarrow (CH_3)_2S=O$	Chlorpromazine
Reductive dehalognation	$C_6H_5CH_2Br \rightarrow C_6H_5CH_3$	Halothane

brane-bound protein to be ubiquitous, occurring in life forms ranging from bacteria to man. Although various isozymes exist, they all catalyze the oxidation of organic substrates by dioxygen (Equation 11.11 and Table 11.4), performing essential roles in biosynthesis, metabolism, the detoxification of harmful substances, and, in some cases, the inadvertent generation of highly active carcinogens.

$$\overset{|}{\underset{/}{C}}-H + O_2 + 2H^+ + 2e^- \longrightarrow \overset{|}{\underset{/}{C}}-OH + H_2O \qquad (11.11)$$

Also:

$$R_3N \longrightarrow R_3NO$$

$$R_2S \longrightarrow R_2SO$$

    (ii) Structural and spectroscopic properties of cytochrome P-450$_{cam}$. The use of spectroscopic methods and model complexes to define the various redox and spin states of cytochrome P-450 is a paradigm in bioinorganic chemistry for the characterization of metalloenzyme-active sites. A rather complete and

Table 11.5
Comparison of EPR spectral data, Fe(III)
form, for cytochrome P-450 and Mb
derivatives

	g_1	g_2	g_3
P-450	2.41	2.26	1.91
Mb-SH⁻	2.4	2.3	1.91
Mb-SMe⁻	2.38	2.25	1.94
Mb—OH⁻	2.61	2.19	1.82
Mb—N₃⁻	2.8	2.25	1.75

accurate picture emerged prior to the availability of X-ray structural information. We shall therefore discuss these results briefly, focusing specifically
on the enzyme from *Pseudomonas putida*. This soluble bacterial enzyme that
hydroxylates camphor was the first P-450 to be purified and to have its
crystal structure determined.

In the so-called resting, substrate-free state, the 45,000 dalton molecular
mass P-450$_{cam}$ enzyme contains a single low-spin ferric heme. This assignment was made on the basis of its characteristic electron-spin resonance
spectrum, with g-values at 2.41, 2.26, and 1.91. When camphor is added,
more than 80 percent of the protein is converted to the high-spin form, with
EPR g-values of 7.8, 3.9, and 1.8. During the catalytic cycle responsible for
the chemistry illustrated by Equation 11.11, electrons are transferred to the
protein, resulting in the formation of a high-spin ferrous intermediate; a
low-spin ferrous form can be obtained from this species by addition of CO.
Thus, four distinct states of the heme group are accessible in cytochrome
P-450, high- and low-spin forms for both Fe(II) and Fe(III).

(a) *The low-spin ferric state.* In ferric MbN$_3$⁻, where azide occupies the
O$_2$-binding site, the EPR spectrum has g-values of 2.8, 2.25, and 1.75, clearly
different from those of resting P-450. When thiolate complexes of Mb were
examined, however, the spectra much more closely resembled those of cytochrome P-450 (Table 11.5). These results, together with the observation that
thiol-group modifying reagents inactivated P-450, strongly suggested that
the two axial ligands in the protein were a cysteine thiolate and possibly a
histidine imidazole group. The cysteine ligand assignment was supported by
model studies of iron(III) porphyrin-thiolate complexes and by EXAFS results
for the protein that identified an Fe—S bond distance of 2.22 Å. The models
indicated, however, that a variety of additional axial ligands could be employed without modifying the spectroscopic properties of the protein, as

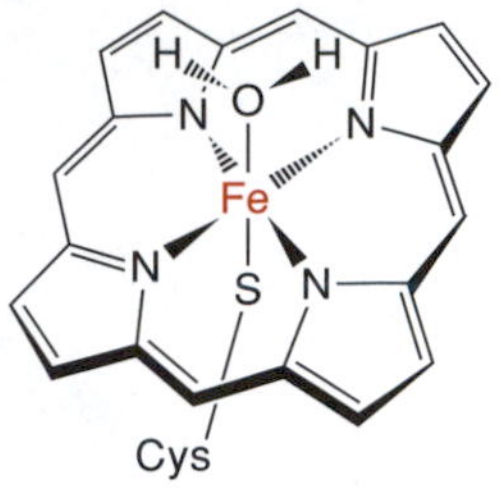

6-coordinate Fe(III) porphyrin-thiolate
low-spin ferric

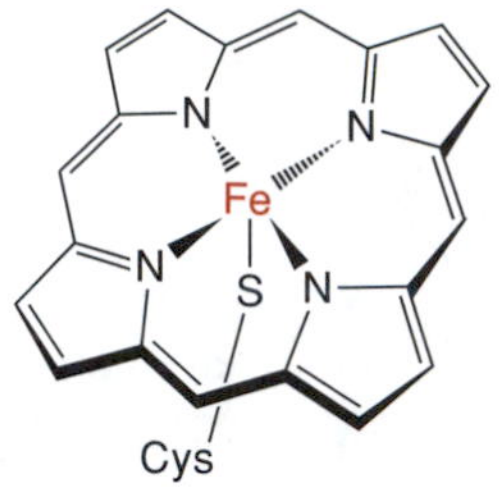

5-coordinate Fe(III) porphyrin-thiolate
high-spin ferric

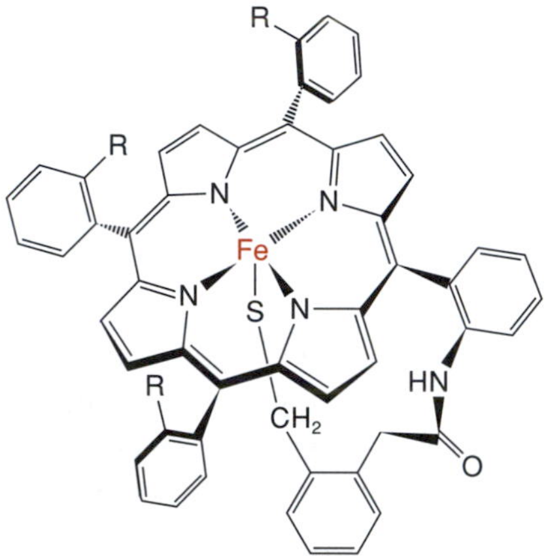

5-coordinate Fe(II) porphyrin thiolate model
high-spin ferrous

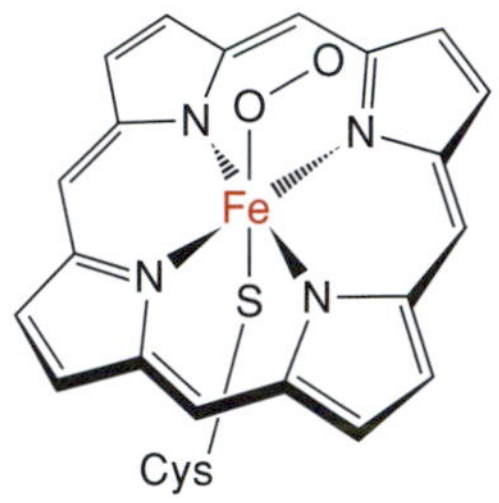

6-coordinate Fe porphyrin species
low-spin

Figure 11.11
Structures of some iron-porphyrin complexes that mimic intermediates known or
believed to occur during cytochrome P-450 turnover.

long as at least one RS$^-$ group was present. The single-crystal X-ray structure
of P-450$_{cam}$ ultimately confirmed this finding, revealing the iron porphyrin to be coordinated to a cysteinate sulfur (Fe–S, 2.20 Å) and a sixth
ligand assigned either as water or hydroxide ion (Figure 11.11). Proton NMR
relaxation studies revealed that the axial water ligand is in rapid exchange
with bulk solvent.

(b) *The high-spin ferric state.* Comparison of the EPR and MCD data of
ferric porphyrin model complexes with those of the camphor-bound cytochrome P-450 strongly indicated a five-coordinate structure with cysteinate
sulfur as the lone axial ligand. From EXAFS studies, the Fe–S bond distance
was determined to be 2.23 Å, comparable to the value of 2.20 ± 0.05 Å
ultimately obtained from the crystal structure. The structure is shown in
Figure 11.11.

(c) *The high-spin ferrous state*. Reduction of the camphor-bound P-450-cam enzyme leads to characteristic UV-visible absorption, ^{1}H NMR, MCD, and EXAFS spectroscopic data that match very well those of high-spin ferrous porphyrin model complexes having a single thiolate axial ligand. Although no crystal structure of the protein is yet available for this state, a ferrous tetraarylporphyrin-thiolate complex has been crystallographically characterized that may be viewed as a close structural analog of this state of the enzyme (Figure 11.11).

(d) *Low-spin ferrous states*. The low-spin, CO-bound ferrous form of cytochrome P-450 that gives it a distinctive 450 nm Soret band contains axial thiolate and carbon monoxide ligands. Extensive UV-visible, Mössbauer, MCD, ^{13}C NMR for the ^{13}CO-bound protein, and EXAFS data for the enzyme and model complexes definitively prove such a structure. As with Mb, a diamagnetic dioxygen derivative of ferrous P-450 can also be synthesized. Resonance Raman spectra reveal the O–O stretching frequency to occur at 1,140 cm^{-1} consistent with a formal Fe^{III}–O$_2$$^-$ electronic structure assignment. The structure of this complex, as revealed by EXAFS and comparison with model complexes, is depicted with the other states in Figure 11.11. This structure has now been confirmed by X-ray crystallography.

(iii) *The catalytic cycle*. Figure 11.12 reveals the currently accepted mechanism for the oxidation of hydrocarbons by cytochrome P-450. This mechanism was deduced from extensive ^{18}O-labeling experiments, the regioselectivity of the hydroxylation reaction, and other information gathered largely through studies of the liver microsome and P-450$_{cam}$ enzymes. In these systems, electrons (e$^-$) are supplied by NADPH and involve coupling to reductase and putidaredoxin systems (Figure 11.13). In the first step of the enzyme mechanism, substrate binds, converting the iron(III) center from the low-spin six-coordinate to the high-spin five-coordinate ferric state. X-ray studies of the camphor adduct of cytochrome P-450$_{cam}$ revealed the substrate to bind in a hydrophobic region of the protein, near the heme iron center, expelling all of the water molecules present in the substrate-binding pocket. Very little rearrangement of polypeptide backbone or side chain atoms occurs upon substrate binding. Interestingly, the presence of bound substrate shifts the redox potential from −300 to −173 mV for P-450$_{cam}$, thus making it easier to reduce.

The next step in the catalytic cycle (Figure 11.12) is the addition of an electron to form the high-spin ferrous, substrate-bound enzyme. The protein is primed for the oxygen-binding step, which leads to a ferrous dioxygen (or ferric superoxide, see above) intermediate that has been isolated and characterized. The next step of the postulated reaction mechanism, introduction

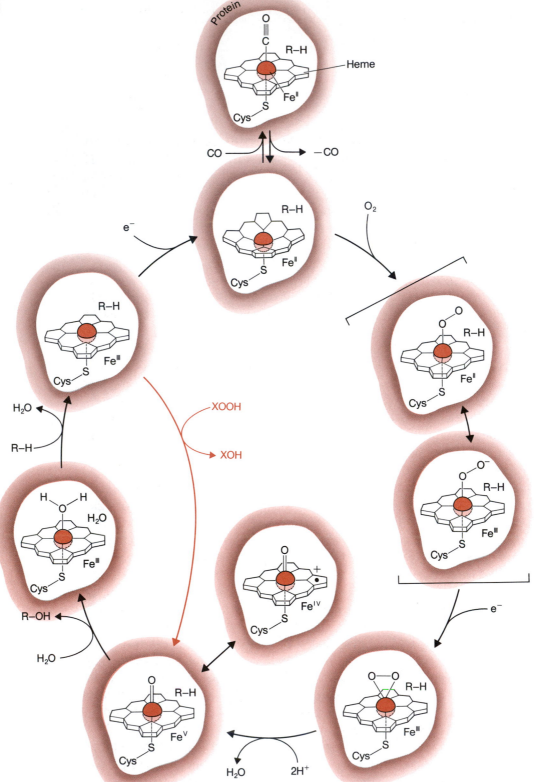

Figure 11.12
Catalytic mechanism proposed for cytochrome P-450. The peroxide shunt is depicted as XOOH going to XOH.

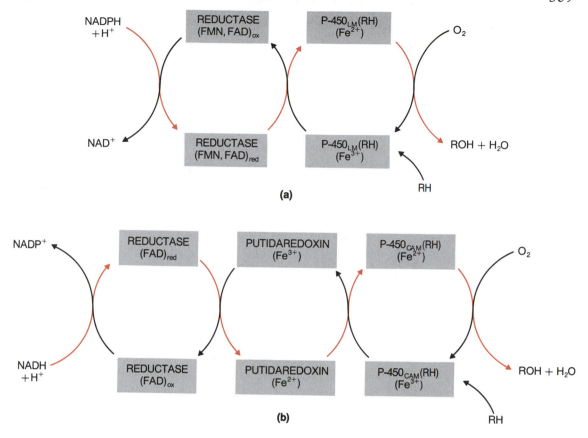

Figure 11.13
Electron transfer schemes for cytochrome P-450 hydroxylation of substrates in (a) liver microsomes and (b) camphor in a bacterial system. The coupling to other proteins involved in electron transfer is depicted for both cases.

of the second reducing equivalent, has not been directly observed for any of the P-450 enzymes. Formally, this intermediate may be written as a ferric-peroxide complex of unknown protonation state.

The cycle now proceeds through the formation of key intermediates and reaction steps that are even less well understood. Importantly, the O–O bond is cleaved to produce, upon addition of protons, water, resulting in an oxygen atom of formal oxidation state zero and an iron center of formal oxidation level +3. This diatomic unit, which can be written as $Fe^{III}-O^0 \leftrightarrow Fe^{IV}-O^- \leftrightarrow Fe^V-O^{2-}$, has been termed an *oxenoid* moiety. Moreover, the π-system of the porphyrin ring can itself be oxidized to a cation radical such that the PFe^V-O^{2-} or $PFe^V=O$ oxo form, where P stands for the porphyrin, can alternatively be written as $P \cdot Fe^{IV}=O$, where $P \cdot$ represents a porphyrin radical cation.

The reaction of the PFe=O oxenoid species with hydrocarbon substrates to produce alcohol and resting protein is one of the most intriguing transformations in bioinorganic chemistry. The fact that camphor, norbornane, and other saturated hydrocarbons partially epimerize, and that large deuterium isotope effects are observed for substrates such as *exo,exo,exo,exo*-tetra- -deuteronorbornane, strongly suggests a mechanism in which the C–H bond is broken homolytically. This reaction is followed by very rapid recombination with the Fe–OH unit to give the product, as indicated in Equation 11.12.

$$R–H + Fe^V{=}O \rightarrow [R\cdot + Fe^{IV}–OH] \rightarrow ROH + Fe^{III} \qquad (11.12)$$

Together these postulated steps have been termed a *rebound mechanism*. A value of $\sim 10^9$ s^{-1} has been estimated for the radical recombination step, fast enough that some radical rearrangements cannot compete with it, for example, ring opening of cyclopropylmethyl radical (k $\sim 10^8$ s^{-1}). Heterolysis or concerted addition of the R–H to the Fe^V=O bond seem less likely, but cannot be entirely ruled out.

The final step of the catalytic cycle (Figure 11.12) is dissociation of ROH product. Presumably, the alcohol is briefly coordinated to the ferric porphyrin site before its release to solution. This species would have a low reduction potential, by analogy to other low-spin Fe^{III} P-450 centers, which would tend to preclude addition of more electrons and a second round of oxidation to form diols. Moreover, it is likely that such a bound alcohol ligand would have to dissociate from the iron center before another substrate molecule could coordinate.

Although it is not known for certain which step in the catalytic cycle is rate-limiting, it is likely to be addition of the second electron or one of the subsequent two reactions. As indicated in Figure 11.12, the reduction and O$_2$-binding steps of the cycle can be bypassed by adding peroxide, XOOH, to P-450 enzymes containing bound substrate. This reaction, called the *peroxide shunt*, leads immediately to the putative oxenoid intermediate and subsequent alkane oxidation chemistry.

(iv) Models for cytochrome P-450 chemistry. Some of the most convincing evidence for aspects of the cytochrome P-450 reaction mechanism shown in Figure 11.12 comes from synthetic model studies. In one such line of work, the 5,10,15,20-tetramesitylporphyrinatoiron(III) complex was used. The methyl groups above and below the porphyrin-ring plane in this molecule provide a stereochemical block to formation of (μ-oxo)diiron(III) species. By analogy to chemistry of the P-450 peroxide shunt, this metalloporphyrin reacts with peroxy acids, such as *p*-nitroperbenzoic acid, or iodosylbenzene

(PhIO), a known oxo transfer reagent, to form an unstable green species capable of doing hydrocarbon oxidations. This green iron-porphyrin-oxo complex could be stabilized for EXAFS and Mössbauer studies at low temperature, however. The former revealed a 1.6 Å Fe=O bond, and the latter could be reasonably well interpreted in terms of an $S = 1$ Fe(IV) center strongly ferromagnetically coupled to a porphyrin cation radical. Large downfield shifts ($\delta \sim 70$ ppm for the meta protons) in the proton NMR spectrum of this species reflect the radical cation character of the porphyrin ring.

With the use of oxo transfer reagents and simple iron porphyrins, the hydrocarbon oxidation chemistry of cytochrome P-450 has been successfully reproduced. Cyclohexane oxidation to cyclohexanol (Equation 11.13) or cyclohexene oxidation to the epoxide and allylic alcohol (Equation 11.14) with tetraphenylporphyrin iron(III) chloride and iodosylbenzene mimic the P-450 reactions.

$$\text{(11.13)}$$

$$\text{(11.14)}$$

11.2.b. Methane Monooxygenase.

(i) *The three proteins of MMO.* The most difficult hydrocarbon substrate to hydroxylate is methane, which has a high C—H bond energy (104 kcal/mol), no dipole moment, and no functionality to assist in binding to or reacting with a protein-active site. Cytochrome P-450 does not hydroxylate methane. It is therefore intriguing that bacteria have evolved that use methane as their sole carbon and energy source. The first reaction catalyzed by these methanotrophic bacteria is methane hydroxylation, and the enzyme system employed is methane monooxygenase. The MMO from *Methylococcus capsulatus* (Bath) [Bath, England, a thermal spa from the waters of which the organism was isolated], consists of three proteins (Figure 11.14), a hydroxylase (previously called protein A), a reductase (protein C), and a regulatory, or coupling, protein B that in some manner modulates electron transfer between proteins A and C. As shown in Figure 11.14, electrons enter the system via the iron-sulfur core and flavin cofactors in the reductase protein. Electron transfer then occurs, with the aid of the coupling protein, to the

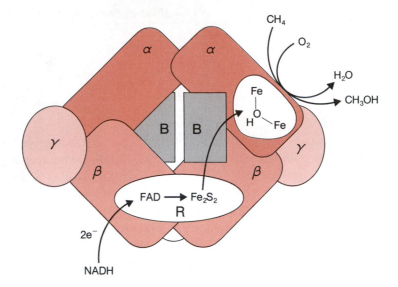

Figure 11.14
Schematic diagram of the $\alpha_2\beta_2\gamma_2$ hydroxylase (H) protein of methane monooxygenase associated with two copies each of the coupling (B) and reductase (R) proteins required for enzymatic activity. Transfer of electrons occurs from NADH via the reductase to the dinuclear iron center in the α subunit of the hydroxylase. The relative positions of the α, β, and γ subunits of H are depicted as recently determined by X-ray crystallography.

hydroxylase enzyme, which is the site of dioxygen activation and substrate binding. This compartmentalization assures that the reducing agent, NADH, never comes in direct contact with the highly reactive hydoxylation unit generated during enzyme turnover. A similar strategy is employed in the cytochrome P-450 enzyme system (see Figure 11.13). Moreover, evidence suggests that electron transfer into the MMO hydroxylase may not occur until after the substrate has bound, and, under these circumstances, two electrons are transferred at the same time. This control of electron transfer, which is effected by the coupling protein, prevents oxidative damage to the hydroxylase and wasteful expenditure of reducing equivalents in the absence of substrate. Again, these gating strategies mimic those adopted by the cytochrome P-450 system.

(ii) Structure and chemical reactivity of the dinuclear iron core in the hydroxylase protein. The active site of the hydroxylase protein of MMO contains a dinuclear nonheme iron unit. As shown by EXAFS investigations, the Fe$\cdots$Fe distance is ~ 3.4 to 3.5 Å in at least the first two of its three redox forms, Fe(III)Fe(III), Fe(III)Fe(II), Fe(II)Fe(II). There is no short μ-oxo bond, and both iron atoms have either five or six oxygen and/or nitrogen donor atoms. From power saturation studies of the EPR spectrum of the mixed valent form, the two iron atoms are known to be antiferromagnetically coupled to one another with a spin-exchange coupling constant $J = -30$ cm^{-1}. Recent proton

ENDOR measurements of this form have provided good evidence for a hydroxo bridge. There is some sequence similarity between the α subunit of the hydroxylase of MMO and protein R2 of ribonucleotide reductase, X-ray structural studies of which have recently shown to have the dinuclear iron core illustrated in Figure 11.15. This information suggested that the diiron core in MMO hydroxylase is most likely bridged by one or two carboxylates as well as a single O atom from hydroxide ion. Evidence for histidine ligation on both iron atoms has been provided by ENDOR and ESEEM studies. Both conclusions were recently confirmed by X-ray crystallography.

As already mentioned, the dinuclear iron center in MMO is capable of catalyzing mixed-function oxidation chemistry analogous to that performed by cytochrome P-450. Much less is known about the detailed reaction mechanism, although it has been established in a single turnover experiment that the reduced protein will react with dioxygen and alkane substrate to produce the corresponding alcohol. Elucidation of the structures of chemical intermediates recently detected and the development of suitable model systems remain important challenges for the bioinorganic chemist. It is interesting to speculate that nature has chosen to use the dimetallic iron center in MMO to parallel the mononuclear iron porphyrin prosthetic group for alkane oxidation because, in both systems, two electrons can be carried without having to invoke a high-valent iron(V) species. As shown in Equations 11.15a and 11.15b, the diiron(III) unit is electronically equivalent to the PFe^{III} center.

$$PFe^{III} + O_2H^- \xrightarrow{\;\;H_2O\;\;} PFe^{V}{=}O \longleftrightarrow \cdot PFe^{IV}{=}O \qquad (11.15a)$$

$$Fe_2^{III} + O_2H^- \xrightarrow{\;\;H_2O\;\;} Fe_2^{IV}{=}O \qquad\qquad (11.15b)$$

On the other hand, attempts to prove the existence of radical intermediates similar to that given in Equation 11.12 have been ambiguous, and there are important differences in the MMO and cytochrome P-450 systems that require clarification. Very recently, the hydroxylase protein has been crystallized, and the 2.2 Å X-ray structure determined. The active site was found to resemble that of the R2 protein (Fig. 11.15), with a hydrophobic pocket for substrate binding. The results of these ongoing investigations should be quite helpful in elucidating details of the enzyme mechanism.

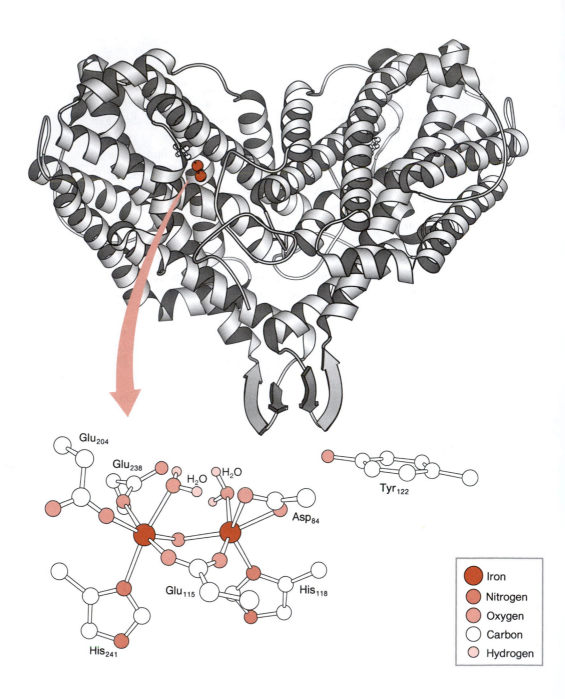

Glu$_{204}$

Glu$_{238}$

H$_2$O   H$_2$O

Tyr$_{122}$

Asp$_{84}$

Glu$_{115}$

His$_{118}$

His$_{241}$

●	Iron
●	Nitrogen
●	Oxygen
○	Carbon
○	Hydrogen

Figure 11.15
Structure of the R2 protein of bacterial ribonucleotide reductase displaying the dinuclear iron center and nearby tyrosyl radical.

11.2.c. Catechol and Other Dioxygenases. Both cytochrome P-450 and MMO are mixed-function oxidases; only one atom of dioxygen is transferred to the alkane, the other being converted into water. This process requires energy in the form of reducing equivalents. Dioxygenase enzymes, on the other hand, incorporate both atoms of the O_2 molecule into substrate. The best-characterized enzyme of this class is protocatechuate 3,4-dioxygenase, which catalyzes the reaction shown in Equation 11.16. This

$$(11.16)$$

transformation involves the cleavage of an aromatic ring to form a dicarboxylic acid. From labeling studies, it was determined that each of the carboxylate groups contains one of the oxygen atoms from the dioxygen substrate.

The enzymes are generally red because of charge-transfer transitions from tyrosinate ligands to active site Fe(III) ions. The crystal structure of protocatechuate 3,4-dioxygenase from *Pseudomonas aeroginosa* has been determined to 2.8 Å resolution. This enzyme is a large oligomer with twelve copies of each of two subunits, a 22-kilodalton α subunit and a 27-kilodalton β subunit; its overall symmetry is T. The active site of this enzyme lies at an interface between the α and β subunits, although all the iron ligands are derived from the β subunit. The structure of the active site is shown in Figure 11.16. The ligands are Tyr-118, Tyr-147, His-160, and His-162, as well as a bound solvent molecule, either a water or a hydroxide ion.

Based on a wide variety of kinetic and substrate variation studies, a reasonably detailed mechanistic scheme has been proposed for this enzyme. This scheme is shown in Figure 11.17. A key feature of this mechanism is the occurrence of significant charge transfer between the metal center and the coordinated catecholate. This electron transfer from the ligand to the iron center imparts significant semiquinone radical-like character to the ligand, activating it for direct attack by the dioxygen molecule. The ligand-dioxygen adduct then undergoes a series of transformations leading to products.

Some aspects of this mechanism have been experimentally probed. In particular, EXAFS studies of the corresponding enzyme from *Brevibacterium fuscum* have been carried out. The EXAFS spectra of the resting state are consistent with the crystal structure results, revealing three oxygen or nitrogen ligands at 1.90 Å and two O/N at 2.08 Å. The shorter distances are assigned to the two tyrosinate ligands and a bound hydroxide ion; the longer distances derive from the two histidines. Addition of a catechol-based sub-

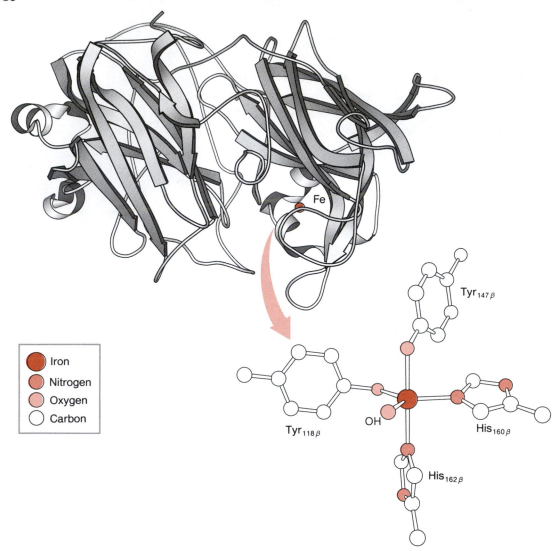

Figure 11.16
Structure of protocatechuate 3,4-dioxygenase and its iron center.

strate, which coordinates to the iron via both of its hydroxyl groups as deduced from a variety of EPR studies, results in very little change in the parameters derived from EXAFS curve fitting. This result suggests that the iron remains five-coordinate, which conclusion requires that two of the ligands present in the resting enzyme be displaced. The hydroxide ion is one obvious candidate, with concomitant protonation via one of the phenolic hydrogen atoms. It is also proposed that one of the histidine or tyrosinate ligands is displaced and protonated by the remaining phenolic hydrogen

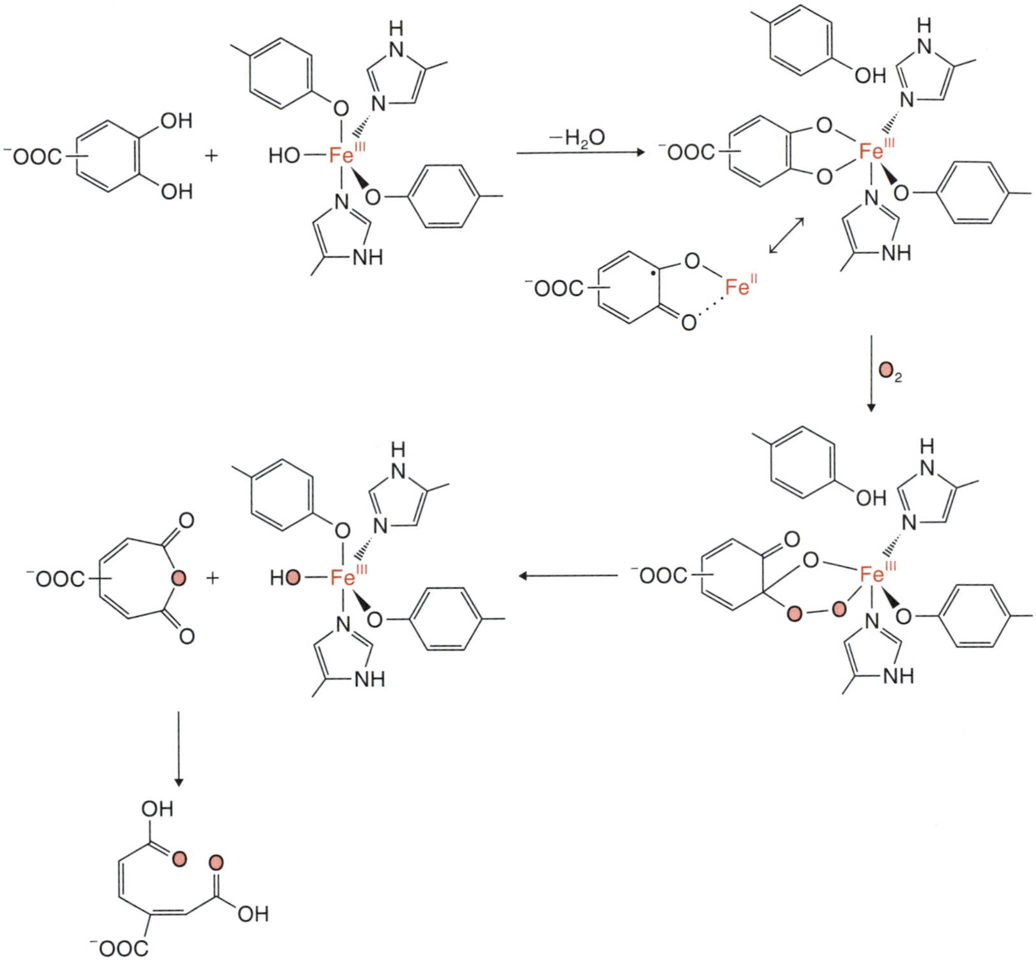

Figure 11.17
Mechanism proposed for the action of protocatechuate 3,4-dioxygenase. The
colored atoms show the fate of oxygen atoms introduced as ^{18}O-labeled dioxygen.

atom. The overall changes are summarized in Figure 11.17. Many of the
details of the oxygen insertion and cleavage reactions remain to be elucidated
definitively.

The iron in this dioxygenase binds and activates substrate for electro-
philic attack by dioxygen through partial charge transfer from the substrate
to the metal center. This role may be contrasted with that of the metal ions
in the enzymes discussed in Chapter 10, which act as Lewis acids. A completely redox-inert metal ion such as zinc would probably not be able to

substitute functionally for iron(III). It will be interesting to see whether this degree of required redox character applies to other dioxygenase enzyme systems.

11.3. Oxygen-Atom-Transfer Reactions: Mo

11.3.a. Molybdenum Oxotransferase Enzymes.   Another large class of enzymes that promote oxygen-atom-transfer reactions are those that contain molybdenum. All of these molybdenum-containing enzymes with the exception of nitrogenase appear to require a common cofactor, the likely structure of which has been elucidated in recent years (see Section 5.4.d. and Figure 5.20). Spectroscopic studies, particularly those involving EPR and X-ray absorption techniques, have revealed that the detailed nature of the molybdenum sites varies significantly from enzyme to enzyme, in a manner analogous to that seen for porphyrin-containing proteins. For the molybdenum proteins, however, the variability seems to be even greater. The enzymes all catalyze reactions, some of which are shown in Table 11.6, that can be written as oxygen-atom-transfer reactions. The strongest evidence that these enzymes share a common cofactor comes from an assay involving a mutant of the mold *Neurospora crassa*. This organism normally produces nitrate-reductase, but a mutant strain, termed nit-1, is devoid of this activity. Treatment of extracts of nit-1 with acid-inactivated nitrate reductase can reconstitute the latent nitrate-reductase activity. Importantly, extracts from all of the other known molybdenum-containing enzymes (except nitrogenase) are capable of reactivating nit-1 extracts as well.

Here we shall concentrate on two of these enzymes, namely, xanthine oxidase and sulfite oxidase. Xanthine oxidase, isolated from cow's milk, has been a much studied enzyme. It exists as a dimer of molecular mass ~ 300 kilodaltons. Each subunit contains one molybdenum, one flavin adenine dinucleotide, and two Fe_2S_2 centers. Some aspects of the structure of the molybdenum site have been determined by EXAFS studies on xanthine oxidase and the closely related xanthine dehydrogenase from liver. The oxidized form has one terminal molybdenum oxo (Mo=O) group, at least two thiolate-type sulfur ligands (probably from the pterin dithiolene side chain), and one terminal sulfido (Mo=S) group. A variety of chemical studies foreshadowed the discovery of this sulfido group by the spectroscopic methods. Treatment of xanthine oxidase with CN^- inactivates the enzyme and releases one equivalent of thiocyanate ion, SCN^-. The reduced form still has one terminal oxo group, but it appears that the terminal sulfido group has been

Table 11.6
Mo-containing oxygen-transfer enzymes

Enzyme	Reaction Catalyzed
Xanthine oxidase	Xanthine + $H_2O \rightarrow$ uric acid + $2N^+ + 2e^-$
Sulfite oxidase	$SO_3^{2-} + H_2O \rightarrow SO_4^{2-} + 2H^+ + 2e^-$
Nitrate reductase	$NO_3^- + 2H^+ + 2e^- \rightarrow NO_2^- + H_2O$
Formate dehydrogenase	$HCOOH \rightarrow CO_2 + 2e^- + 2H^+$
CO dehydrogenase	$CO + H_2O \rightarrow CO_2 + 2e^- + 2H^+$
Dimethylsulfoxide reductase	$2e^- + 2H^+ + (CH_3)_2SO \rightarrow H_2O + (CH_3)_2S$
Trimethylamine N-oxide reductase	$(CH_3)_3NO + 2e^- + 2H^+ \rightarrow (CH_3)_3N + H_2O$
Biotin sulfoxide reductase	Biotin sulfoxide + $2e^- + 2H^+ \rightarrow$ biotin + H_2O

protonated to form a hydrosulfide (SH^-) ligand. Cyanide-treated enzyme has two terminal oxo groups in the oxidized form.

A key experiment has been reported that bears strongly on the mechanism of xanthine oxidation. Reduced xanthine oxidase was oxidized with an ^{18}O-labeled amine oxide derivative followed by a substoichiometric amount of xanthine. The isolated uric acid was found to contain at least 79 percent of the oxygen label. The converse experiment using ^{18}O water and unlabeled enzyme revealed little incorporation of label into the product. These results indicate that the enzyme-catalyzed reactions proceed via transfer of an oxygen atom or its equivalent to substrate rather than by molybdenum-promoted attack of solvent water or hydroxide.

A variety of EPR spectra has been observed for Mo(V) forms of xanthine oxidase with the use of rapid freeze methods for sample preparation. Since the molybdenum center shuttles between Mo(VI) and Mo(IV) redox states during the catalytic cycle, the relationship between species giving rise to these EPR signals and those along the main reaction pathway is somewhat unclear. Because the enzyme also contains other redox-active centers, however, Mo(IV) species generated in the course of the reaction could be in equilibrium with EPR-active Mo(V) species via intramolecular electron-transfer reactions. In addition, Mo(V) can be generated in the course of reoxidizing the enzyme to the Mo(VI) state. Thus such EPR studies have been crucial in providing information about the reaction mechanism and the structures of potential intermediates. One possible mechanistic scheme is

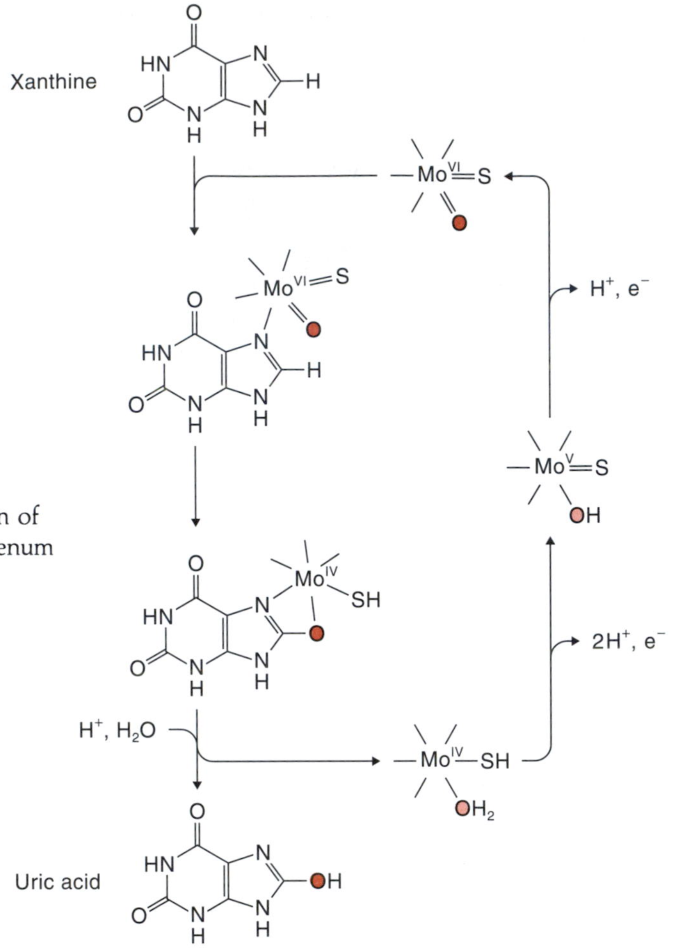

Figure 11.18
Mechanism proposed for conversion of xanthine to uric acid at the molybdenum site of xanthine oxidase.

shown in Figure 11.18. An oxygen atom is transferred from the molybdenum center to xanthine, with the terminally bound sulfido group acting as a base for hydrogen that is removed from the substrate. The oxo group is then restored at the Mo center from a water molecule via a series of coupled deprotonation and electron transfer steps.

Sulfite oxidase is a somewhat simpler enzyme than is xanthine oxidase, in that it contains only a single prosthetic group, namely, a heme, in addition to the molybdenum cofactor. Furthermore, the two prosthetic groups in the enzyme from chicken liver are separable from one another by limited proteolysis. Thus, the holoenzyme is a dimer with a molecular mass of 120 kDa. Treatment with trypsin yields two equivalents of a monomeric fragment of 9.5 kDa that contains the heme group and a dimeric unit of 94 kDa that contains the molybdenum center. The fact that the sum of the molecular weights of the two fragments is less than that of the holoenzyme reflects uncertainties that arise in the experimental determination of protein molecular weights. The ability to cleave the enzyme in this manner has allowed separation of the two half-reactions as well as spectroscopic charac-

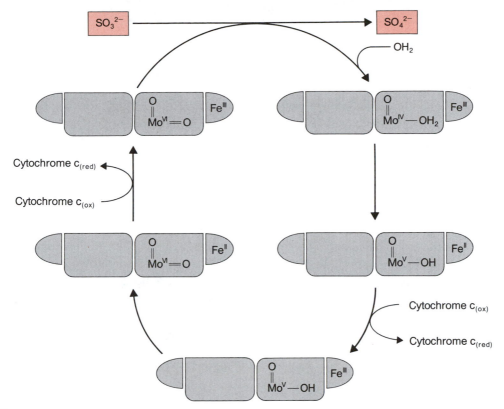

Figure 11.19
Scheme depicting the mechanism postulated for the conversion of sulfite to sulfate by sulfite oxidase, showing both the oxygen atom transfer and internal electron transfer steps.

terization of the molybdenum center without interference from other strongly absorbing species. In particular, whereas the intact enzyme can oxidize sulfite by using oxidized cytochrome *c*, ferricyanide, or dioxygen as electron acceptors, the isolated molybdenum domain has lost the ability to transfer electrons to cytochrome *c* but retains its activity with the other reagents. This observation shows that sulfite oxidation occurs at the molybdenum center, with the heme center serving to couple this two-electron oxidation to the reduction of two molecules of cytochrome *c*.

As for xanthine oxidase, much of the structural information about the molybdenum site in sulfite oxidase has come from EXAFS studies. The oxidized form of the enzyme has two terminal oxo groups as well as two or three thiolate sulfur atoms. Upon reduction, one of the oxo groups is lost, and the number of thiolate sulfur atoms can be more confidently assigned to be three. The *cis*-dioxomolybdenum(VI) unit is quite common in molybdenum chemistry. Such species with thiolate sulfur ligation are well known to undergo oxygen-atom-transfer reactions with oxo-group acceptors such as phosphines. Based on such chemical observations as well as kinetic studies of the enzyme, the catalytic cycle shown in Figure 11.19 has been proposed.

The substrate oxidation reaction occurs via direct oxygen atom transfer from molybdenum to sulfite to form sulfate. The monooxomolybdenum(IV) species is then reoxidized to Mo(VI) in two one-electron steps coupled with deprotonation of a bound water molecule.

These two examples illustrate the role of oxomolybdenum centers in executing oxygen-atom-transfer reactions. The stability of the Mo(VI) and Mo(IV) oxidation levels with an appropriate number of terminally bound ligands facilitates such reactions without the generation of substrate radical species. The role of the special pterin that acts as one of the ligands for molybdenum in these centers in not known. It could be involved in electron-transfer processes to substrate, although there are no data that require such a function. A more likely possibility is that it facilitates the electron-transfer pathways that couple the molybdenum-centered reactions to the other redox-active groups that are almost invariably present in this class of enzymes.

11.3.b. Models for Molybdenum Oxotransferases.

Several model complexes for the active sites of molybdenum oxotransferases have been synthesized. These models generally do not attempt to reproduce the pterin moiety found in the enzymes, although they do include reasonable approximations to the donor atoms bound to molybdenum as revealed by EXAFS analysis. Most importantly, it has been possible to mimic the reactivity of the enzymes, particularly the oxygen-atom-transfer step. The best-studied reactions of this kind involve the reduction of sulfoxides to sulfides and of amine oxides to amines.

The system that has been most extensively characterized utilizes the pyridine-thiolate ligand L shown below. A complex having the formula

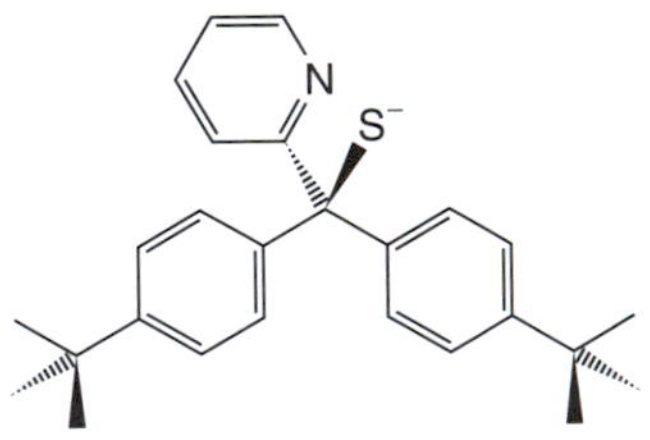

[MoO$_2$(L)$_2$] containing the dioxomolybdenum(VI) moiety has been prepared, the structure of which is shown in Figure 11.20. Reaction of this complex with phosphines such as Et$_3$P results in oxygen-atom transfer from molybdenum to phosphorus to yield the corresponding phosphine oxide and the oxomolybdenum(IV) complex [MoO(L)$_2$], the structure of which is also depicted in Figure 11.20. Note that the bulky *tert*-butyl groups were incorporated into the ligand to preclude the reaction between [MoO$_2$(L)$_2$]

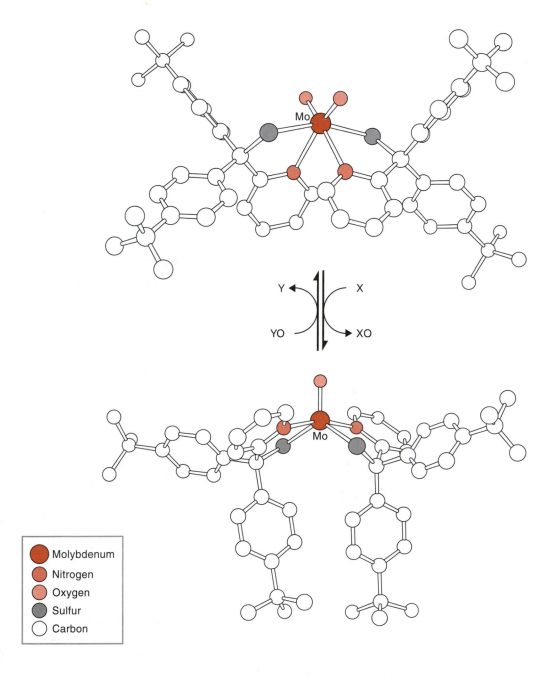

Figure 11.20
A model for molybdenum oxo-transferase enzymes. The structures of both the oxidized (top) and reduced (bottom) forms have been determined by X-ray crystallography.

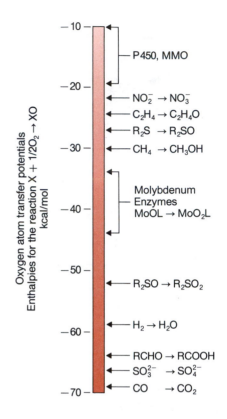

Figure 11.21
Relative oxygen-atom-transfer enthalpies.
The molybdenum- and iron-containing
enzyme-based couples are well placed
thermodynamically to execute the biochemical reactions that they catalyze.

and $[MoO(L)_2]$, which produces the oxo-bridged Mo(V) complex $[(L)_2MoOMo(L)_2]$. In much the same way as occurs for the myoglobin and hemoglobin models discussed in Section 11.1.a.iii, these μ-oxo species represent thermodynamically stable dead ends that limit the utility of complexes of many ligands as enzyme models. The reduced $[MoO(L)_2]$ complex will react with a variety of oxygen-atom donors, including several enzyme substrates, to regenerate $[MoO_2(L)_2]$ with concomitant substrate reduction. For example, dimethylsulfoxide, a substrate for several molybdenum enzymes, is quantitatively reduced to dimethylsulfide by $[MoO(L)_2]$.

One of the most important observations from studies of these model complexes is the demonstration that the oxygen-atom-transfer enthalpy, defined in Figure 11.21, for the dioxomolybdenum(VI)/oxomolybdenum(IV) couple is a key feature of the system. Some molybdenum enzymes, such as sulfite oxidase, transfer oxygen atoms to substrate, whereas others, including nitrate reductase, transfer the oxygen atoms from the substrate. The molybdenum centers in these enzymes are ideally positioned thermodynamically to execute these different transformations. This feature is illustrated in Figure 11.21, which orders the relative oxygen-atom-transfer enthalpies for some of these reactions as well as a sampling of those catalyzed by cytochrome P-450 and MMO.

11.4. Other Reactions of Dioxygen and Its Byproducts

11.4.a. Protective Metalloenzymes: The Cu–Zn Superoxide Dismu- $Superoxide\ O_2^-$
tase.   Superoxide dismutase (SOD) metalloenzymes function to disproportionate the superoxide ion according to Equation 11.17. As can be seen from
this reaction, hydrogen peroxide, itself a potentially harmful substance, is one

$$2\,O_2^- + 2\,H^+ \rightleftharpoons H_2O_2 + O_2 \qquad (11.17)$$

of the products. The hydrogen peroxide is scavenged by another metalloenzyme, catalase, that is discussed in more detail in the next section. Thus,
working in concert, SOD and catalase serve to protect organisms that utilize
dioxygen from the potentially toxic byproducts of O_2 metabolism. Superoxide dismutases having redox-active manganese, iron, or copper centers
have been characterized. Here we focus on the last, a copper-zinc superoxide
dismutase sometimes referred to as the brass enzyme after its heterometallic
core.

(i) *Copper superoxide dismutases and their chemistry.* In the late 1960s, it
was discovered that copper-containing proteins, previously isolated from
various sources and known as cupreins (erythrocuprein from the red blood
cell, hepatocuprein from liver, and so forth), were capable of catalyzing the
disproportionation of superoxide ion (Equation 11.17). These enzymes are all
composed of two identical subunits, each of which contains one Cu and
one Zn atom. The discovery of a function led to intensive studies of the
Cu_2Zn_2SOD enzymes. By using rapid-reaction techniques, it was soon established that the catalytic mechanism consisted of two diffusion-controlled
steps, illustrated in Equations 11.18a and 11.18b. These reactions are

$$ECu^{II} + O_2^- \rightleftharpoons ECu^I + O_2 \qquad (11.18a)$$

$$ECu^I + O_2^- + 2\,H^+ \rightleftharpoons ECu^{II} + H_2O_2 \qquad (11.18b)$$

sometimes referred to as a "ping-pong" mechanism, because of the shuttling
back and forth of copper between its two common oxidation states. Protonation of the superoxide ion in the second step renders both of the reactions thermodynamically feasible. As indicated in Table 1.2, the redox potential for the O_2/O_2^- couple is -0.33 V and for the O_2^-/H_2O_2 couple $+0.89$
V. Thus, any metal atom capable of one-electron redox chemistry between
M^{n+} and $M^{(n+1)+}$ states that has a potential between the limits $-0.3 \le E^\circ \le$
$+0.9$ V will be thermodynamically capable of serving as a superoxide dismutase. It is therefore not surprising that, in addition to the Cu_2Zn_2SODs,

there are both Fe and Mn SODs in nature. These latter enzymes have only a single metal atom at their active sites, however.

The disproportionation of superoxide ion depicted by Equation 11.17 occurs rapidly even in the absence of a catalyst, the second-order rate constant for the uncatalyzed reaction being $\sim 4 \times 10^5$ M^{-1} s^{-1}. The enzyme-catalyzed reactions have rate constants $\sim 2 \times 10^9$ M^{-1} s^{-1}, however, such that the overall rate is limited only by diffusion of the O_2^- ion to the protein-active site. In attempting to account for the need for SOD enzymes, it has been argued that the concentration of superoxide ion *in vivo* is sufficiently low that, were it to disproportionate by the uncatalyzed process of Equation 11.17, it would first react with and damage biological tissue. Thus the role of the enzyme is to disproportionate superoxide ion in a putative protective function. Although free metal ions also efficiently catalyze Equation 11.17, the rate constant for the Cu-catalyzed reaction being $\sim 4 \times 10^9$ M^{-1} s^{-1}, the concentrations of unbound cupric or ferric cations are believed to be far too low for this alternative means of superoxide dismutase activity to be important in living systems.

(ii) Structure of copper-zinc superoxide dismutase. Bovine erythrocyte superoxide dismutase has been crystallized and its structure determined by X-ray diffraction studies to 2 Å resolution. The polypeptide backbone comprises a barrel of eight antiparallel chains of β-pleated sheet, a common motif in protein structures that has become known as the β-barrel. Two such polypeptide chains form a dimer in the Cu_2Zn_2SOD protein. The metal-binding site is depicted in Figure 11.22. The Cu(II) ion is coordinated by four histidine side chains, His 44, His 46, His 118, and His 61. In addition, there is evidence for a fifth axial water ligand. Interestingly, His 44 and His 46 are in *trans* positions of the distorted square-planar CuN_4 coordination sphere, despite their proximity in the sequence. The tripeptide His 44-Val 45-His 46 completely blocks access to the copper from one side of the CuN_4 plane. The other side is accessible to solvent by means of a conical channel ~ 4 Å wide at the copper and opening to solution at the protein surface. Lining this channel are positively charged amino-acid side chain residues, including a sequence-conserved Lys 134 ε-amino group located 13 Å from Cu at the far end of the channel. Possibly this residue is positioned to attract O_2^- ions, which diffuse over the protein surface, down into the channel from which they can never return as a toxic species. An adjacent, conserved Glu 131 residue partly neutralizes the charge and helps define an electrostatic gradient that drives the O_2^- ion toward the positively charged Cu^{2+}/Zn^{2+}-active site. Other charged residues in the channel include Glu 119 and Glu 130, Lys 120, and Arg 141.

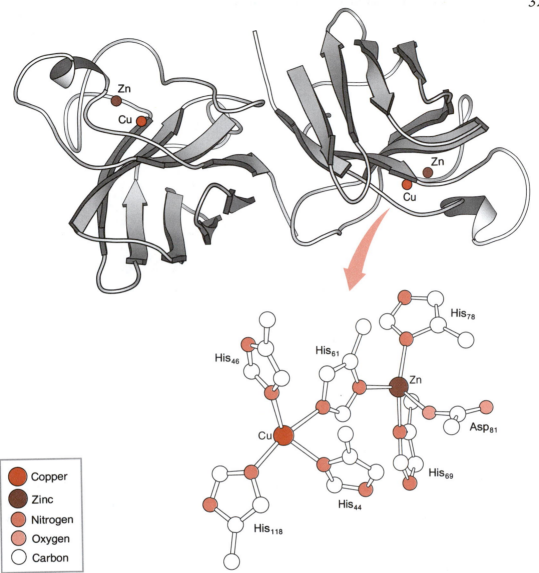

Figure 11.22
Structure of copper-zinc superoxide dismutase and its catalytic dimetallic core.

In addition to the cupric ion at the active site, the floor of the cavity also contains Zn(II) (Figure 11.22). Coordinating this ion in a tetrahedral fashion are three histidine residues, His 69, 78, and 61, the last bridging to the copper ion, and the carboxylate group of Asp 81. The imidazolate-bridged dimetallic center in Cu_2Zn_2SOD is a novel structural feature that, prior to its discovery in the enzyme, had not been previously encountered in coordination chemistry.

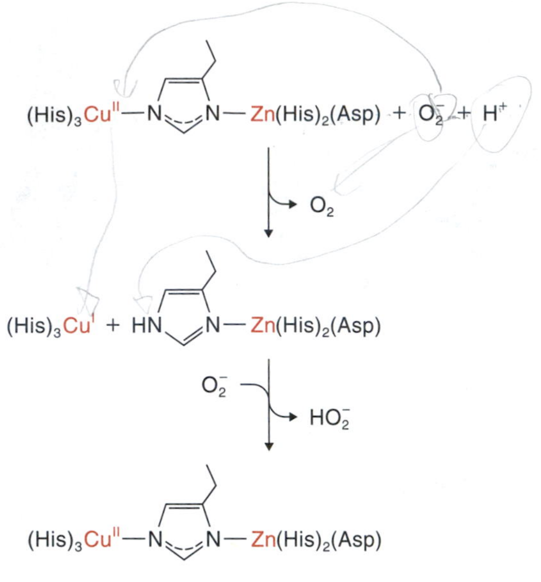

Figure 11.23
Possible bridge-breaking mechanism for copper zinc
superoxide dismutase.

(iii) Role of the zinc. The presence of the zinc in Cu_2Zn_2SOD suggested
that a second metal ion might be required for superoxide dismutase activity.
One specific proposal is shown in Figure 11.23. In this mechanism, the
imidazolate bridge is broken and reformed during each catalytic cycle. The
rate at which this cleavage can occur has been shown to be as fast as the
turnover rate of the enzyme. Coordination of a histidine ring nitrogen atom
to zinc(II) lowers the pK_a of the other nitrogen atom from 14 to ~ 7 to 10.
This second nitrogen atom will prefer to bind to a proton rather than Cu(I),
a preference that leads to bridge cleavage following the first step of the
enzyme mechanism (Equation 11.18a and Figure 11.23). Reoxidation of Cu(I)
to Cu(II) (Equation 11.18b) reforms the bridge, with transfer of the proton
to peroxide ion, which is released as H_2O_2. Such a mechanism creates an
open site at the Cu(I) center for binding to O_2^-, obviating the need for a
potentially high-energy five-coordinate transition state.

As attractive as the mechanism of Figure 11.23 might be, subsequent
studies revealed that Cu_2E_2SOD, in which the zinc site is empty (E), is just as
active as the holoenzyme. What then might be the true role of Zn^{2+} in SOD?
One possibility, mentioned above, is that Zn(II) helps to create the electric
field gradient that attracts O_2^- into the Venus flytrap conical cavity of SOD.
Both O_2 and H_2O_2, the products of the disproportionation reaction, are
neutral and hence can readily diffuse out, whereas O_2^-, an anion, cannot.
Another possible role for zinc is to confer thermal stability to the protein.

Table 11.7
Reactions of heme-containing peroxidases and catalase

Enzyme	Reaction
Cytochrome c peroxidase	$ROOH + 2H^+ + 2e^- \rightarrow ROH + H_2O$
Horseradish and spinach peroxidase	$A\text{-}H_2 + H_2O_2 \rightarrow A + 2H_2O$
Chloroperoxidase	$A\text{-}H + X^- + H^+ + H_2O_2 \rightarrow A\text{-}X + 2H_2O$
Catalase	$2H_2O_2 \rightarrow 2H_2O + O_2$

Copper-zinc SODs are remarkably stable to heat, their spectral properties and activity being unchanged after 7 min incubation in 10 mM pH 7.2 phosphate buffer at 75 °C. By contrast, when the metals are removed from the enzyme, the thermal stability is markedly diminished.

11.4.b. A Protective Metalloenzyme, Catalase, and Peroxidases. Another reduced form of dioxygen of importance is hydrogen peroxide. A variety of enzymes has been discovered that utilize hydrogen peroxide as a substrate, either simply destroying it by disproportionation to dioxygen and water or by employing it as an oxidant. The enzymes to be considered in this section and the reactions they catalyze are summarized in Table 11.7. All of these enzymes contain heme groups at their active sites, although the nature of the axial ligand and the details of the reaction mechanisms are quite different.

Cytochrome c peroxidase from yeast consists of a single chain of 294 amino acids, the structure of which has been determined crystallographically to 1.7 Å resolution, as discussed briefly in Chapter 9. One axial site is occupied by histidine in a manner similar to that seen in hemoglobin; the other side is unoccupied in the resting, Fe(III), state. Crystal structures of horseradish peroxidase and chloroperoxidase are not yet available. A variety of spectroscopic investigations, however, including detailed EXAFS studies, has revealed that the axial ligand is again histidine in the former. In the latter, it is, cysteine thiolate, making it similar to the active site of the cytochromes P-450. The heme ligation in these active sites is depicted in Figure 11.24.

These various enzymes share a generalized common reaction mechanism, although the details are quite different, as we shall discuss below. The common outline is given in Equation 11.19. The initial step is the two-electron

$$EPFe(III) + H_2O_2 \xrightleftharpoons{-H_2O} EPFe(-2e^-) \xrightleftharpoons{X} X(-2e^-) + EPFe(III) \quad (11.19)$$

oxidation of the enzyme, designated EPFe(III), by hydrogen peroxide with the release of one molecule of water. The nature of the two-electron oxidized

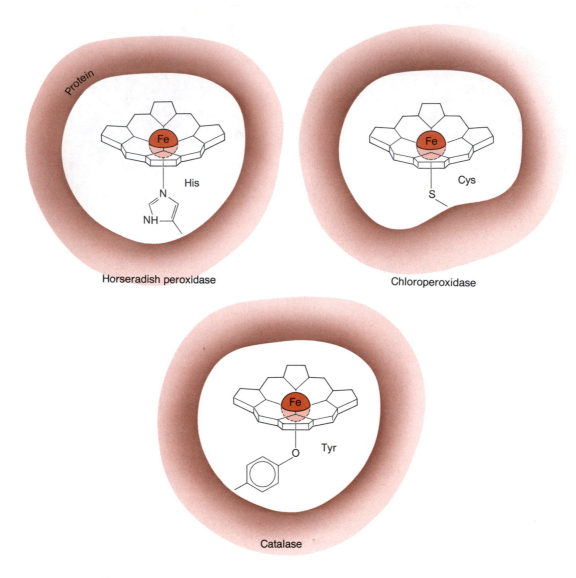

Figure 11.24
Ligation of heme in the active sites of two peroxidases and a catalase.

forms, EPFe($-2e^-$), of the enzymes has been the subject of intense investigation. For all the enzymes this form includes an Fe(IV)=O unit at the iron center. This iron-oxo unit is responsible for one of the two electrons that has been removed. For catalase, horseradish peroxidase, and chloroperoxidase, the other electron comes from the porphyrin ring to yield a cation radical species. This state is referred to as Compound I in horseradish peroxidase, and is analogous to the postulated active oxidant in the cytochrome P-450 reaction cycle discussed in Section 11.2.a. For cytochrome c peroxidase, however, the electron comes from a protein side chain. Recent work using site-directed mutagenesis and detailed ENDOR studies has demonstrated that the free radical is centered at Trp 191. The indole ring of this amino acid is oriented perpendicular to the porphyrin ring, parallel to and in contact with the axial histidine ligand. Furthermore, these two aromatic rings are each hydrogen bonded to Asp 225.

Once the two-electron oxidized species of the enzymes are formed, they can undergo a variety of reactions with species X in returning to the EPFe(III) state. For catalase or chloroperoxidase, this step is achieved by reaction with a second equivalent of hydrogen peroxide. In this reaction, the peroxide is oxidized rather than cleaved to yield dioxygen. The oxo group from the heme center is released as water. The reaction can also occur one step at a time, which, for cytochrome c peroxidase, results in the stepwise oxidation of two molecules of cytochrome c. For horseradish peroxidase, similar reactions occur, with a number of electron donors such as alkylamines or sulfides. Again, the oxo group is released as water; it is not transferred to substrate. Finally, for chloroperoxidase, the two-electron oxidized form can react with chloride to yield hypochlorite, which then can transfer halogen to an organic substrate with release of hydroxide.

Catalase enzymes have evolved to effect the disproportionation of hydrogen peroxide according to Equation 11.20. Catalases are multisubunit enzymes of M_r 250,000 daltons that contain a heme group at their active

$$2\ H_2O_2 \rightleftharpoons O_2 + 2\ H_2O \qquad (11.20)$$

sites. The structure of catalase from bovine liver has been determined crystallographically. This protein is depicted in Figure 11.25. Coordinated to one of the axial positions is the deprotonated phenolic oxygen atom of Tyr 357, a unique feature among heme proteins. The other axial position is free to bind substrate. Situated in this distal site is the phenyl ring of Phe 160, which is parallel to and stacked on the plane of one of the porphyrin pyrrole rings, and hydrogen-bonding components His 74 and Asn 174, which are essential residues for the catalytic function of the enzyme.

The catalytic mechanism of the enzyme, established by means of kinetics investigations, is similar to that of SOD. The disproportionation of hydrogen peroxide occurs in two steps (Equations 11.21a and 11.21b) that have

$$EFe^{III} + H_2O_2 \rightleftharpoons Compound\ I + H_2O \qquad (11.21a)$$

$$Compound\ I + H_2O_2 \rightleftharpoons EFe^{III} + H_2O + O_2 \qquad (11.21b)$$

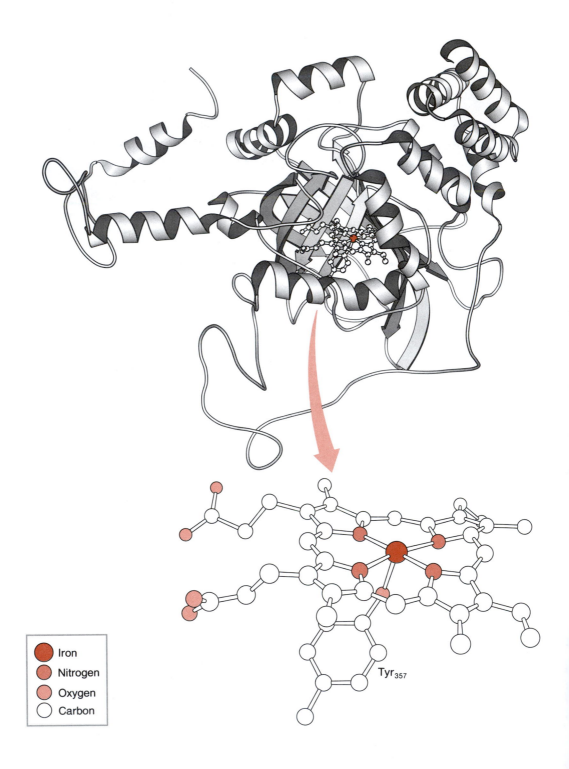

Iron
Nitrogen
Oxygen
Carbon

Tyr₃₅₇

Figure 11.25
Structure of catalase from bovine liver, highlighting its active site heme group.

obvious parallels to Equation 11.18. In the first step, the substrate is reduced to water with concomitant oxidation of the enzyme. The product of this oxidation, Compound I, has a heme redox level that is the same as that in the oxidized state of cytochrome P-450. It is best written as a PFe=O oxenoid species. Not surprisingly, Compound I from catalase can oxidize formate, nitrite, and ethanol as well as hydrogen peroxide. Oxidation of H_2O_2 to form dioxygen and water completes the catalytic cycle.

11.4.c. Dioxygen Radical-Generating Systems.

In the reactions of redox-active metal ions with dioxygen, there are several steps where reactive species containing unpaired electrons are generated. In particular, superoxide ion, O_2^-, and hydroxyl radical, $OH\cdot$, as well as the O_2 molecule itself, are all inorganic radicals. These species are all utilized in bioinorganic systems, each combining coupled electron and atom-transfer chemistry to achieve a particular function. Three examples will be discussed here to illustrate the common themes that connect these seemingly disparate topics in bioinorganic chemistry. They are the enzyme ribonucleotide reductase, the antitumor antibiotic bleomycin, and a system employed to footprint or map nucleic acid structures, namely, Fe/EDTA.

(i) *Ribonucleotide reductase (RR).* This enzyme catalyzes the conversion of ribo- to deoxyribonucleotides (Equation 11.6) in the first committed step in the biosynthesis of DNA. It has therefore been a target for antitumor and antiviral agents developed with the strategic purpose of selectively inhibiting the enzyme in diseased tissue. Herpes virus, for example, encodes an essential RR required for the progression of the disease. There are two components of RR enzymes. One is a metal-containing unit that functions to generate radicals; the other binds substrate and catalyzes the dehydroxylation of the ribose 2'-hydroxyl group by a mechanism initiated by removal of the 3'-H atom of the ring. There are two kinds of RR proteins that differ in their radical-generating unit. One class employs coenzyme B_{12} as its cofactor, utilizing chemistry discussed in more detail in Section 11.5 below. The other contains a carboxylate-bridged diiron unit similar to those found in Hr and MMO.

The structure of the iron-containing R2 subunit of RR from *E. coli* is depicted in Figure 11.15. Of particular interest is the presence of a stable tyrosyl radical situated ~5 Å away from one of the iron atoms in the dinuclear $\{Fe_2O(O_2CR)\}^{3+}$ center. This neutral radical is relatively stable in its protein environment and gives rise to a characteristic rhombic EPR spectrum. Replacement of this tyrosine with phenylalanine in a site-specic mutagenesis study inactivates the enzyme, demonstrating the importance of the tyrosyl radical in the catalytic mechanism.

The tyrosyl radical and enzymatic activity can be destroyed by addition of reagents such as hydroxylamines, which correspondingly have antiviral activity. The enzyme may be reactivated by reducing the diiron center to the Fe(II)Fe(II) form and adding dioxygen. The latter reaction is believed to occur as shown in Equation 11.22. The similarities between this reaction and the dioxygen-iron chemistry utilized by cytochrome P-450, MMO, and

$$EFe^{II}Fe^{II} + Tyr\text{-}OH + O_2 + e^- + H^+ \rightarrow EFe^{III}OFe^{III} + Tyr\text{-}O\cdot + H_2O$$
$$(11.22)$$

catalase (Equations 11.11 and 11.10 and Figure 11.12) are striking and illustrate that electron transfer coupled with dioxygen cleavage can also be used to generate a mechanistically important tyrosyl radical in a bioinorganic system. The chemistry written in Equation 11.22 is still a subject of active investigation; for example, the source of the electron shown in the equation is currently unknown.

(ii) Small molecule DNA cleaving agents: Bleomycin and iron-EDTA. Several small molecule-metal ion complexes that produce radical-generating systems capable of cleaving DNA have been discovered. One of the first of such molecules to be extensively investigated was the natural product bleomycin, the structural formula of which is shown in Figure 3.23. In the presence of iron(II) and dioxygen or iron(III) and hydrogen peroxide, bleomycin induces single- and double-strand cleavages in DNA. Despite a considerable amount of effort, the structure(s) of the iron complexes of bleomycin are still somewhat controversial, although the peptide region on the left-hand side of the molecule, rather than the dithiazole unit, appears to be the site of metal binding. Upon activation, a ferric peroxide or a high-valent species, perhaps an Fe(V)-oxo unit, that abstracts hydrogen atoms from the deoxyribose rings of DNA is formed. The subsequent chemistry that results in DNA cleavage has been extensively and elegantly investigated with the development of many tools useful for studying a wide range of radical-based reactions of DNA. A simple scheme that outlines some of the relevant reactions is shown in Figure 11.26.

The hydroxyl radical has also found great use in studying the structural properties of DNA and DNA-protein complexes. As shown in Figure 11.27, this species can be efficiently generated via the Fenton reaction from Fe(II)-EDTA in the presence of hydrogen peroxide and a reducing agent such as ascorbate. This OH· radical can diffuse and abstract hydrogen atoms from DNA, a process that again results in cleavage following subsequent chemistry. These and other reagents that facilitate DNA or RNA cleavage via hydrogen atom abstraction and radical chemistry have been and are being

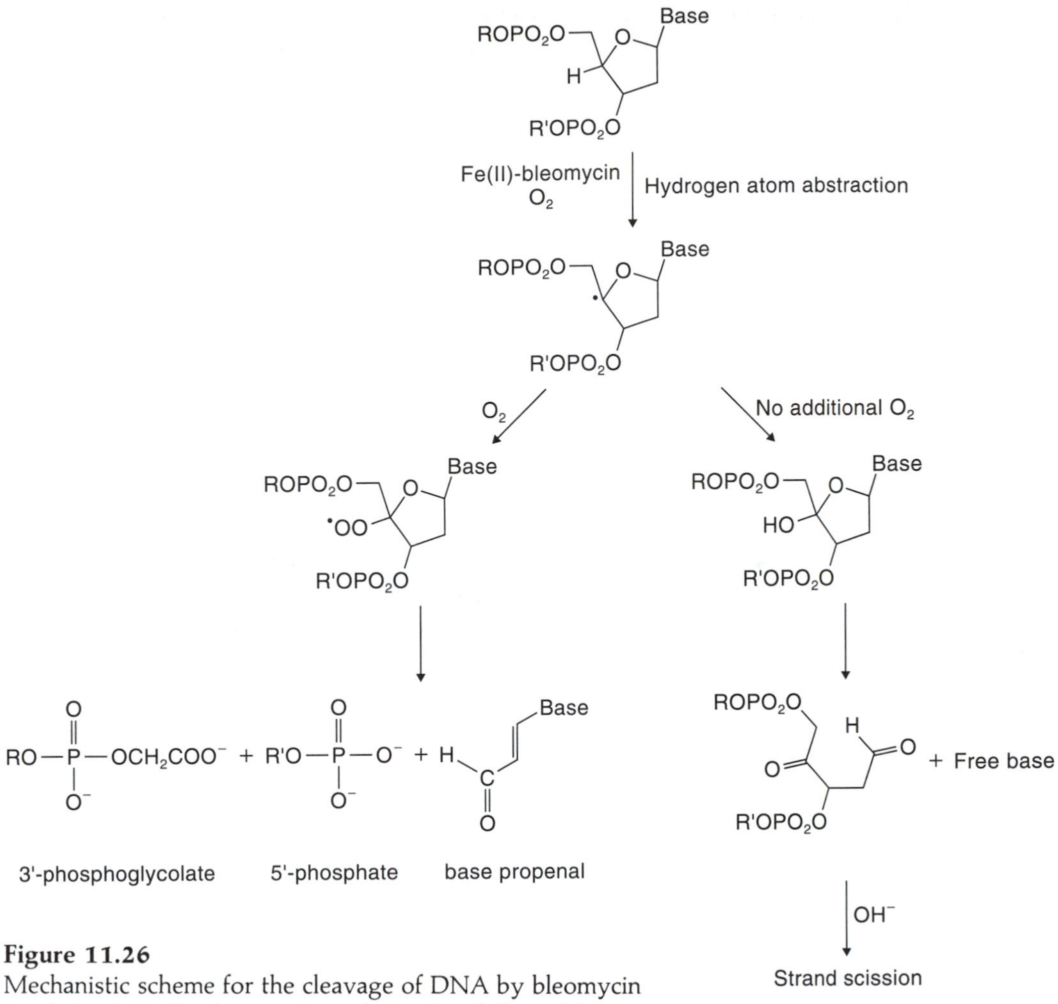

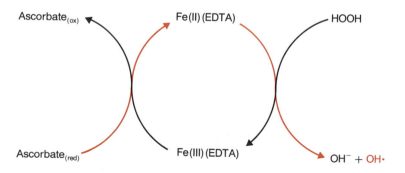

Figure 11.26
Mechanistic scheme for the cleavage of DNA by bleomycin involving initial hydrogen atom abstraction followed by additional dioxygen-dependent or dioxygen-independent steps.

Figure 11.27
Fenton reaction in which ferrous ion reacts with hydrogen peroxide to generate hydroxyl radical and hydroxide ion. The presence of a reductant such as ascorbate allows the reaction to proceed catalytically.

developed. The various systems differ in that the hydrogen-abstracting agent may be a metal complex, often bound to the DNA (as in bleomycin), or a freely diffusible species such as hydroxyl radical.

11.5. Coenzyme B-12-Dependent Reactions

Organometallic compounds are defined by the presence of metal-to-carbon bonds. Such compounds are of fundamental interest to inorganic chemists and are of tremendous and increasing importance in organic synthesis and catalysis. Known examples in biological systems are rare, however. The notable exception is vitamin B_{12}, a coenzyme required for several enzymatic transformations. Its chief functional unit is a Co—C bond that is cleaved during the course of these reactions. This landmark compound has played an important role in several fields of research and continues to be the subject of investigation. Its inclusion in this chapter derives from its use by nature to generate radicals needed in group-transfer reactions that parallel those already seen to arise from metal/dioxygen reactions. The production of radicals is afforded by homolytic cleavage of the Co—C bond. The occurrence of both B_{12}- and Fe/O_2-dependent ribonucleotide reductase enzymes nicely illustrates the natural connection between these functionally related bioinorganic systems.

11.5.a. Historical Perspective. In 1926 it was discovered that the disease pernicious anemia could be effectively treated by feeding patients large amounts of liver. The active ingredient purified from liver was a cobalt complex with the approximate empirical formula $C_{61-64}H_{84-90}N_{14}O_{13-14}PCo$. This compound was crystallized in forms suitable for X-ray diffraction analysis, but at the time (1948), no structure of this magnitude and complexity had ever been solved.

Through a heroic combination of chemical and crystallographic approaches, however, the structure was determined by 1956. The result is shown in Figure 1.6. The molecule consists of a central corrin ring with a pendent dimethylbenzimidazole unit. The corrin ring is similar to the porphyrin ring found in the heme proteins, as discussed previously (see Figures 1.3 and 1.6 and Chapter 3), but lacks one carbon atom linking the A and D rings. Furthermore, whereas the porphyrin ring contains a full complement of double bonds, the corrin system is quite reduced. All the outer carbon centers of the corrin ring are sp^3 hybridized and tetrahedral, whereas the corresponding centers in a porphyrin are sp^2 hybridized and trigonal planar. A consequence of this difference is that a porphyrin has no chiral centers, whereas the corrin ring of vitamin B_{12} has nine. At the center of the structure is an octahedrally coordinated cobalt(III) ion. The corrin ring provides four equatorial nitrogen donor ligands, and the axial ligands are contributed by the pendant benzimidazole ring and by cyanide ion. The latter was present in the derivative originally isolated and crystallized, designated cyanocobalamin. The determination of its structure by crystallographic methods demonstrated the power of this technique for complex molecules. The axial cyanide group in cyanocobalamin is an artifact of the isolation procedures. The species utilized in enzyme-catalyzed reactions is coenzyme B_{12}. In this compound, the sixth ligand is a 5'-deoxyadenosyl group, as illustrated in Figure 1.6. The structure of this compound was established in 1962. This organometallic compound has a metal-alkyl bond with a Co−C distance of 2.05 ± 0.05 Å and a Co−C−C angle of approximately 130°.

Coenzyme B_{12} is synthesized biologically via reduction of vitamin derivatives from Co(III) through Co(II) eventually to Co(I) redox levels. In this last form, the metal center is quite electron-rich and, hence, nucleophilic. It attacks the 5' carbon atom of ATP to displace the triphosphate group and form the Co−C bond. This reaction amounts to an oxidative addition at the cobalt center, oxidizing it to the Co(III) state. This chemistry illustrates the importance of the availability of multiple cobalt oxidation states for the functioning of vitamin B_{12} and has obvious parallels to the iron redox chemistry discussed previously in this chapter.

11.5.b. Enzymatic Reactions and the Catalytic Mechanism.

Some of the reactions catalyzed by coenzyme-B_{12}-dependent enzymes are shown in Figure 11.28. Whereas several coenzyme-B_{12}-promoted reactions constitute true oxidation-reduction chemistry, for example, the reduction of ribonucleoside to deoxyribonucleoside triphosphates, many are simply rearrangements that can be represented by the general form indicated in Equation 11.23. As can be seen, these reactions all involve a 1,2-carbon shift, in which the

$$\overset{\diagdown}{\underset{X}{C_1}}-\overset{\diagup}{\underset{H}{C_2}}-\quad\longrightarrow\quad\overset{\diagdown}{\underset{H}{C_1}}-\overset{\diagup}{\underset{X}{C_2}}-\tag{11.23}$$

substituent X is interchanged with a hydrogen atom on the adjacent carbon center. A mechanistic scheme that accounts for such a rearrangement

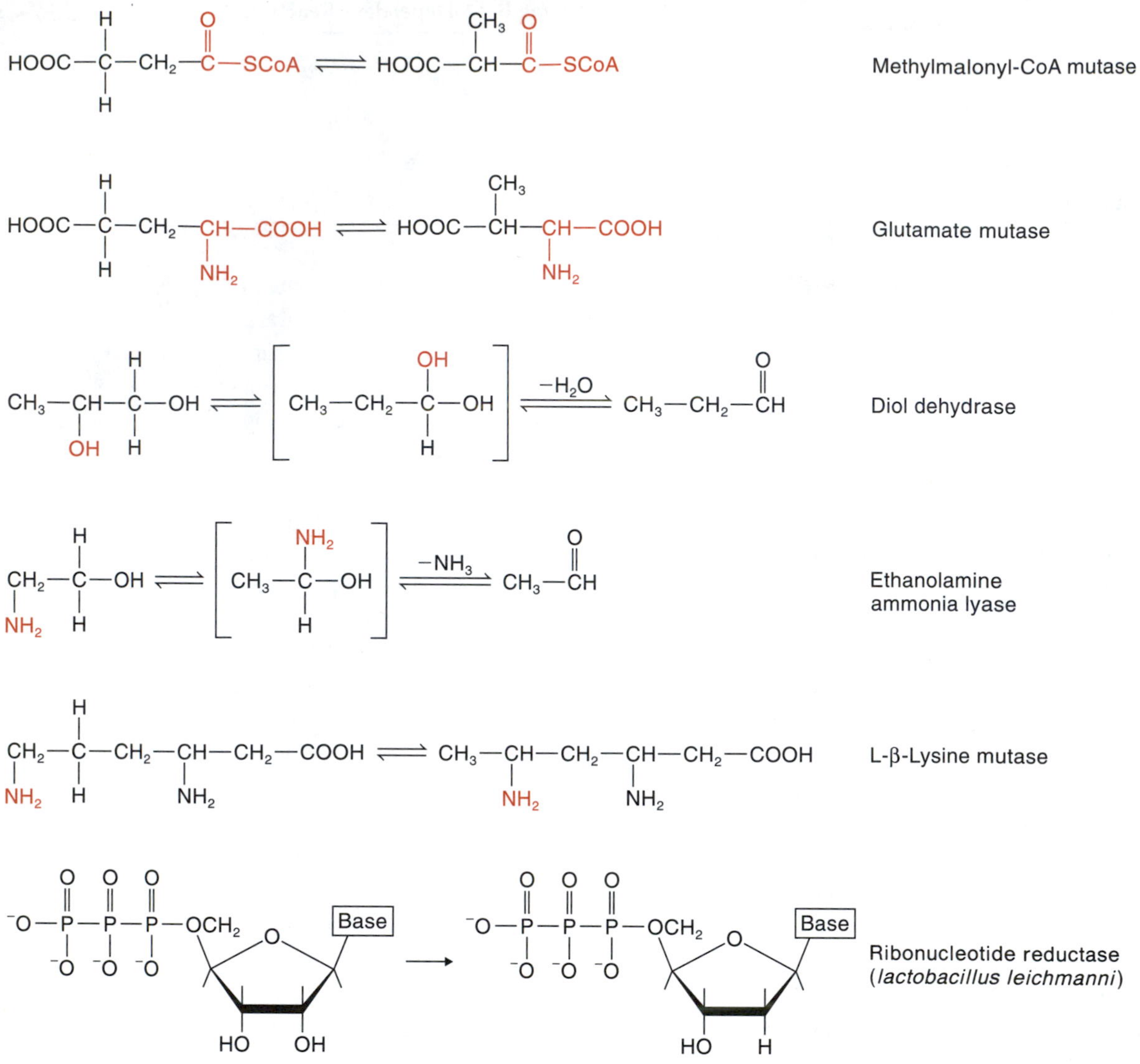

Figure 11.28
Reactions catalyzed by coenzyme B_{12}-dependent enzymes.

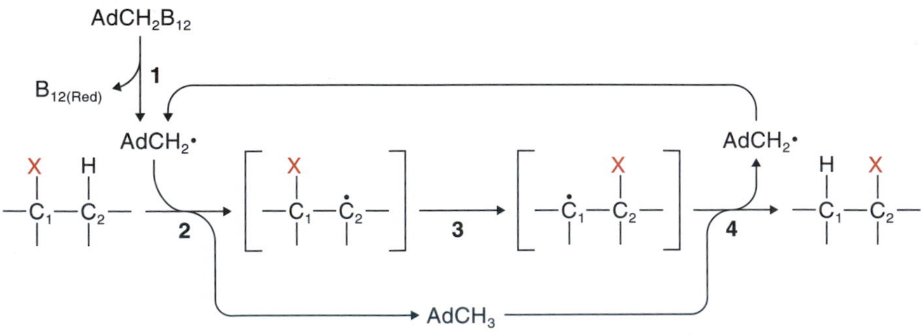

Figure 11.29
Mechanistic scheme for the 1,2-shift that occurs in many coenzyme B_{12}-activated enzymes.

is shown in Figure 11.29. The first step involves homolytic cleavage of the Co–C bond to form a Co(II) species plus a deoxyadenosyl radical. The radical then abstracts a hydrogen atom from substrate to produce a substrate radical, which then rearranges. A variety of mechanisms is possible for this step, and different ones may apply to different enzymatic reactions. The possibilities include direct rearrangement of the radical to form product; electron transfer from the radical to the cobalt center, producing a Co(I) species and a substrate carbocation, which could then rearrange; electron transfer from cobalt to the substrate radical, forming a Co(III) species and a substrate carbanion, which could then rearrange; or coupling of the substrate radical with cobalt to form additional organometallic species that mediate the rearrangement. Regardless of the intimate details of this rearrangement step, a product radical must form that abstracts hydrogen back from the deoxyadenosine moiety to afford product and regenerate the initial deoxyadenosyl radical. Product release and substrate binding allow continuation of a chain reaction without additional initiation via Co–C bond cleavage.

There is substantial evidence to support this scheme. Radical intermediates, a low-spin d^7 Co(II) complex and a carbon-based free radical, have been directly observed by EPR spectroscopy. Moreover, they are formed at rates kinetically consistent with the overall rates of the enzymatic processes. In one series of experiments, an aliquot of diol dehydrase plus five equivalents of coenzyme B_{12} were mixed with 1,2-propanediol and, within 10 milliseconds, the sample was frozen in isopentane at $-140\ °C$. The EPR spectrum of this sample (Figure 11.30) was essentially identical to that obtained by mixing enzyme, coenzyme, and substrate and allowing them to react for 15 seconds. Thus, $t_{1/2} < 3$ milliseconds for formation of the radical, at least two or three times faster than the overall rate of the enzymatically catalyzed reaction. Double integration of the spectrum revealed a total of 1.2 spins per mole of enzyme, consistent with significant conversion to the Co(II) + R· form, and accounting for up to two free spins. Further evidence for the deoxyadenosyl radical intermediate was afforded by isotopic labeling studies. Coenzyme B_{12} was synthesized with the methylene group bound to cobalt labeled nonstereospecifically with a single tritium atom. Use of this labeled coenzyme with diol dehydrase revealed transfer of all the tritium to the product propionaldehyde. This result indicates that hydrogens are transferred nonstereospecifically to product. The two methylene hydrogen atoms must therefore become equivalent during catalysis, a fact easily reconciled by the formation of a freely rotating methyl group on deoxyadenosine after cleavage of the Co-methylene bond, and hydrogen-atom abstraction from substrate by the deoxyadenosyl radical.

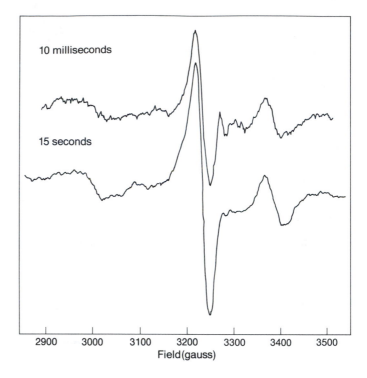

Figure 11.30
Electron paramagnetic resonance signals formed when diol dehydrase and five equivalents of coenzyme B_{12} were incubated with 1,2-propanediol, showing the generation and persistence of the deoxyadenosyl radical. (Adapted from J. E. Valinsky et al., *Journal of the American Chemical Society* **96**, 4709–4710 (1974)).

11.5.c. Thermodynamic Considerations and the Role of the Enzyme.
The data discussed above indicate that coenzyme B_{12} functions as a free-radical source. The Co–C bond is sufficiently labile that homolytic cleavage yielding the deoxyadenosyl radical and the Co(II) corrin complex apparently occurs on the enzymes at rates consistent with the overall rates required for catalysis. This observation has prompted studies of the strengths of Co–C bond energies and of the factors that determine these energies. The dissociation strengths were determined by kinetic methods in the following manner. Consider the scheme given in Equation 11.24. With the use of the cobalt

$$\text{Ado-B12(Co}^{\text{III}}) \underset{k_{-1}}{\overset{k_1}{\rightleftharpoons} } \text{Ado} \cdot + \text{B12(Co}^{\text{II}})$$
$$\text{Ado} \cdot + \text{trap} \xrightarrow[k_2]{\text{fast}} \text{Ado-trap}$$

(11.24)

complex Co(DH)_2 as a trapping agent, where DH represents the dimethyl-glyoximato ligand, the reaction proceeded at reasonable rates near 100 °C in water. From the temperature dependence of k_1, the activation parameters for deoxyadenosyl radical formation were determined to be $\Delta H^{\ddagger} = 28.6$ kcal/mol and $\Delta S^{\ddagger} = 2$ cal/mol-deg. Given that the recombination reactions, corresponding to k_{-1} and k_2, appear to be diffusion-controlled, and correcting $\Delta H^{\ddagger}$ accordingly by ~ 2 kcal/mol, this result was used to estimate a Co−C bond dissociation energy of approximately 26 kcal/mol. A slightly higher value of 31.5 kcal/mol was obtained by using a stable nitroxide molecule as a radical trap in ethylene glycol solution.

Two points about this bond-dissociation energy are important. First, it is relatively low. Carbon-carbon single-bond-dissociation energies are typically 80 to 90 kcal/mol. Second, it is still much too high to be useful in the catalytic reaction without other influences, such as additional Co−C bond weakening afforded by interactions of the coenzyme with the enzyme. The values of $\Delta H^{\ddagger}$ and $\Delta S^{\ddagger}$ determined experimentally can be used to estimate values for $\Delta G^{\ddagger}$ at 37 °C. The calculated energies of 28−30 kcal/mol correspond to rate constants $k_1 \sim 10^{-7} - 10^{-8}$ sec^{-1}. Enzymes such as diol dehydrase turn over with rates near 10^2 sec^{-1}. Thus, enzymes that bind coenzyme B_{12} must further lower the bond-dissociation energy by approximately a factor of two in order to generate a free-radical source that is capable of operating at useful rates.

At least two mechanisms for decreasing this dissociation energy seem plausible and have been tested to some extent by using model systems. The first involves electronic influences by the axial ligand *trans* to the bond that generates the radical. Since bond dissociation is accompanied by reduction of the cobalt center from Co(III) to Co(II), it is expected that decreasing the electron-donating ability (basicity) of the axial ligand should stabilize the Co(II) form and, hence, decrease the bond-dissociation energy. This concept has been tested with the use of a series of isosteric pyridine bases and the bis(dimethylglyoximato)cobalt model system discussed above. A linear correlation between the pK_a of the pyridine ligand and the kinetically determined bond-dissociation energy was observed.

A second factor that may influence the strength of the cobalt-carbon bond involves steric interactions between the corrin ring and the deoxyadenosyl unit. As noted in the initial structure determination of coenzyme B_{12}, the Co−C−C angle is significantly larger than 109.5°. Examination of the structure of the coenzyme reveals several relatively close contacts between the deoxyadenosyl group and the corrin ring and its substituents. This result suggests a mechanism by which an enzyme could weaken the Co−C bond. By deforming the corrin ring appropriately, steric interactions could be introduced between the ring substituents and the deoxyadenosyl

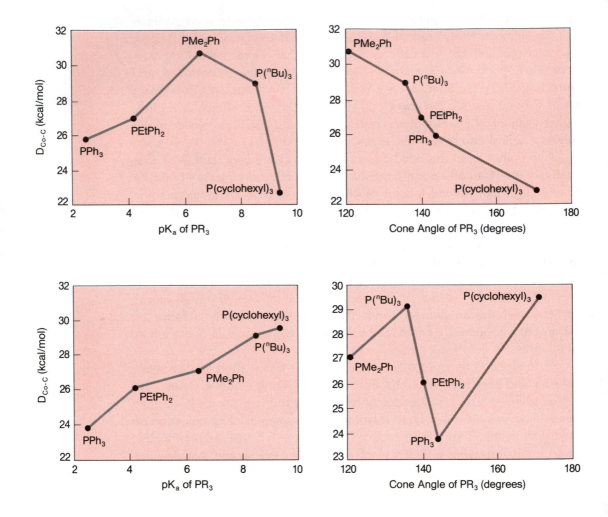

Figure 11.31
Dependence of Co–C bond dissociation energy on pK_a values (left) and cone angles (right) of various phosphine ligands. The top two panels are for dimethylglyoxime cobalt benzyl complexes and the bottom two are for the corresponding cobalt octaethylporphyrin benzyl derivatives. (Adapted from M. K. Geno and J. Halpern, *Journal of the American Chemical Society* **109**, 1238–1240 (1987)).

group that would lead to strain in the coenzyme that could be relieved by Co−C bond cleavage. The consequences of such steric interactions have also been probed by using the bis(dimethylglyoximato)cobalt model system with various tertiary phosphines as axial ligands. In addition, comparative studies have been made with cobalt octaethylporphyrin derivatives with interesting results. As shown in Figure 11.31 for the dimethylglyoxime complexes, the bond-dissociation energies correlate well with the steric properties of the phosphine ligands, as measured by the cone angle, and not with their basicity. Thus, a large but highly basic phosphine such as tricyclohexylphosphine yields a complex with a quite low bond-dissociation energy. In contrast, for the corresponding octaethylporphyrin derivatives, the bond-dissociation energies correlate well with phosphine basicity and not with steric parameters (Figure 11.31).

These results can be interpreted in terms of the differing flexibility of the two systems. The bis(dimethylglyoxime) complexes are more flexible than the porphyrin complexes such that binding of a sterically demanding axial ligand can be transmitted to the alkyl group coordinated on the other side of the equatorial plane. The corrin system is more saturated and, hence, more flexible than a corresponding porphyrin system. This explanation suggests a reason why nature has selected this corrin macrocycle for use in generating and storing carbon free-radical equivalents.

Finally, we return to the rearrangement processes themselves. As with many reactions, the intimate details of coenzyme-B_{12}-dependent transformations have been quite difficult to unravel completely. Stereochemical studies with various enzymes have revealed a wide range of behaviors for the rearrangements. Consequently, model systems have been required to test the chemical and kinetic feasibility of some of the proposed mechanisms. For example, analogs of some of the proposed organometallic intermediates that would be formed via coupling of the initially formed radical with the Co(II) center for diol dehydrase have been prepared and studied. These complexes did not undergo the chemistry that would be expected if they were, in fact, accurate representations of intermediates in the enzyme-catalyzed processes. In contrast, analysis of the chemistry of specifically generated free-radical species corresponding to possible intermediates in the methylmalonyl-CoA mutase reaction revealed that they did undergo appropriate rearrangements at nearly the required rates. Future chemical and enzymatic investigations will be required to solidify and extend these studies, but at this point, it seems clear that the fundamental function of cobalt in coenzyme-B_{12}-utilizing enzymes is to provide a site for free-radical storage and release, and not to serve as the catalytic center for substrate rearrangement.

Study Problems

1. As indicated in this chapter, a number of Co(II) complexes bind dioxygen. For example, $[Co(CN)_5]^{3-}$ reacts with O_2 to produce a binuclear complex of the form $[(CN)_5Co-O_2-Co(CN)_5]^{6-}$. Although the starting material is paramagnetic, the product is diamagnetic. Propose a formulation for the electronic structure (oxidation states) in the product. How would you test your hypothesis?

2. The cytochromes P-450 are regarded as oxidizing enzymes; yet they consume one equivalent of the reducing agent NADH for each catalytic cycle. Why is NADH required? What atoms are formally reduced in the overall reaction?

3. Is it possible to produce derivatives of superoxide dismutase that have either copper in all the metal sites (that is, copper replacing zinc) or zinc in all the metal sites (zinc replacing copper)? Would you expect either of these derivatives to be active? What would you expect for the magnetic properties of the two derivatives?

4. Some molybdenum-containing enzymes are capable of reducing sulfoxides (R_2SO) to the corresponding sulfides (R_2S). Attempts to use them to convert the corresponding sulfones (R_2SO_2) to either the sulfoxide or the sulfide have been unsuccessful. Propose an explanation for this observation.

Bibliography

Dioxygen Transport

J. P. Collman, T. R. Halpern, and K. S. Suslick. 1980. "O_2 Binding to Heme Proteins and Their Synthetic Analogs," in T. G. Spiro, ed., *Metal-Ion Activation of Dioxygen*. Wiley-Interscience, New York, 1–72.

R. E. Dickerson and I. Geis. 1983. *Hemoglobin: Structure, Function, Evolution, and Pathology*. Benjamin/Cummings, Menlo Park, CA.

H. D. Ellerton, N. F. Ellerton, and H. A. Robinson 1983. "Hemocyanin—A Current Perspective." *Prog. Biophys. Molec. Biol.* 41, 143–248.

N. Kitajima, K. Fujisawa, C. Fujimoto, Y. Moro-oka, S. Hashimoto, T. Kitagawa, K. Toriumi, K. Tatsumi, and A. Nakamura. 1992. "A New Model for Dioxygen Binding in Hemocyanin: Synthesis, Characterization, and Molecular Structure of μ-η^2,η^2 Peroxo Dinuclear Copper(II) Complexes, $[Cu(HB(3,5-R_2pz)_3]_2(O_2)$ (R = i-Pr and Ph)." *J. Am. Chem. Soc.* 114, 1277–1291.

D. M. Kurtz, Jr. 1990. "Oxo- and Hydroxo-Bridged Diiron Complexes: A Chemical Perspective on a Biological Unit." *Chem. Rev.* 90, 585–606.

S. J. Lippard. 1988. "Oxo-Bridged Polyiron Centers in Biology and Chemistry." *Angew. Chem., Int. Ed. Engl.* 27, 344–361.

M. F. Perutz, G. Fermi, B. Luisi, B. Shaanon, and R. C. Liddington. 1987. "Stereochemistry of Cooperative Mechanisms in Hemoglobin." *Acc. Chem. Res.* 20, 309–321.

L. Que, Jr., and A. E. True. 1990. "Dinuclear Iron- and Manganese-Oxo Sites in Biology," in S. J. Lippard, ed., *Progress in Inorganic Chemistry*, Volume 38. Wiley-Interscience, New York, 97–200.

E. I. Solomon, M. J. Baldwin, and M. D. Lowery. 1992. "Electronic Structures of Active Sites in Copper Proteins: Contributions to Reactivity." *Chem. Rev.* 92, 521–542.

Z. Tyeklár and K. D. Karlin. 1989. "Copper-dioxygen Chemistry: A Bioinorganic Challenge." *Acc. Chem. Res.* 22, 241–248.

P. C. Wilkins and R. G. Wilkins. 1987. "The Coordination Chemistry of the Binuclear Iron Site in Hemerythrin." *Coord. Chem. Rev.* 79, 195–214.

Oxygen Atom Transfer Reactions—Fe

L. A. Andersson and J. H. Dawson. 1991. "EXAFS Spectroscopy of Heme-Containing Oxygenases and Peroxidases." *Struct. and Bond.* 74, 1–40.

M. J. Coon and R. E. White. 1980. "Cytochrome P-450, a Versatile Catalyst in Monooxygenation Reactions," in T. G. Spiro, ed., *Metal Ion Activation of Dioxygen*. Wiley-Interscience, New York, 73–123.

J. H. Dawson. 1988. "Probing Structure-Function Relations in Heme Containing Oxygenases and Peroxidases." *Science* 240, 433–439.

P. G. Debrunner, I. C. Gunsalus, S. G. Sligar, and G. C. Wagner. 1978. "Monooxygenase Hemoproteins: Cytochromes P-450," in H. Sigel ed., *Iron in Model and Natural Compounds. Metal Ions in Biological Systems*, Volume 7. Marcel Dekker, New York, 242–277.

J. T. Groves. 1980. "Mechanisms of Metal-Catalyzed Oxygen Insertion," in T. G. Spiro, ed., *Metal Ion Activation of Dioxygen*. Wiley-Interscience, New York, 125–162.

F. P. Guengerich. 1991. "Reactions and Significance of Cytochrome P-450 Enzymes." *J. Biol. Chem.* 266, 10019–10022.

F. P. Guengerich and T. L. Macdonald. 1990. "Mechanisms of Cytochrome P-450 Catalysis." *FASEB J.* 4, 2453–2459.

H. G. Jang, D. D. Cox, and L. Que, Jr. 1991. "A Highly Reactive Functional Model for the Catechol Dioxygenases: Structure and Properties of $[Fe(TPA)DBC]BPh_4$." *J. Am. Chem. Soc.* 113, 9200–9204.

M. Nozaki and O. Hayashi. 1984. "Dioxygenases and Monooxygenases," in J. V. Bannister and W. H. Bannister, eds., *The Biology and Chemistry of Active Oxygen*. Elsevier, New York, 68–104.

P. R. Ortiz de Montellano. 1986. *Cytochrome P-450: Structure, Mechanism, and Biochemistry*. Plenum Press, New York.

T. D. Porter and M. J. Coon. 1991. "Cytochrome P-450: Multiplicity of Isoforms, Substrates, and Catalytic and Regulatory Mechanisms." *J. Biol. Chem.* 266, 13469–13472.

A. C. Rosenzweig, C. A. Frederick, S. J. Lippard, and P. Nordlund. 1993. "Crystal Structure of a Bacterial Non-haem Iron Hydroxylase that Catalyzes the Biological Oxidation of Methane." *Nature* 366, 537–543.

Oxygen Atom Transfer Reactions—Mo

M. W. W. Adams and L. E. Mortenson. 1985. "Mo Reductase: Nitrate Reductase and Formate Dehydrogenase," in T. G. Spiro, ed., *Molybdenum Enzymes*. Wiley-Interscience, New York, 519–593.

M. J. Barber, M. P. Coughlan, K. V. Rajagopalan, and L. M. Siegel. 1982. "Properties of the Prosthetic Groups of Rabbit Liver Aldehyde Oxidase: A Comparison of Molybdenum Hydroxylase Enzymes." *Biochemistry* 21, 3561–3568.

R. C. Bray. 1988. "The Inorganic Biochemistry of Molybdoenzymes." *Quart. Rev. Biophys.* 21, 299–329.

S. J. N. Burgmayer and E. I. Stiefel. 1985. "Molybdenum Enzymes, Cofactors, and Model Systems, The Chemical Uniqueness of Molybdenum." *J. Chem. Ed.* 62, 943–953.

S. P. Cramer and E. I. Stiefel. 1985. "Chemistry and Biology of the Molybdenum Cofactor," in T. G. Spiro, ed., *Molybdenum Enzymes*. Wiley-Interscience, New York, 411–441.

S. F. Gheller, B. E. Schultz, M. J. Scott, and R. H. Holm. 1992. "A Broad-Substrate Analogue Reaction System of the Molybdenum Oxotransferases," *J. Am. Chem. Soc.* 114, 6934–6935.

E. W. Harlan, J. M. Berg, and R. H. Holm. 1986. "Thermodynamic Fitness of Molybdenum(IV,VI) Complexes for Oxygen Atom Transfer Reactions, Including Those with Enzymatic Substrates." *J. Am. Chem. Soc.* 108, 6992–7000.

T. R. Hawkes and R. C. Bray. 1984. "Quantitative Transfer of the Molybdenum Cofactor from Xanthine Oxidase and from Sulphite Oxidase to the Deficient Enzyme of the *Nit-1* Mutant of *Neurospora crassa* to Yield Active Nitrate Reductase." *Biochem. J.* 219, 481–493.

S. M. Hinton and D. Dean. 1990. "Biogenesis of Molybdenum Cofactors." *CRC Crit. Rev. Microbiol.* 17, 169–188.

R. H. Holm and J. M. Berg. 1986. "Toward Functional Models of Metalloenzyme Active Sites: Analogue Reaction Systems of the Molybdenum Oxo Transferases." *Acc. Chem. Rev.* 19, 363–370.

B. E. Schultz, S. F. Gheller, M. C. Muetterties, M. J. Scott, and R. H. Holm. 1993. "Molybdenum-Mediated Oxygen Atom Transfer: An Improved Analogue Reaction System of the Molybdenum Oxotransferases." *J. Am. Chem. Soc.* 115, 2714–2722.

Other Reactions of Dioxygen and Its Byproducts

J. M. Bollinger, Jr., D. E. Edmondson, B. H. Hyunh, J. Filley, J. R. Norton, and J. Stubbe. 1991. "Mechanism of Assembly of the Tyrosyl Radical-Dinuclear Iron Cluster Cofactor of Ribonucleotide Reductase." *Science* 253, 292–298.

G. Cohen and R. A. Greenwald. 1983. *Oxy Radicals and Their Scavenger Systems*, Volumes 1 and 2. Elsevier Biomedical, New York.

J. A. Fee. 1980. "Superoxide, Superoxide Dismutases and Oxygen Toxicity," in T. G. Spiro, ed., *Metal Ion Activation of Oxygen*. Wiley-Interscience, New York, 209–237.

I. Fridovich. 1989. "Superoxide Dismutases." *J. Biol. Chem.* 264, 7761–7764.

E. D. Getzoff, J. A. Tainer, P. K. Weiner, P. A. Kollman, J. S. Richardson, and D. C. Richardson. 1983. "Electrostatic Recognition Between Superoxide and Copper, Zinc Superoxide Dismutase." *Nature* 306, 287–290.

S. M. Hecht. 1986. "The Chemistry of Activated Bleomycin." *Acc. Chem. Res.* 19, 383–391.

A. M. Michelson, J. M. McCord, and I. Fridovich. 1977. *Superoxide and Superoxide Dismutases.* Academic Press, London.

A. M. Pyle and J. K. Barton. 1990. "Probing Nucleic Acids with Transition Metal Complexes," in S. J. Lippard, ed., *Progress in Inorganic Chemistry*, Volume 38. Wiley-Interscience, New York, 413–475.

T. J. Reid III, M. R. N. Murthy, A. Sicignano, N. Tanaka, W. D. L. Musick, and M. G. Rossman. 1981. "Structure and Heme Environment of Beef Liver Catalase at 2.5 Å Resolution." *Proc. Natl. Acad. Sci. USA* 78, 4767–4771.

G. Rotilio, ed. 1986. *Superoxide and Superoxide Dismutase in Chemistry, Biology, and Medicine.* Elsevier, Amsterdam.

D. S. Sigman. 1986. "Nuclease Activity of 1,10 Phenanthroline—Copper Ion." *Acc. Chem. Res.* 19, 180–186.

J. Stubbe. 1990. "Ribonucleotide Reductases: Amazing and Confusing." *J. Biol. Chem.* 265, 5329–5332.

J. Stubbe and J. W. Kozarich. 1987. "Mechanisms of Bleomycin-Induced DNA Degradation." *Chem. Rev.* 87, 1107–1136.

J. A. Tainer, E. D. Getzoff, J. S. Richardson, and D. C. Richardson. 1983. "Structure and Mechanism of Copper, Zinc Superoxide Dismutase." *Nature* 306, 284–287.

T. D. Tullius. 1987. "Chemical 'Snapshots' of DNA: Using the Hydroxyl Radical to Study the Structure of DNA and DNA-Protein Complexes." *TIBS* 12, 297–300.

T. D. Tullius, ed. 1989. *Metal-DNA Chemistry.* ACS Symposium Series, Number 402. American Chemical Society, Washington, DC.

J. S. Valentine and M. W. Pantoliano. 1981. "Protein-Metal Ion Interactions in Cuprozinc Protein (Superoxide Dismutase)," in T. G. Spiro, ed., *Copper Proteins.* Wiley-Interscience, New York, 291–358.

Vitamin B_{12}-Dependent Reactions

G. Choi, S-C. Choi, A. Galan, B. Wilk, and P. Dowd. 1990. "Vitamin B_{12s}-Promoted Model Rearrangement of Methylmalonate to Succinate Is Not a Free Radical Reaction." *Proc. Natl. Acad. Sci. USA* 87, 3714–3716.

D. Dolphin. 1982. B_{12}, Volume 1. Wiley, New York.

R. G. Finke and B. P. Hay. 1984. "Thermolysis of Adenosylcobalamin: A Product, Kinetic, and Co-C5′ Bond Dissociation Energy Study." *Inorg. Chem.* 23, 3041–3043.

R. G. Finke and D. A. Schiraldi. 1983. "Model Studies of Coenzyme B_{12} Dependent Diol Dehydrarase, 2: A Kinetic and Mechanistic Study Focusing Upon the Cobalt Participation or Nonparticipation Question." *J. Am. Chem. Soc.* 105, 7605–7617.

M. K. Geno and J. Halpern. 1987. "Why Does Nature Not Use the Porphyrin Ligand in Vitamin B_{12}?" *J. Am. Chem. Soc.* 109, 1238–1240.

J. Halpern. 1985. "Mechanisms of Coenzyme B_{12}-Dependent Rearrangements." *Science* 227, 869–875.

D. C. Hodgkin, J. Kamper, M. Mackay, J. Pickworth, K. N. Trueblood, and J. G. White. 1956. "Structure of Vitamin B_{12}." *Nature* 178, 64–66.

J. M. Pratt. 1972. *Inorganic Chemistry of Vitamin B_{12}.* Academic Press, London.

CHAPTER 12

Protein Tuning of Metal Properties to Achieve Specific Functions

---

Principles: Similar metal-based prosthetic groups can perform very different functions, depending on the protein environment. Changes in the number of coordination sites at the metal center can be used to control whether electron transfer, substrate activation, or ligand binding occurs. Redox potentials can be tuned by the choice of donor atoms from amino-acid side chains, by distortions of the coordination geometry, as well as by more indirect mechanisms, such as modulating the hydrophobicity, hydrogen bonding, and local dielectric constant of the active site. Control of electron transfer can be achieved by substrate binding without coordination through modulation of metal ligation that accompanies conformational changes at the active site. Substrate specificity is dictated by the positioning of amino-acid side chains to construct a pocket for specific delivery of a coordinated reactant. Proteins containing more than one prosthetic group can couple multiple processes to accelerate specific reactions or facilitate complex reactions that would not otherwise readily occur.

12.1. The Basic Concept

Our discussions thus far have demonstrated that metal ions carry out a variety of biological functions. In this chapter we illustrate the important concept that the same metallic unit can perform very different functions, depending on its chemical context in a biological environment. The tuning of the metal core of a metalloprotein is one of the most fascinating topics in bioinorganic chemistry. In electron-transfer reactions, the redox potential for a given change in metal-oxidation level can be modified both by the particular ligand donor set supplied by the protein and by the noncoordinated, surrounding amino acids that define the electrostatic properties of the local medium. By incorporating more than one metal prosthetic group, a protein can couple multiple processes to accelerate specific reactions or facilitate complex chemistry that would otherwise be difficult to accomplish. For reactions involving atom or group transfer, the active site of a metalloenzyme can be exquisitely sculptured by the folding of nearby protein chains to afford shape selectivity, hydrophobicity, hydrophilicity, and transition-state geometries that lead to substantial rate increases. The properties of the metal ion occupying the coordination site are also important, as has been explored by addition of different metal ions to the apoprotein. These features are evident to the biomimetic chemist trying to duplicate the reaction rates and specificities of simple enzymatic processes such as ester or peptide hydrolysis.

A related principle is that substrate binding can itself lead to activation of a metal ion required for catalysis, for example, by changing its redox potential. This phenomenon allows for maximal control of the timing of a sequence of chemical steps in the overall transformation; in the example cited, the control of redox potential might prevent premature electron transfer. Moreover, redox changes at a metal site are often accompanied by geometric rearrangements that induce subtle structural movements in the protein chain supplying the coordinating amino acids. These changes can, in turn, modulate other functional properties of the protein. In a multisubunit protein such as hemoglobin, the geometric and electronic changes that occur upon dioxygen binding to the heme group affect the O_2 binding in other subunits. This behavior is one of the most elegant bioinorganic control mechanisms devised by nature. For some of the more difficult transformations in biology, such as the six-electron reduction of dinitrogen to ammonia or the four-electron oxidative coupling of water to form dioxygen, nature has assembled an array of metallobiomolecules that work in concert. These proteins are often positioned strategically with respect to one another in a membrane or as part of a multiprotein complex, where tuning might be initiated via interactions at the protein-protein interfaces of the resulting macromolecular species that are transmitted through peptide conformational changes to the active metal core.

The idea that proteins might tune the properties of bound metal centers is not new to bioinorganic chemistry. Given known differences in the spectroscopic and other properties of metallobiomolecules in comparison with those of known small molecules, it was proposed in the late 1960s that metalloenzymes might represent *entatic states*, states that are "closer to that of a unimolecular transition state than to that of a conventional, stable molecule." This idea has stimulated much discussion and research. Subsequent experimentation has revealed that many systems having unusual spectroscopic features contain metal centers that do not appear to be strained or poised, but were simply not represented in the known repertoire of small molecules at the time. Examples may be found among blue copper or ferredoxin proteins. The synthesis by bioinorganic chemists of models that accurately mimic the properties of such metalloprotein cores demonstrates that relatively simple ligands can be used to reproduce their spectra and, in favorable circumstances, reactivity. Thus, although some appear to meet the criteria for the entatic state as indicated above, many seem to involve selection of a metal center that is well suited to carry out the function at hand. When more than one function is possible, tuning by the protein surroundings provides the requisite control. In this chapter we focus on such systems.

12.2. Control of Function by Protein Side Chains: Opening Coordination Sites

One of the most fundamental aspects of a bound metal ion or metal-containing cofactor that a protein can influence is the number of open coordination sites. Control of this parameter can be achieved by limiting the number of potential ligation sites that are not occupied by protein-derived ligands. In this manner, a bioinorganic chip such as a metal porphyrin or iron-sulfur cluster can be used in one of two fundamentally different ways, either as a simple outer-sphere electron carrier or as a site for substrate binding and, potentially, activation. Several examples of this phenomenon are described in this section.

12.2.a. Heme Proteins. In Chapter 9 we discussed cytochrome c, in which the protein supplies two axial ligands, a histidine and a methionine, to complete octahedral coordination in both the reduced and oxidized states of an iron porphyrin. Cytochrome c appears to function as a very effective one-electron carrier without any ligand-binding or catalytic properties. A similar situation holds for mammalian cytochrome b_5, a protein that also contains heme coordinated by two axial ligands, both histidines, in both the oxidized and reduced states. In an early application of site-directed mutagenesis to the study of heme proteins, one of these histidines (His-39) was converted to methionine with the goal of producing a protein with an iron site similar to that of cytochrome c. Spectroscopic and reactivity studies of the resulting mutant protein revealed that the methionine group either does not coordinate to the iron or is easily displaced by other ligands. In the ferrous state the mutated protein forms a stable complex with CO, with histidine serving as the sixth ligand. In the ferric state the protein catalyzes the hydrogen peroxide-dependent demethylation of dimethylaniline, a reaction not effected at all by the native protein. These observations provide direct evidence of the profound effects that creating an available coordination site can have on a metalloporphyrin unit. For use in a protein that serves only as a redox partner, the heme should have no open coordination site, because structural changes that accompany changes in redox state increase the reorganizational energy at the site and hence inhibit or at least retard the rate of electron transfer. In contrast, for a protein involved in ligand binding or catalysis, an open coordination site is an absolute requirement.

12.2.b. Iron-Sulfur Proteins: Aconitase. The use of iron-sulfur clusters in electron-transfer proteins was also extensively discussed in Chapter 9. Both Fe_2S_2 and Fe_4S_4 units can undergo rapid, reversible redox reactions that have wide application in biology. Recently, a class of enzymes has emerged in which iron-sulfur clusters are utilized as catalysts for nonredox reactions, specifically, to perform hydratase or dehydratase functions. The best-characterized enzyme in this class is aconitase, which catalyzes the stereospecific interconversion of citrate and isocitrate via the intermediate *cis*-aconitate. This chemistry, which represents an important step in the Krebs cycle, is illustrated in Equation 12.1. A key to controlling the reactivity of the Fe_4S_4

$$
HO-\underset{\substack{| \\ COO^- \\ | \\ COO^-}}{\overset{\substack{COO^- \\ | }}{|}}\quad\underset{-H_2O}{\overset{-H_2O}{\rightleftharpoons}}\quad \underset{\substack{| \\ COO^- \\ | \\ COO^-}}{\overset{\substack{COO^- \\ |}}{|}}\quad\underset{-H_2O}{\overset{H_2O}{\rightleftharpoons}}\quad HO-\underset{\substack{| \\ COO^- \\ | \\ COO^-}}{\overset{\substack{COO^- \\ |}}{|}}\qquad (12.1)
$$

   citrate   *cis*-aconitate   *iso*-citrate

center, which can serve alternatively to effect electron transfer or hydratase/dehydratase chemistry, is the availability of a ligand-binding site at one corner of the iron-sulfur cube.

Recent X-ray crystal structure determinations of aconitase with bound isocitrate and nitroisocitrate, an inhibitor, have provided important insights into key features of the catalytic mechanism that illustrate this principle. Unlike the Fe_4S_4 ferredoxins, only three of the four iron atoms in the cube are coordinated by a cysteine thiolate ligand (Figure 12.1). In the resting state of the enzyme, the remaining iron atom contains a bound hydroxide ion that serves as a hydrogen-bond acceptor from a nearby histidine residue (His-101) and a donor to Asp-165. Like the other iron atoms in the cluster, this center has a distorted tetrahedral, four-coordinate geometry. Upon binding the substrate, the hydroxide ion becomes protonated to form a water ligand, and, importantly, the iron expands its coordination number to six through the formation of a chelate ring with two of the oxygen atoms of the substrate. The system is thus set up to be dehydrated through attack by Ser-642 alk-

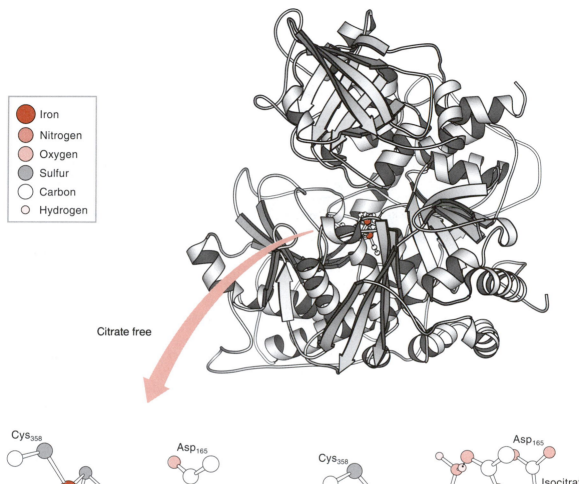

Iron
Nitrogen
Oxygen
Sulfur
Carbon
Hydrogen

Citrate free

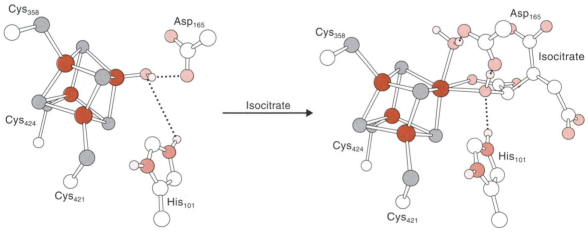

Cys₃₅₈

Asp₁₆₅

Cys₄₂₄

Cys₄₂₁

His₁₀₁

Isocitrate

Cys₃₅₈

Asp₁₆₅

Isocitrate

Cys₄₂₄

His₁₀₁

Cys₄₂₁

Figure 12.1
Crystal structure of aconitase showing its iron-sulfur cluster in the substrate-free
form and with bound isocitrate.

oxide on the activated isocitrate $C\beta$-H bond with subsequent cleavage of the $C\alpha$-OH bond. Additional details of the postulated mechanism are depicted in Figure 12.2.

The catalytically active iron in aconitase undergoes no redox changes during the enzyme cycle and, in many respects, functions in a manner entirely analogous to that described in Chapter 10 for many zinc enzymes. Water or hydroxide ion is bound to the metal in the absence of substrate. There is extensive use of hydrogen-bonding units, here three histidine and one aspartic acid residues, to form the active site and to bind and orient the substrate. Coordination of the hydroxyl and carboxylate oxygen atoms to iron(III) at the cube corner activates the substrate for proton abstraction by a suitably positioned basic residue on the protein, leading ultimately to the evolution of the product molecule. One difference between the chemistry at the metal centers is that, in aconitase, the iron cycles in coordination number between four in the resting enzyme and six in the presence of bound substrate, whereas zinc usually remains four- or five-coordinate. Perhaps this greater flexibility is one reason why nature has used a corner of the iron-sulfur cube in the aconitase system.

Finally, recall (Section 6.2.b) that the iron-responsive element-binding protein (IRE-BP) turned out to be a cytosolic aconitase. The ability of iron to bind reversibly to the apo form of the protein and serve as part of a cellular mechanism to regulate the levels of iron further illustrates the control of protein function by choice of protein side chain. If all of the labile iron atoms were coordinated to cysteine thiolate ligands, it is unlikely that either the IRE-BP or aconitase functions could be carried out.

12.2.c. Zinc Proteins. As discussed in previous chapters, zinc can play important structural as well as catalytic roles in proteins. One major feature that distinguishes these two types of functions is that an open coordination position is always present at a catalytic zinc site but absent at a purely structural site. Apart from this distinction, there are other substantial differences among the types of ligands within each class. For example, in Chapter 7, we discussed structural zinc sites within nucleic-acid-binding proteins having compositions $[Zn(Cys)_4]$, $[Zn(Cys)_3(His)]$, $[Zn(Cys)_2(His)_2]$, and $[Zn_2(Cys)_6]$. In Chapter 10 we encountered enzymatic sites that had $[Zn(His)_2(Glu)(OH_2)]$, $[Zn(His)_3(OH_2)]$, and $[Zn(Cys)_2(His)(OH_2)]$ compositions. For the structural sites, the tetrahedral coordination spheres of each metal are saturated with protein-derived ligands. The presence of two or more cysteinate ligands strongly reduces the likelihood that the coordination sphere might be expanded beyond four, thus precluding the potential use of the centers for substrate binding or activation. In contrast, for catalytic zinc, one of the coordination sites in the resting enzyme is occupied by a water or

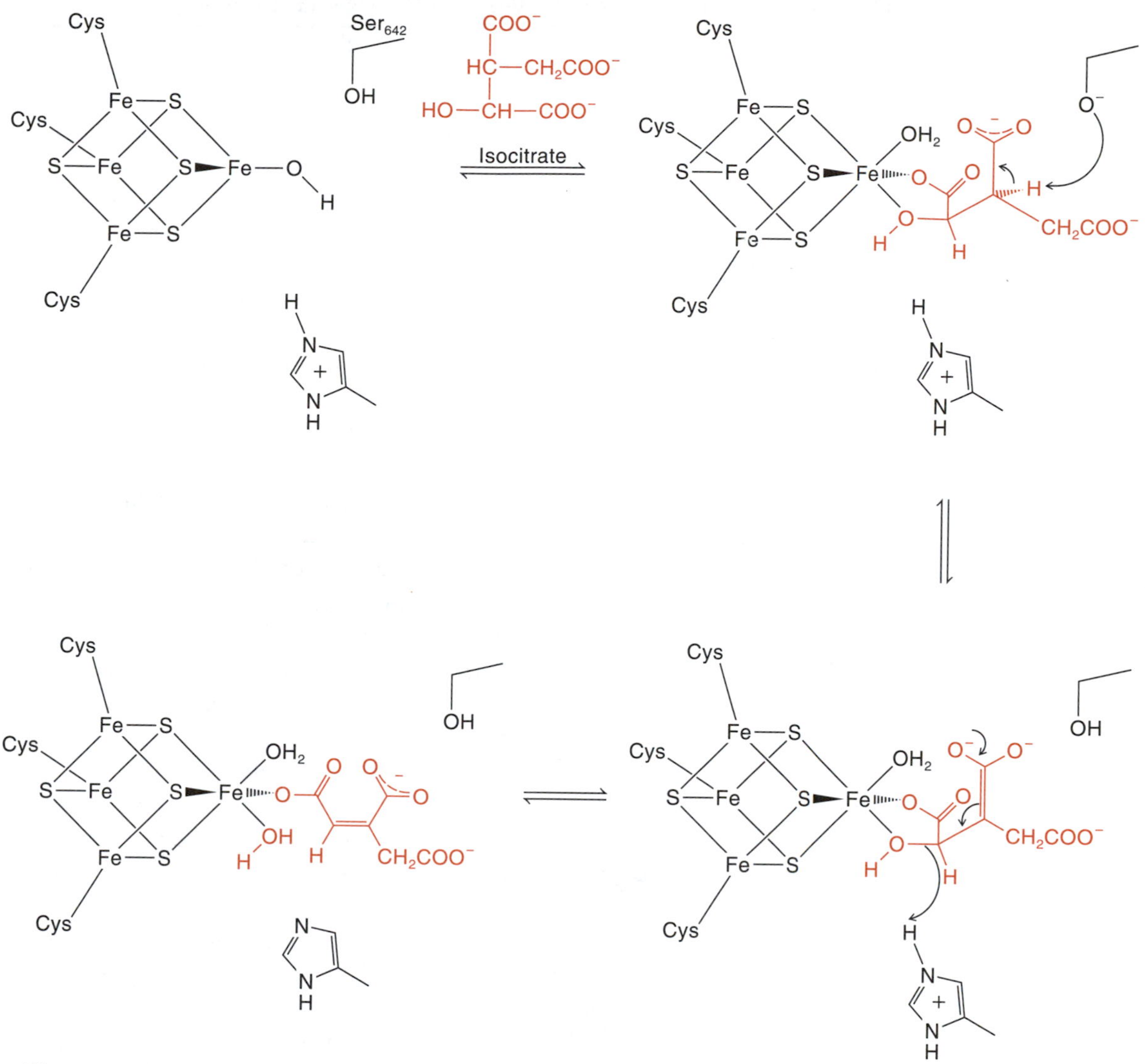

Figure 12.2
Mechanistic scheme for the conversion of isocitrate to *cis*-aconitate at the active site of aconitase, showing the expansion of the iron coordination sphere from four- to six-coordinate.

hydroxide ligand, which either functions directly in the catalytic reaction or is displaced by substrate, or both.

12.3. Control of Function by Ligand Type

As discussed in Chapter 3, the active sites of many metalloproteins contain complexes between metals and organic cofactors or metal clusters that act as bioinorganic chips. These are semiautonomous units for which many of the properties are determined by the units themselves. Since they are held in place by one or more additional donors supplied to the metal ion by the protein side chains, however, the nature of these ligands provides a simple means for tuning the properties of the center to optimize a particular function.

12.3.a. Heme Proteins.   In Chapter 11, we encountered several examples in which hemes are coordinated to a single protein-derived ligand. Included were myoglobin, hemoglobin, and cytochrome c peroxidase, for which the protein-derived ligand is histidine, cytochrome P-450 and chloroperoxidase, for which the ligand is cysteine, and catalase, which employs a tyrosine. Although there are many differences between these proteins apart from their axial ligand, one property that appears to be directly associated with the nature of the axial ligand is the stability of their ferrous forms. Thus, for myoglobin and hemoglobin, which must avoid oxidation to the thermodynamically preferred, inactive ferric (met) form, a neutral histidine ligand is utilized such that the porphyrin-iron-axial ligand assembly is uncharged when the metal is in the $+2$ oxidation state. For enzymes in which the resting state is, and should mechanistically be, ferric, anionic tyrosinate or cysteinate ligands are present, so that neutrality is achieved when the iron has a $+3$ charge.

Two experimental results that bear on this concept have been described. First, by using site-directed mutagenesis, the histidine that is bound to the iron in myoglobin was changed to tyrosine. The expressed mutated protein formed a complex that had spectroscopic properties consistent with tyrosinate coordination to iron. The mutant protein was highly susceptible to autoxidation. Exposure to dioxygen at room temperature resulted in oxidation to the inactive, met (ferric) state rather than production of a dioxygen complex.

The second example involves conversion of the axial methionine ligand to cysteine in cytochrome c. For this protein, it is possible to accomplish the same consequences afforded by site-directed mutagenesis with the use of a technique known as *semisynthesis*. Treatment of horse-heart cytochrome c with cyanogen bromide results in cleavage of the protein at methionines 65 and 80. The large fragment, containing amino acids 1 to 65 and the bound heme, is then isolated. The cleavage reaction converts methionine 65 to a homoserine lactone residue. The modified fragment can then be combined with a synthetic peptide corresponding to residues 66–104 with any desired changes in sequence. The amino group of this added peptide opens the lactone ring to reform the peptide bond! A scheme summarizing this methodology is shown in Figure 12.3. This approach has been used to produce a

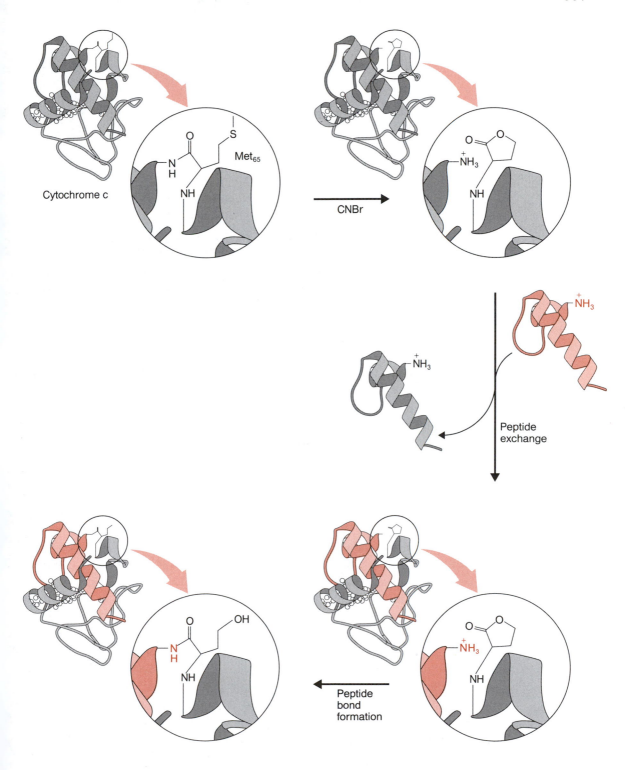

Figure 12.3
Scheme for the preparation of cytochrome *c* derivatives by semisynthesis.

protein with a cysteine replacing the methionine. The oxidized Fe(III) form of the resulting protein has absorption spectroscopic characteristics quite similar to those of ferric cytochrome P-450. Upon reduction, however, the spectral features resemble those of ferrous cytochrome c rather than reduced cytochrome P-450, indicating that the iron-thiolate bond is cleaved or possibly protonated. The reduction potential for this process is -390 mV compared with $+262$ mV for the native protein. Both the large change in reduction potential and the instability of the ferrous-thiolate bond reflect the importance of site neutrality and illustrate how dramatically the metal properties can be tuned by changes in axial ligation.

12.3.b. Iron-Sulfur Proteins: Rieske Centers.

As discussed in Chapters 5 and 9, iron-sulfur clusters are generally ligated to proteins by terminal cysteinate ligands. Thus, for example, cores of the Fe_2S_2 type are bound through four cysteinates to form $[Fe_2S_2(S\text{-}Cys)_4]$ clusters. These units undergo redox reactions from an oxidized state containing two ferric ions and, therefore, an overall cluster charge on $[Fe_2S_2(S\text{-}Cys)_4]$ of -2, to a reduced form with one ferric and one ferrous ion and an overall charge of -3. The potentials for these reactions are in the range -0.23 to -0.46 V. A novel type of center, called the Rieske center, has been discovered in both mitochondrial and photosynthetic electron-transport chains. These centers have much higher redox potentials, falling in the -0.15 to $+0.35$ V range but generally closer to the higher value. Despite the large difference in redox potential, spectroscopic data indicated that these centers contain Fe_2S_2 cores. Although no crystallographic information is yet available for any protein containing a Rieske center, detailed studies on the enzyme phthalate deoxygenase from *Pseudomonas cepacia* have been carried out. By using, among other techniques, ENDOR spectroscopy on specifically [14]N- and ^{15}N-labeled proteins, it was demonstrated that the center has two cysteinates coordinated to one iron atom and two histidines coordinated to the other, resulting in the structure shown in Figure 12.4. The presence of the two histidine ligands accounts for the higher redox potentials observed for the Fe_2S_2 core in the Rieske center; nitrogen-donor ligands stabilize the Fe(II), compared with the Fe(III), oxidation level, and the diminished charge on the histidine-coordinated center further stabilizes the lower oxidation state. The proteins cycle from oxidized forms containing two ferric ions and, hence, a neutral cluster of the form $[(Cys)_2Fe_2S_2(His)_2]$, to a reduced form containing one ferric and one ferrous ion, resulting in an overall cluster charge of -1. Thus, both for the heme and for these dinuclear iron-sulfur centers, the choice

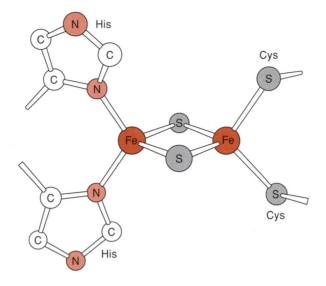

Figure 12.4
Structure of the Rieske center showing the iron-sulfur cluster
in which coordination occurs by histidine as well as the usual
cysteine residues.

of the ancillary ligands provides for coarse tuning of the redox potential
into a functionally useful region. Finer tuning can be accomplished by other
means, as we shall see.

12.4. Control of Function by Coordination Geometry: Blue Copper Proteins

In addition to determining the choice of ligands for a given metal site, proteins can also dictate the coordination geometry to some extent. This control
feature requires that the protein framework be sufficiently stabilized by structural elements other than the coordination preferences of the metal ion. By
far the best-characterized example of this phenomenon is provided by the
blue copper protein plastocyanin. As noted above, blue copper proteins were
suspected early on to be examples of such entatic states. This conjecture was
based upon the unusual EPR and optical spectroscopic features exhibited by
these proteins, which, at the time, were different from those of any known
simple copper complex. As discussed in Section 9.1.b, however, many of
the spectroscopic features can be attributed to the coordination of a cysteine
thiolate to copper(II). Nonetheless, other evidence suggests that these sites
probably represent entatic states, very much as envisioned in the original

hypothesis. Most four-coordinate complexes of copper(II) are square-planar, whereas the corresponding copper(I) complexes are tetrahedral. Thus the transition state for a redox reaction involving interconversion of these two forms would be expected to have an intermediate geometry in order that the reversible electron-transfer reaction rate not be retarded by a substantial structural change. Examination of the structure of plastocyanin in its oxidized and reduced forms indicates that, indeed, only small structural changes accompany the redox changes, as discussed in Chapter 9, with a distorted geometry being present in each.

This observation can be accounted for in two ways. One possibility is that, with this set of ligands, copper might actually prefer such an intermediate geometry in both oxidation states. This question cannot be directly addressed because of the great difficulty in preparing complexes of this type with unconstrained or, indeed, any other ligands, owing to the ease of oxidation of the thiolate. The alternative explanation is that the protein imposes this geometry on the metal. Such a hypothesis is strongly supported by the observation that the structures of the apoprotein and the mercury-substituted protein are essentially identical to those of the copper-containing forms, as shown in Figure 12.5. Thus, plastocyanin appears to have a coordination geometry largely determined by the protein. Such tuning of the copper stereochemistry at this site minimizes the local reorganizational energy upon redox reactions and probably optimizes the potential for the requisite biological functions of the protein as well.

12.5. Control of Function by Indirect Effects

The above examples demonstrate how ligands bound directly to a metal ion in a protein can be used to tune its redox and coordination properties. For a number of systems, however, the composition and gross geometry of the metal center and its protein-derived ligands are identical; yet the properties are significantly different. In these instances, other characteristics of the folded protein chain must tune the properties of the metal to allow it to fulfill its biological function most effectively.

12.5.a. [Fe$_4$S$_4$(Cys)$_4$] Centers: Hydrophobicity and Hydrogen-Bonding Effects. One of the most interesting examples of this type of tuning involves iron-sulfur proteins. The bacterial ferredoxins undergo redox reactions at quite low potentials, near -400 mV, whereas the redox reaction of

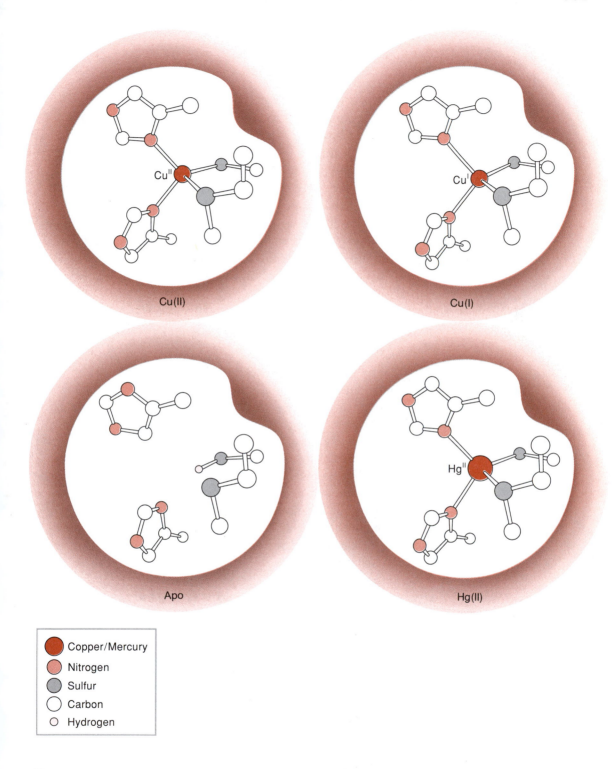

361

Figure 12.5
Structures of the active sites in oxidized (upper left), reduced (upper right), apo (lower left), and Hg(II)-derivatized (lower right) plastocyanin.

high-potential iron proteins (HiPIPs) occur at much higher potentials, near +350 mV. Early crystallographic analysis revealed the remarkable fact that these two proteins each contain [$Fe_4S_4(Cys)_4$] centers, as discussed in Chapters 5 and 9. This finding presented an apparent paradox, namely, how can two proteins with the same active sites have such different redox potentials? This question was resolved with the realization, supported by comparison of the spectroscopic properties of the proteins with those of well-characterized model complexes, that two different redox couples are used, as summarized in Equation 12.2. Although this discovery removed the need to explain how two proteins with the same center exhibit a 750-mV difference in potential, it

$$Fe_4S_4(Cys)_4{}^{3-} \rightleftharpoons Fe_4S_4(Cys)_4{}^{2-} \rightleftharpoons Fe_4S_4(Cys)_4{}^{-}$$

$$\text{Ferredoxin} \qquad\qquad \text{HiPIP}$$

$$(12.2)$$

did not explain how a given protein selects one or the other redox couple. More detailed analyses of the crystal structures indicate several differences between the environments of the clusters that may contribute to tuning the potential and, hence, determining which couple is used physiologically.

The major differences involve the overall hydrophobicity of the sites and the number of peptide NH-to-sulfur hydrogen bonds. A hydrophobic environment will favor the less highly charged [$Fe_4S_4(Cys)_4$]$^-$ center and hence support the HiPIP redox couple. In contrast, peptide NH-to-sulfur (sulfide or cysteinate) hydrogen bonds help to neutralize the negative charge on the cluster, as discussed in Section 8.2. These interactions will be most important for the more negatively charged forms of the cluster and, thus, will help to stabilize the reduced ferredoxin state. Such expectations are consistent with the observed structural differences. The environment in the ferredoxin is less hydrophobic, and the cluster is more nearly exposed to solvent. Furthermore, these sites have eight to nine NH-to-sulfur hydrogen bonds per cluster, as compared with only five such H-bonds for HiPIP. The hydrogen bonds for one of the clusters from bacterial ferredoxin are illustrated in Figure 12.6. The relative importance of the hydrophobic and hydrogen-bonding effects has not yet been established. It is clear, however, that such effects can tune the redox potential enough to move one couple or the other into the range required for the protein to function as an electron carrier under physiologically required conditions.

12.5.b. Cytochromes: Electrostatic Potential.
In Chapter 9 we learned that the redox potential at a metal center in a protein is strongly dependent on the composition and state of protonation of its ligands. An alternative

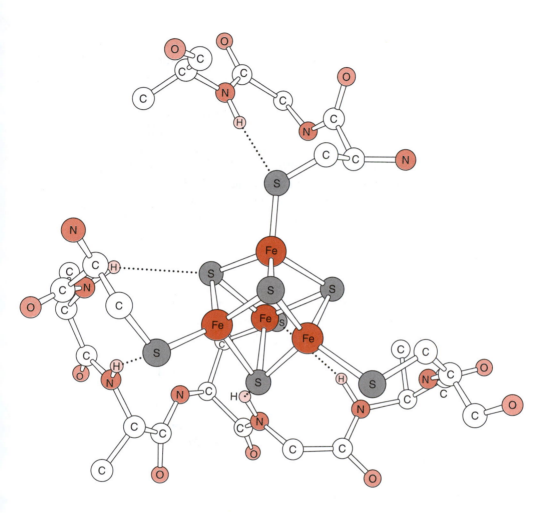

Figure 12.6
Some of the peptide NH-to-sulfur hydrogen bonds
around an iron-sulfur cluster in ferredoxin.

mechanism for tuning the redox potential of an electron carrier involves the positioning of charged amino acids on the protein surface. The charges of these surface residues will interact with any residual charge on the metal according to Coulomb's law. The energy of such an interaction depends on the inverse of the distance between the redox center and the charged group and on the effective dielectric constant of the medium that separates them. The value of the latter depends both on the contribution of the protein residues and on the solvent. Without the shielding due to the high dielectric constant of the aqueous solvent, the tuning effects due to surface charges would be quite large. As it is, the effects are significantly smaller than those due to ligands bound directly to the metal, discussed above, but they can be important for the fine tuning required to match the potentials of two electron-transfer proteins.

The role of charged groups in modulating the redox potential has been demonstrated by using mammalian cytochrome b_5 generated by genetically altered DNAs encoding the protein. The redox potentials for the native, or wild-type, protein and a series of mutants were directly measured by means of a cysteine-modified gold electrode. The mutations involved neutralization of acidic residues via conversion of glutamic acid to glutamine and of aspartic acid to asparagine as well as by a few other types of changes. The potential of the wild-type protein is -7 mV versus NHE, whereas those of the mutants ranged from -13 mV (for a serine to aspartic acid mutation) to $+13$ mV (for a triple mutant involving neutralization of three acidic residues). The data could be fit by using a computer-generated model of the protein that incorporated its crystallographically determined shape and a low dielectric constant ($\varepsilon = 2-4$) surrounded by solvent of high dielectric ($\varepsilon = 80$). As shown in Figure 12.7, the quality of the fit suggests that these surface charges should be effective in tuning redox potentials via this electrostatic mechanism.

12.6. Substrate Specificity

The above discussion illustrates how the properties of the metal site itself can be tuned to generate a center that is appropriate for a given function. Another level of tuning involves active-site changes that do not affect the metal ion itself but instead modulate the interactions between the metalloprotein and other molecules. This feature allows active sites constructed with the same metal and ligand components to execute different functions. One of the

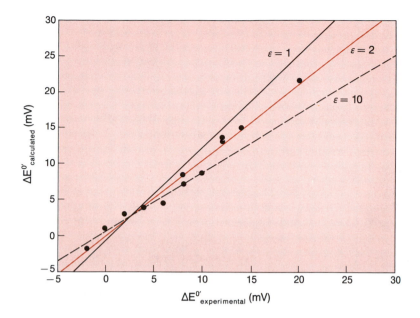

Figure 12.7
Experimentally determined redox potentials for cytochrome b_5 mutants compared to calculated values assuming different dielectric constants for the protein interior. Linear fits to the data suggest that dielectric constants in the range from 2 to 4 best reproduce the experimental results. (Adapted from K. K. Rodgers and S. G. Sligar, *Journal of the American Chemical Society* **113**, 9419–9421 (1991)).

best illustrations of this phenomenon involves the specificity of metalloenzymes for their substrates. Here, different enzymes having essentially identical metal centers can bind and turn over different substrates and, hence, perform different functions.

12.6.a. Proteases. In Chapter 10, we saw that the zinc peptidases carboxypeptidase A and thermolysin have active sites that are remarkably similar, even though these two enzymes do not appear to be structurally or evolutionarily related. Moreover, the two enzymes catalyze the hydrolysis of different types of substrates. Carboxypeptidase A is an exopeptidase cleaving the C-terminal amino acid of the end of a polypeptide chain. Furthermore, this enzyme is most active when the C-terminal residue is an aromatic or large hydrophobic residue such as phenylalanine or leucine. In contrast, thermolysin is an endopeptidase. It cleaves amide linkages in the middle of a polypeptide chain with differing degrees of specificity for positions on both sides of the scissile bond.

The crystal structures of the two proteins provide information as to how this specificity is accomplished. For carboxypeptidase A, there is a binding pocket for the terminal carboxylate group that features the positively charged side chain from Arg-145 (Figure 10.2). In addition, there is a hydrophobic binding pocket for the side chain of the terminal residue that accounts for the preference for hydrophobic amino acids. Thermolysin, on the other hand, has a binding pocket that extends several residues beyond the scissile bond. Studies with a series of inhibitors containing the sequence Gly---Leu-X, where --- indicates the bond to be cleaved, revealed greater binding by two orders of magnitude for X = Leu than for X = $-NH_2$.

An even more direct comparison is provided by carboxypeptidase B. This enzyme is highly analogous to carboxypeptidase A, being approximately 50 percent identical in amino-acid sequence and very similar in its tertiary structure. The substrate preference is different, however. Carboxypeptidase B favors basic amino acids rather than hydrophobic residues in the terminal site, a difference that is nicely explained by comparison of the structures. In the center of the region corresponding to the hydrophobic pocket in carboxypeptidase A lies the Asp-255 side chain of carboxypeptidase B, as shown in Figure 12.8. This residue is presumed to interact with positively charged side chains to favor substrate binding and subsequent cleavage.

12.6.b. Cytochromes P-450. A second example of substrate tuning in the active-site pocket is provided by the cytochromes P-450. As discussed in Chapter 11, these enzymes catalyze the hydroxylation of a number of hydrophobic substrates. The family of enzymes is very large, with over 50 different known sequences. It has been estimated that the human genome has genes for somewhere between 100 and 200 different P-450 enzymes. A crystal structure is available for a bacterial cytochrome P-450 that hydroxylates camphor. The mammalian enzymes are membrane-bound and thus have been harder to purify and crystallize. Sequence comparisons indicate, however, that the mammalian enzymes have the same basic structure as that of the bacterial enzyme, except that they have additional sequences at the amino terminus that appear to be responsible for membrane binding. The sequences of members of this family of enzymes differ considerably in detail, however. This variability allows for a great deal of tailoring of the substrate-binding site around the conserved iron-porphyrin catalytic site, so that each of the enzymes is capable of converting its own set of substrates. Considerable research is currently under way in an attempt to understand the relationships between amino-acid sequence, binding-site structure, and substrate speci-

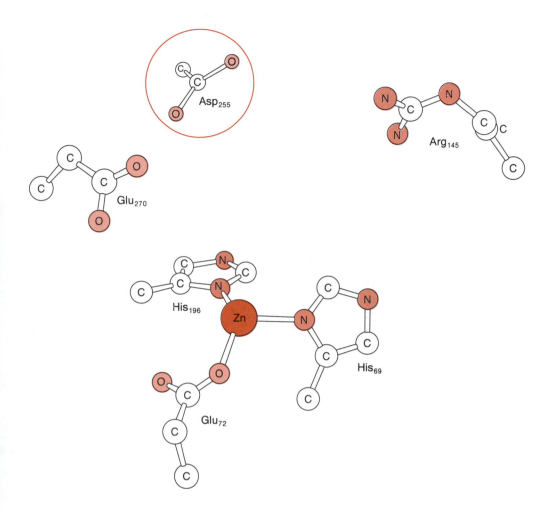

Figure 12.8
Active site region of the carboxypeptidase B structure, showing the charged aspartate residue in position 255 that leads to preferential cleavage of peptide bonds adjacent to basic residues.

ficity. The positions of some of the residues that are thought to be involved in determining substrate specificity, based on the known structure of the camphor enzyme, are shown in Figure 12.9. If the features that control substrate specificity can be identified, the P-450 enzymes will provide an excellent opportunity for protein engineering, with the goal of producing enzymes that oxidize hydrocarbons regiospecifically.

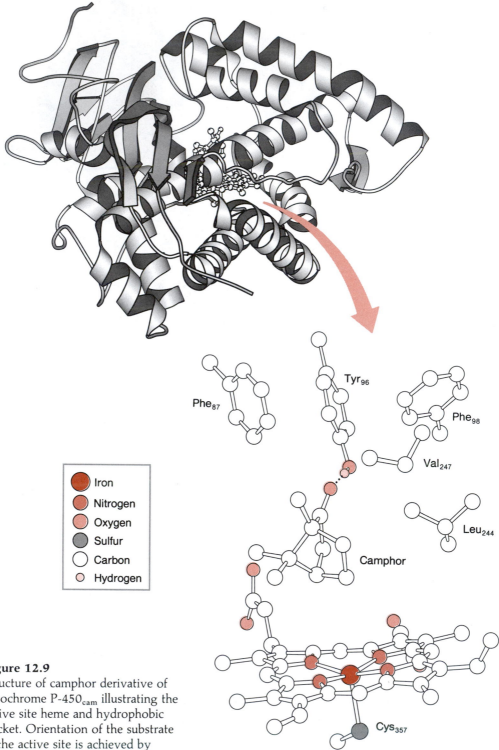

Figure 12.9
Structure of camphor derivative of cytochrome P-450$_{cam}$ illustrating the active site heme and hydrophobic pocket. Orientation of the substrate in the active site is achieved by hydrogen bonding to Tyr$_{96}$, as indicated by the dotted line.

12.7. Coupled Processes

One of the most important manifestations of protein tuning involves the ability to couple two or more processes. Such coupling is crucial to the execution of multielectron redox reactions, the transduction of different forms of chemical energy, and the minimization of undesired side reactions due to inappropriate generation of highly reactive species, among other phenomena. Indeed, without such coupling, many of the most fundamental processes in biology would be impossible. The detailed mechanisms responsible for coupling are known for only a few systems; however, these systems reveal some of the strategies that are most likely utilized in other, less well characterized, systems.

12.7.a. Electron Transfer Coupling—Sulfite Oxidase.   As discussed in Chapter 11, sulfite oxidase is an enzyme that contains a molybdenum cofactor and a b-type heme. It catalyzes the reaction shown in Equation 12.3. Other electron acceptors such as ferricyanide or dioxygen can be substituted

$$SO_3^{2-} + 2 \text{ cytochrome } c_{(ox)} + H_2O \rightleftharpoons SO_4^{2-} + 2\text{cytochrome } c_{(red)} + 2H^+$$

$$(12.3)$$

for cytochrome c. The protein from chicken liver is a dimer with a molecular mass of 120 kilodaltons that contains one Mo and one heme per subunit. Treatment of the enzyme with proteases produces a dimeric Mo-containing domain with a subunit molecular mass of 47 kilodaltons and a monomeric heme domain of 9.5 kilodaltons. The isolated Mo domain is still capable of oxidizing sulfite, with ferricyanide or dioxygen as the electron acceptor, but it has lost the ability to use cytochrome c. Even combining the isolated Mo and heme domains, or using the unfractionated proteolysis reaction mixture, does not restore the cytochrome-c-dependent activity. This result reveals that the relationship between the two domains, as defined by the intact polypeptide chain, is necessary for coupling the two-electron sulfite oxidation reaction that takes place at the Mo center with the two one-electron cytochrome c reduction reactions that occur at the heme centers. The overall mechanism, including this coupling phenomenon, was summarized previously in Figure 11.19.

12.7.b. Chemical and Electron-Transfer Coupling: Nitrogenase. The enzyme nitrogenase reduces dinitrogen to ammonia. The full reaction catalyzed by this two protein-enzyme system is given in Equation 12.4.

$$N_2 + 8H^+ + 16Mg\text{-}ATP + 8e^- \rightleftharpoons 2NH_3 + H_2 + 16Mg\text{-}ADP + 16P_i \quad (12.4)$$

Two molecules of ATP, the Mg complex of which is the actual substrate (Chapter 10), are hydrolyzed for every electron that is transferred. The chemical details of the mechanism for coupling ATP hydrolysis to dinitrogen reduction remain to be elucidated. Several general features have been established, however. First, Mg-ATP binds to the Fe protein and, by so doing, lowers the reduction potential by approximately 100 mV. The ability of substrate binding to modulate the redox potential of a metalloenzyme is a strategy previously encountered for cytochrome P-450 and MMO (Chapter 11). This decrease in redox potential is necessary to allow electron transfer from the Fe protein to the MoFe protein. ATP hydrolysis appears to occur concomitantly with this electron transfer. The two ATP molecules hydrolyzed for each electron transferred reflect the dimeric nature of the Fe protein.

The nitrogenase system also illustrates a different sort of coupling. In the reduction of each molecule of dinitrogen, the enzyme consumes a total of eight electrons. Two of the electrons are used to produce one molecule of dihydrogen; the remaining electrons are actually utilized in the production of ammonia. The rate-determining step in the nitrogenase reaction involves dissociation of the Fe protein from the MoFe protein following electron transfer. The MoFe protein must store up electrons, added one at a time, until substrate reduction can occur. Again, the mechanisms that allow this coupling of a series of one-electron-transfer reactions to produce a multielectron reduction remain to be determined, but they must involve coordinated tuning by the protein chains of the properties of several novel metal clusters present in this fascinating enzyme.

12.7.c. Substrate Binding: Cytochrome P-450. The cytochromes P-450 provide one of the best-characterized examples of chemical coupling. As mentioned in Chapter 11, reduction of iron from Fe(III) to Fe(II), the form that binds dioxygen, does not occur prior to substrate binding. Such control is crucial, since this system must generate a very reactive intermediate; if this species were formed in the absence of substrate, it might either react with and damage the enzyme itself or be short-circuited and waste reducing equivalents. Through studies of the native bacterial enzyme P-450$_{cam}$ with its natural (camphor) or unnatural substrates, and by investigating site-directed mutants, the molecular mechanisms responsible for this coupling have been elucidated.

The crystal structure of P-450$_{cam}$ in the absence of substrate revealed that the substrate binding site is filled with water molecules, including one that is directly coordinated to the metal to produce an octahedral low-spin Fe(III) center. Upon camphor binding, all of these waters are displaced, and the coordination number of the iron is reduced to five. The camphor is contacted extensively by the protein via van der Waals interactions and one hydrogen bond to the camphor oxygen. The change in coordination number is accompanied by conversion of iron to the high-spin state. These observations suggest how the reduction of the iron is coupled to substrate binding. The extensive changes in the coordination number, spin state, and polarity of the iron site increase the redox potential of the iron and thus the ease with which it is reduced to the ferrous state. This reduction step then allows dioxygen binding to occur, since both an open coordination site and a reduced iron center are produced (see Section 11.2.a). Once the dioxygen is bound, probably in a manner similar to that observed in hemoglobin and myoglobin, the center is reduced by addition of a second electron to produce the initial highly reactive form. Under normal circumstances, the O−O bond in this intermediate is cleaved heterolytically to release water and generate the iron-oxo species that hydroxylates the substrate. Displacement of the oxidized substrate by water completes the catalytic cycle. The economy of this chemistry requires proper placement of the substrate in the active site and elimination of all possible side reactions to ensure that the uptake of electrons is fully coupled to substrate hydroxylation.

The roles of the substrate and the enzyme in mediating this tight coupling have been directly probed. Let us first consider the substrate. Camphane, an analog of camphor that simply lacks the carbonyl group, can also be processed by the enzyme. In this reaction, 90 percent of the hydroxylated product is the 5-exo isomer, a result analogous to that obtained with the normal substrate. Whereas the oxidation of camphor is essentially 100 percent efficient, however, with all the consumed dioxygen and reducing equivalents ending up in product, the efficiency of the oxidation of camphane is only 8 percent. The reason for this difference was revealed by crystallographic analysis of the complex of the enzyme with camphane, which differs from the camphor derivative in several important respects. First, water molecules remain in the active site, including one bound to the iron, even though the substrate is present. Second, the mobility of the substrate is significantly higher than that observed for camphor. These differences allow the reactive oxygen species to be discharged through processes other than substrate oxidation, such as protonation of the reduced dioxygen to release hydrogen peroxide and reductive protonation of the iron-oxo species to produce water. A mechanistic scheme that includes these reactions is shown in Figure 12.10.

Similar uncoupling phenomena can be affected by changes in the enzyme. Replacement of a conserved threonine residue with alanine resulted in a mutant enzyme that was uncoupled even with camphor as substrate. Crystallographic studies revealed that this residue participates in several interactions that stabilize and rigidify the active site, particularly when both dioxygen and substrate are bound. These results indicate that extensive fine tuning has occurred in cytochrome P-450 to allow highly efficient channeling of a reactive intermediate into a single, desired reaction pathway.

12.7.d. Membrane Transport: Photosynthetic Reaction Center.

Our final example of coupling involves the bacterial photosynthetic reaction center, which couples the absorption of light to the generation of a potential across a membrane. In Chapter 9, we saw that the reaction center consists of an elaborate array of prosthetic groups arranged in a complex, membrane-bound protein assembly. Absorption of a photon by the special pair of chlorophylls results in the ejection of an electron. This electron is very rapidly transferred from one prosthetic group to the next until it arrives at a quinone, Q_B, via the scheme shown in Figure 12.11. The back transfer of the electron is blocked by the rapid shuttling of an electron from the bound cytochrome to the oxidized special pair. Once two electrons have been transferred to Q_B, it is released from the reaction center in the protonated hydroquinone form. The transfer of electrons across the membrane and the uptake of protons to produce hydroquinone contribute to the generation of a proton electrochemical gradient that can be utilized by other processes. For this system, the disposition of the reaction center in the membrane is the basis for coupling; without the presence of the intervening membrane, the system would function only to cycle electrons to and from the quinone pool. Further characterization of other membrane-bound systems, such as those involved in the photosynthetic apparatus of higher plants and in the oxidative phosphorylation pathway, will be necessary to appreciate fully other strategies that nature has evolved to couple reactions across bilayer membranes. These challenging problems are among the many frontiers of bioinorganic chemistry to be discussed in the final chapter.

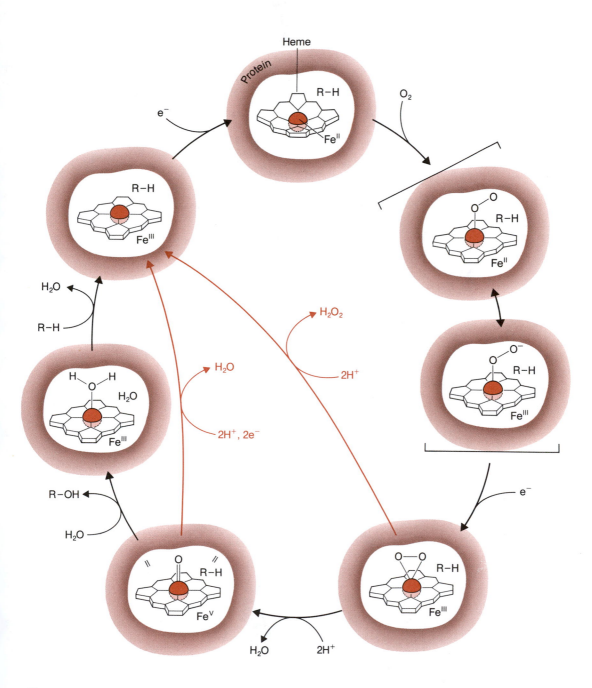

Figure 12.10
Mechanistic scheme for cytochrome P-450$_{cam}$. Pathways for uncoupled reduction of dioxygen without oxidation of substrate are shown in color.

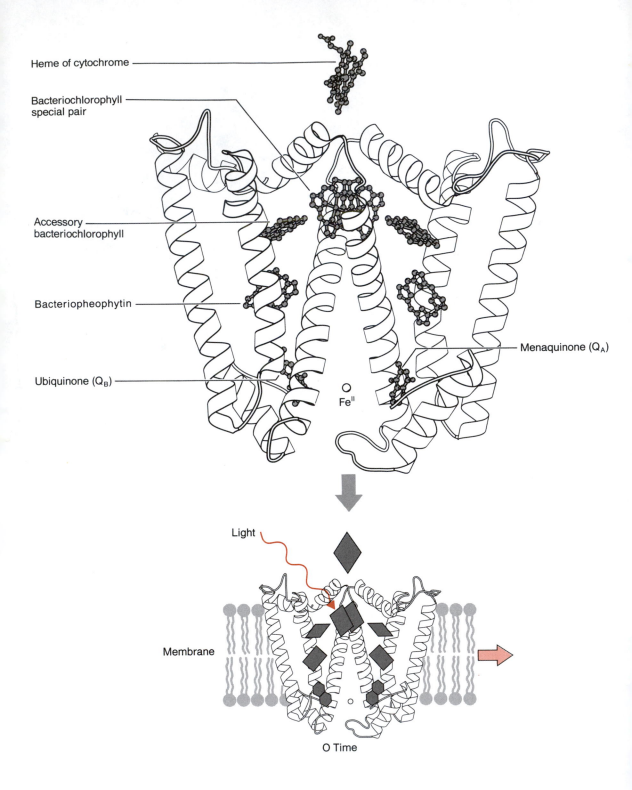

Heme of cytochrome

Bacteriochlorophyll
special pair

Accessory
bacteriochlorophyll

Bacteriopheophytin

Menaquinone (Q$_A$)

Ubiquinone (Q$_B$)

Fe^{II}

Light

Membrane

O Time

Figure 12.11
Scheme for electron transfer through the photosynthetic reaction center resulting in charge separation across the membrane. Oxidized cytochrome is formed on one side of the membrane while reduced quinone is produced on the other.

374

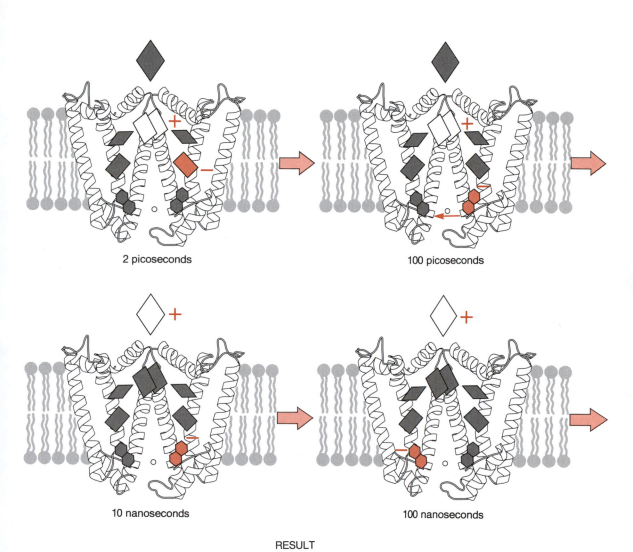

2 picoseconds

100 picoseconds

10 nanoseconds

100 nanoseconds

RESULT

+ Charge (oxidized cytochrome)

− Charge (reduced quinone)

Study Problems

1. The active-site geometry of ribonucleotide reductase is shown below. The diiron center is believed to function as a source of regenerating the tyrosyl radical. The center is first reduced to the diiron(II) form, which then reacts with dioxygen, the Tyr_{122}, and another electron source to form the site shown. Discuss what modifications in the coordination of the iron atoms must occur at the center upon reduction in order for the regeneration chemistry to take place. Suggest specific changes.

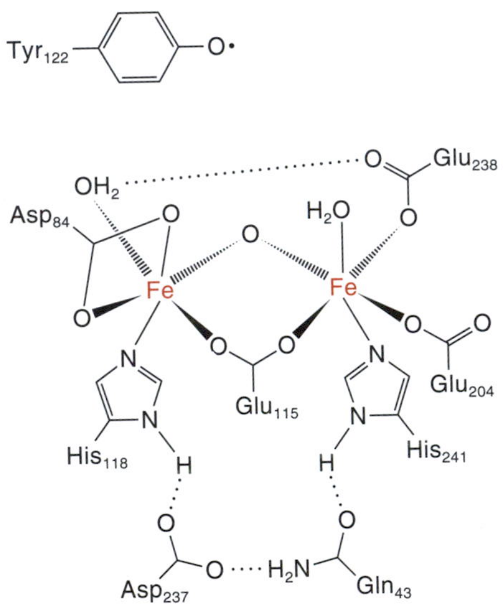

2. Electron-transfer rates in proteins containing blue copper centers such as plastocyanin can be high, as exemplified by the rate constant of electron self-exchange of 1.3×10^6 $M^{-1}s^{-1}$ for azurin. These sites have relatively small ligand reorganization energies, since the copper geometry of the oxidized and reduced forms is very similar, and this feature has been suggested as one reason why the rates are so high. Recently, CuL_2^{n+} complexes, $n = 1$ or 2, were prepared where L is a bidentate bis(imidazole) ligand. The structures are both tetrahedral with average Cu–N bond lengths of 1.96 Å for the Cu(II) form and 2.03 Å for the Cu(I) form. Studies of the self-exchange electron-transfer rate for this couple revealed a self-exchange rate constant of $< 10^2$ $M^{-1}s^{-1}$. Discuss this result in comparison to the azurin value.

3. Early in the X-ray structure determination of metazido hemerythrin, the electron-density maps were erroneously interpreted in terms of a structural model in which two Fe(III) centers were joined by only a single oxo bridge, with three histidines coordinated at one terminal position and two histidines and an azide ligand at the other. Discuss how this structure is less satisfying than the actual structure of met-azidoHr, depicted in Figure 11.4 , in terms of the principles developed in this chapter.

4. In an attempt to create a minizinc enzyme, a zinc-finger-like peptide that lacked the final histidine residue was synthesized. This peptide was shown to bind metal ions such as zinc(II) and cobalt(II) to produce complexes with $M^{II}(Cys)_2(His)(OH_2)$ sites. The bound water molecule could be displaced by other ligands. Attempts to use such complexes to catalyze even simple reactions, such as the hydrolysis of activated esters, proved unsuccessful, however. Suggest reasons why no catalysis was observed and approaches to improve the likelihood of generating an active catalyst.

Bibliography

Control of Function by Protein Side Chains: Opening Coordination Sites
H. Lauble, M. C. Kennedy, H. Beinert, and C. D. Stout. 1992. "Crystal Structures of Aconitase with Isocitrate and Nitroisocitrate Bound." *Biochemistry* 31, 2735–2748.
B. L. Vallee and D. S. Auld. 1990. "Active-Site Zinc Ligands and Activated H_2O of Zinc Enzymes." *Proc. Natl. Acad. Sci. USA* 87, 220–224.
L. Zheng, M. C. Kennedy, H. Beinert, and H. Zalkin. 1992. "Mutational Analysis of Active Site Residues in Pig Heart Aconitase." *J. Biol. Chem.* 267, 7895–7903.

Control of Function by Ligand Type
P. J. Boon, G. I. Tesser, and R. J. F. Nivard. 1979. "Semisynthetic Horse Heart [65-Homoserine] Cytochrome *c* from Three Fragments." *Proc. Natl. Acad. Sci. USA* 76, 61–65.
A. L. Raphael and H. B. Gray. 1991. "Semisynthesis of Axial-Ligand (Position 80) Mutants of Cytochrome c." *J. Am. Chem. Soc.* 113, 1038–1040.
S. Sligar, K. D. Egeberg, J. T. Sage, D. Morikis, and P. M. Champion. 1987. "Alteration of Heme Axial Ligands by Site-Directed Mutagenesis: A Cytochrome Becomes a Catalytic Demethylase." *J. Am. Chem. Soc.* 109, 7896–7897.

Control of Function by Coordination Geometry: Blue Copper Proteins
T. P. J. Garret, D. J. Clingeleffer, J. M. Guss, S. J. Rogers, and H. C. Freeman. 1984. "The Crystal Structure of Poplar Apoplastocyanin at 1.8-Å Resolution." *J. Biol. Chem.* 259, 2822–2825.
J. M. Guss and H. C. Freeman. 1983. "Structure of Oxidized Poplar Plastocyanin at 1.6 Å Resolution." *J. Mol. Biol.* 169, 521–563.
J. M. Guss, P. R. Harrowell, M. Murata, V. A. Norris, and H. C. Freeman. 1986. "Crystal Structure Analyses of Reduced (Cv[I]) Poplar Plastocyanin at Six pH Values." *J. Mol. Biol.* 192, 361–387.

Control of Function by Indirect Effects
R. Langen, G. M. Jensen, U. Jacob, P. J. Stephens, and A. Warshel. 1992. "Protein Control of Iron-Sulfur Cluster Redox Potentials." *J. Biol. Chem.* 267, 25625–25627.

G. R. Moore. 1983. "Control of Redox Properties of Cytochrome c by Special Electrostatic Interactions." *FEBS Lett.* 161, 171–175.

K.K. Rodgers and S.G. Sligar. 1991. "Surface Electrostatics, Reduction Potentials, and the Internal Dielectric Constant of Proteins." *J. Am. Chem. Soc.* 113, 9419–9421.

M. A. Walters, J. C. Dewan, C. Min, and S. Pinto. 1991. "Models of Amide-Cysteine Hydrogen Bonding in Rubredoxin: Hydrogen Bonding Between Amide and Benzenethiolate in $[(CH_3)_3NCH_2CONH_2]_2[Co(SC_6H_5)_4] \cdot 1/2CH_3CN$ and $[(CH_3)_3NCH_2CONH_2][SC_6H_5]$." *Inorg. Chem.* 30, 2656–2662.

Coupled Processes

R. H. Burris. 1991. "Nitrogenases." *J. Biol. Chem.* 266, 9339–9342.

J. L. Johnson and K. V. Rajogopalan. 1977. "Tryptic Cleavage of Rat Liver Sulfite Oxidase—Isolation and Characterization of Molybdenum and Heme Domains." *J. Biol. Chem.* 252, 2017–2025.

R. Raag and T. L. Poulos. 1991. "Crystal Structures of Cytochrome P-450$_{CAM}$ Complexed with Camphane, Thiocamphor, and Adamantane: Factors Controlling P-450 Substrate Hydroxylation." *Biochemistry* 30, 2674–2684.

D. C. Youvan and B. L. Marrs. 1987. "Molecular Mechanisms of Photosynthesis." *Sci. Amer.* 256, 42–48.

CHAPTER 13

The Frontiers of
Bioinorganic Chemistry

Throughout our discussion in Chapters 5–12 of this book, we have tried to illustrate the principles of bioinorganic chemistry with examples where maximal information was available through direct and model studies of a particular system. Some of the cases discussed represent nearly closed areas of research, in which the remaining goals, if any, are not addressing the bioinorganic chemical aspects of the problem. In other instances, considerable work remains to be done to understand in molecular detail the underlying bioinorganic chemistry. In looking back over this discussion, we might ask what are the solved problems in the field. Among the most convincing examples are (1) the way that nature binds dioxygen reversibly; (2) how the toxic by-products, particularly superoxide ion, are eliminated; (3) the constitution of electron-transfer proteins, especially those containing blue copper or iron-sulfur centers; and (4) the role of zinc in hydrolytic enzymes such as carboxypeptidase. Major progress has also been made in understanding the homolytic Co–C bond chemistry of the B_{12} coenzyme, in sorting out the DNA-binding properties of platinum anticancer drugs, in deducing the ways that nature recruits, coordinates, and stores iron, in solving the structures of metalloregulatory proteins such as the zinc-finger proteins and calmodulin, and in sorting out the details of the cytochrome P-450 hydroxylation chemistry. In each of these, however, there remain important bioinorganic issues to be addressed in future experimental studies. Moreover, for every one of these examples, there are many others for which considerably less information is available. These unsolved problems represent the frontier of bioinorganic chemistry as it exists in the mid-1990s. Interestingly, many of these frontier areas were identified two decades ago, when the first major treatise in bioinorganic chemistry was published. Included are such important and incompletely understood systems as nitrogenase, long-range electron-transfer proteins, and cytochrome oxidase, as well as more recently identified problems, such as the role of nickel in hydrogenases, methanogens, and urease, the functioning of manganese in the oxygen-evolving system of photosystem II, and molybdenum oxo transfer proteins. In this chapter we shall briefly describe these frontier areas of bioinorganic chemistry, partly in the hope of stimulating research directed to the resolution of the currently identifiable issues and partly to expose the reader to the tremendous scope of the field. Since it is our thesis in this book that the principles described in Chapters 5–12 will long outlive the individual examples discussed, we expect that knowledge gained in pursuit of these frontier research problems will fall into the various categories delineated in our discussion. Accordingly, we structure our discussion here to mirror the organizational structure of these earlier chapters.

13.1. Choice and Uptake of Metal Ions

The question of how metal ions are taken up and maintained at critical concentration levels in cells has been answered in some molecular detail only for iron. The discovery of enterobactin discussed in Chapter 5, however, offers the hope that related ligands might have evolved to chelate copper, manganese, or indeed virtually all the metal ions used in biology. The identification and characterization of such alternative systems remains a significant challenge. For example, the recent isolation and analysis of the gene responsible for Menkes disease, which results from a copper-transport abnormality, reveals the predicted protein to have membrane-spanning domains and a copper-binding site. This discovery augurs well for the ultimate contribution of bioinorganic chemistry to the treatment of human disease. Moreover, it is likely that metal ions in addition to those listed in Table 1.1 will be found to be essential elements in biology, leading candidates among the transition metals being Ti, which has been identified in a number of marine organisms, and Cr, which has been associated with the glucose tolerance factor and diabetes, although its postulated role is quite controversial. As knowledge about these metals accumulates, studies of their functions and homeostasis will ensue.

There has been much discussion about metalloporphyrins and iron-sulfur clusters as bioinorganic chips in metalloproteins. Yet nature must manage the assembly of many other systems about which we know only a little. For example, how do Mo and Fe and mixed metal clusters, such as that in the Fe/Mo cofactor of nitrogenase, assemble? What is the mechanism of assembly of the Mo pterin cofactor complex involved in oxo-transfer proteins? The bioinorganic chemist interested in the construction of models for metallo-protein-active sites is likely to play a significant role in sorting out the answer to these questions, for these units are amenable to synthesis and study as spectroscopic and functional models for the protein cofactors.

Neurobiology constitutes a major frontier in modern science. Although alkali metal ions have been shown to play an important role in signal trans-duction, it is likely that other metals ions will participate in processes required for neurological functions. A recent report suggests the presence of magnetic particles in the human brain, similar to the magnetite crystallites that occur in magnetotactic bacteria and the homing mechanisms of honeybees and birds. A related area is the use of metal ions as pharmaceuticals to control neurolog-ical behavior; understanding the molecular mechanisms by which lithium ion functions to manage the manic-depressive condition is one frontier of consid-erable interest and importance. Bioinorganic chemists will also help unravel the molecular mechanisms of neurotoxicity caused by metal ions, develop new chelating agents to manage or reverse such toxicity, and make major contributions to areas of medicine where metal ions are used as probes and drugs. In the last area, we call attention especially to the need for better metal complexes to detect diseased tissue by magnetic resonance imaging and as radiopharmaceuticals. Moreover, as mechanistic information about cisplatin, auranofin, and other metal-based drugs continues to accumulate, it should be possible to design more effective ways to screen new compounds for activity *in vitro*.

13.2. Control and Utilization of Metal-Ion Concentrations

Over the past several years great progress has been made in characterizing the mechanisms by which metal-ion concentrations are controlled within cells. As discussed in Chapters 5 and 6, much information has accumulated about iron both in prokaryotic and eukaryotic systems; yet significant details, such as the actual structure of the ferritin core and mechanistic information about its loading and unloading, are lacking. Evidence is mounting that simi-lar systems exist for a wide variety of other metal ions. Many of the initial leads are coming from genetics studies. By selecting for mutants that are unable to tolerate either unusually high or low concentrations of metal ions, it is possible to get initial access to proteins that are necessary for metal-ion uptake, storage, and detoxification. Investigation of these phenomena, both at the level of system organization and at the detailed molecular level, should prove to be an exciting and fruitful research area. Significant preliminary results have been obtained for metals such as molybdenum, nickel, zinc, and copper. These studies can be expected to yield additional examples of the existing strategies used by nature for metal-ion uptake and detoxification as well as some entirely new mechanisms.

A second major research area involves detailed structural characterization of ion-transport systems and ion channels. The determination of the crystal structure of the photosynthetic reaction center had a profound influence on the study of the mechanism of photosynthesis. The structure confirmed some previous ideas and provided a framework for focusing new biophysical experiments. In addition, it identified new questions that could not have been conceived without knowledge of the structure. Similar results can be expected once the first crystal structures of ion transporters and ion channels are determined. Information obtained from lower-resolution structural methods such as electron microscopy, combined with the results of site-directed mutagenesis studies, has been used to produce schematic models that have proved to be useful. Only with structural results at or near atomic resolution, however, will hypotheses concerning phenomena such as metal-ion selectivity be addressable. These are challenging problems because of the difficulty in obtaining adequate quantities of pure material and of producing suitable crystals. The rewards, however, make such projects impossible to postpone.

Finally, the issue of how metal ions participate in intra- and intercellular communication has not in any sense been fully addressed. Calcium is by far the best-understood and studied example. As noted in Chapter 7, the potential for zinc to act as a cellular signal or messenger in some situations is suggested by a variety of circumstantial data but has not yet been established. Moreover, systems are being discovered in which metal ions are central players in intracellular communication networks even if they are not themselves the messengers. A striking example involves nitric oxide, NO. This molecule, which was once believed to be only highly toxic, has been discovered to be an important neurotransmitter, a relaxant of vascular smooth muscle, and a factor involved in platelet deaggregation. Its synthesis is regulated by calcium/calmodulin, whereas its receptor is an iron porphyrin within the enzyme guanylate cyclase that converts guanine triphosphate to cyclic guanine monophosphate. This information opens a treasure chest of interesting projects for study by bioinorganic chemists.

13.3. Metal Folding and Cross-Linking

The ability of metal ions to fold biopolymers has been long recognized; however, the extent to which this phenomenon occurs in biology has been appreciated only in recent years. New classes of proteins that contain metal-organized domains, analogous to the zinc-finger proteins, are being discovered with great rapidity. This progress can be expected to continue and be extended to less-well-characterized systems, such as membrane-bound proteins.

Even more rapid accumulation of knowledge can be expected with regard to other biopolymers. The exciting discoveries of the roles of metal ions, for example, potassium and sodium, in controlling the formation of structures such as those found in telomeres foreshadows the elucidation of other metal-based unusual DNA structures. The structure of RNA is an even more open field. The presence of nonspecific and specific binding sites for metal ions in the complex folds of RNA molecules is becoming the rule rather than a curiosity. Detailed structural characterization of these sites by crystallographic or NMR methods for studying RNA should soon reveal some general principles.

Carbohydrates also contain numerous potential metal-binding sites. The detailed three-dimensional structures of carbohydrates, and the role of metal ions in determining and regulating these structures, constitute essentially uncharted territory, representing perhaps the greatest frontier of bioinorganic chemistry in the next decade. Some important leads are being uncovered, however. For example, certain lectins, carbohydrate-binding proteins, have been shown to require metal ions such as calcium for binding. A recent crystal-structure determination revealed that calcium bridges the protein and carbohydrate moieties.

Finally, we come to biominerals and biomineralization. In Chapter 6, we briefly discussed one example, namely, the iron core of ferritin. Several other examples are known, however, including the magnetite cores of magneto-tactic bacteria, cadmium-sulfide particles produced by certain microorganisms, and, of course, the calcium-phosphate-based materials found in bone and teeth. Such materials can play a variety of roles dependent on the unique properties of inorganic solids. The calcium-based materials are crucial for the structural integrity of multicellular organisms. Furthermore, these materials are important components of metal-ion homeostasis systems. The assembly and disassembly of these biominerals are complex and involve not only the inorganic chemistry component but also a variety of interactions with proteins. The field of "solid-state bioinorganic chemistry" is a challenging one that should expand greatly in the near future.

13.4. Binding of Metal Ions to Biomolecules

The selection of the "correct" metal ion is crucial for metallobiomolecule biosynthesis. Much emphasis has been placed on the role of thermodynamics in controlling such reactions; however, the identification of apparent paradoxes has shown that other factors must also be important. For example, *E. coli* produces two distinct polypeptides with superoxide dismutase activity, one of which contains iron, the other manganese. The structures of the proteins are very similar, making it difficult to see how thermodynamic factors alone could allow one metal to bind to one polypeptide and another to another. This paradox assumes that the proteins are binding metal ions from the same pool, however. Recent studies have revealed that the different polypeptides are biosynthesized under different conditions. Thus, iron may be found in the iron superoxide dismutase polypeptide because this polypeptide is produced when iron is fairly abundant. The interplay between such metal-responsive gene-regulatory systems and metal-ion incorporation thermodynamic and kinetics should be a fruitful area for research.

A related issue is the potential involvement of ancillary proteins in metal incorporation. As mentioned in Section 13.2, many interesting leads have been provided by genetics studies. The reconstitution of some of these systems entirely *in vitro* should provide insights into the mechanisms by which metal ions are inserted into proteins for cases for which the incorporation does not occur spontaneously.

The design of metalloproteins is an additional area of study that is beginning to develop. Various investigators have tried to incorporate metal ions into existing protein structures by adding to or modifying amino-acid side chains, by using unnatural amino acids, or by attempting to design novel metalloproteins *de novo*. Some exciting successes have been reported, although a great deal remains to be done. One example from this field involves building a metal-binding site into the combining site of an antibody. Although metal binding has been demonstrated, it remains to be shown whether the metal is actually coordinated to the designated ligands. Furthermore, it is unclear whether or not it will be possible for such metal sites to react in a manner similar to that observed for metalloenzymes.

13.5. Electron-Transfer Proteins

Although much has been learned about the structural changes that occur within single proteins upon undergoing a redox reaction, the mechanisms by which electrons are transferred from one protein to another remains quite controversial. Specifically, the roles of distance, orientation, driving force, and, particularly, the nature of the intervening medium in determining electron-transfer rates remain to be fully elucidated. A great deal of theoretical work has been done to address these issues, and a number of ingenious experimental approaches has been devised to test them. Unfortunately, it has been extremely difficult to isolate any one of these factors while keeping the others sufficiently constant that they need not be considered further. In addition, it is clear that protein-protein electron transfer involves not only the structure of the protein-protein complex but also its dynamics. In other words, the electron transfer occurs from one set of configurations within a wider ensemble of protein-protein complexes. Although the question of how an electron gets from one protein to another is a simple one to phrase, it is difficult to answer clearly. Much more work will be required in the next few years before a consensus can be expected to be reached. It will be very exciting when this field has advanced to the point that a truly predictive understanding of the mechanisms is achieved. Possibly the important breakthrough will require the development of new physical or theoretical methods for studying electron-transfer phenomena.

13.6. Substrate Binding and Activation

As discussed in Chapter 10, the ability of metal ions to bind and activate substrates has been worked out in considerable detail for certain zinc hydrolytic enzymes. This field should progress considerably as information about other systems accumulates through X-ray structural studies together with the use of site-specific mutagenesis methods to identify the role of protein side chains in the catalytic mechanism. Moreover, bioinorganic chemists have shown renewed interest in preparing functional models of the zinc cores in several of these enzymes, notably carbonic anhydrase. This research activity should further advance the area in the coming years. Apart from zinc, various other metal ions are known to function as centers for binding and activating substrates. One example is the putative dinuclear nickel center in urease, which catalyzes the hydrolysis of urea. Mechanistic studies of the protein and model compounds currently in progress are likely to provide additional examples to illustrate the principles set forth in Chapter 10.

Less well studied but likely to emerge as a significant component of this subdiscipline is the investigation of zinc, magnesium, and probably manganese as catalysts or cofactors in systems that process DNA or RNA hydrolytically. Included here are such important enzymes as phosphatases and kinases, which are essential for controlling cell growth and have been implicated in cancer, as well as catalytic RNAs (ribozymes), exo- and endonucleases, DNA and RNA polymerases, topoisomerases, and enzymes involved in the general response of cells to external stress. A recent example is protein kinase C, an enzyme involved in signal transduction, which binds four non-bridged zinc ions in ligand environments composed of approximately one nitrogen and three sulfur atoms. Another is *E. coli* DNA polymerase I, in which a carboxylate-bridged dimetal(II) center is proposed to occur in the transition state involved in the 3'-5' exonuclease catalytic mechanism. An interesting, related discovery is the demonstration that the *Tetrahymena* ribozyme is a metalloenzyme having mechanistic features in common with related protein enzymes. The development of model complexes that mimic such functions should accompany these research activities. A related goal is the design and synthesis of artificial nucleases, metal complexes that recognize DNA or RNA sequences, shapes, and structures, and cleave the phosphodiester backbone with the aid of hydrolytic, rather than redox, chemistry. The value of such reagents is that the fragments they produce have the potential to be religated, perhaps enzymatically, to reform polynucleotides having native structures. In this respect, the complexes serve as artificial restriction enzymes. Such chemistry could make a powerful contribution to biotechnology and would be a significant improvement over currently existing reagents that use redox chemistry where the cleavage reactions, like those afforded by bleomycin, produce only nonligatable nicks in the nucleic acid.

13.7. Atom- and Group-Transfer Chemistry

Some of the major triumphs of bioinorganic chemistry involve reactions in which both electron transfer and Lewis acid or base reactions occur. As described in Chapter 11, proteins that function in this manner include hemoglobin and myoglobin, hemocyanin, and superoxide dismutase, where the dioxygen molecule or one of its metabolites is processed at a metal center, as well as mono- and dioxygenases. Despite the successes in unraveling the detailed mechanisms by which these proteins function, however, very important unanswered challenges remain in this area of bioinorganic chemistry.

One long-standing system is the MoFe protein of nitrogenase, discussed in Section 12.7.b. A soluble cofactor containing approximately seven Fe, one Mo, and eight inorganic S^{2-} ions as well as the homocitrate ion has been extracted from this protein, and attempts are in progress to characterize its structure. The MoFe protein has itself also been crystallized, and progress has recently been made on the X-ray structure, leading to models for the cofactor site composition and geometry. Intensive studies of this enzyme, as well in synthetic model complexes, promise to continue as the structure becomes refined and the nature of the dinitrogen activation center is clarified. The intriguing discovery that some nitrogenases have a cofactor in which the Mo atom has been replaced by V or even Fe has further stimulated research in this area. Nitrogenase is certain to remain on center stage in bioinorganic chemistry pending the outcome of research in these areas as well as concomitant mechanistic studies of the steps involved in nitrogen fixation, both with the natural systems and with synthetic models for the MoFe protein.

Another enzyme of comparable, central interest is cytochrome c oxidase. Like nitrogenase, cytochrome c oxidase uses multiple metal centers to carry out a multielectron-transfer reaction, namely, the four-electron reduction of dioxygen to water. Although considerable information is available about the nature of the two copper and two heme iron centers in this enzyme, including knowledge about specific ligands as determined through site-directed mutagenesis studies of the bacterial systems, the detailed structure is still uncertain. This protein is likely to yield to the barrage of physical, genetics, and molecular modeling studies that are currently in progress to define its geometry, especially the relative positioning of the metal centers and its catalytic mechanism. Of additional interest is to learn how the energy released during turnover generates a proton electrochemical gradient that subsequently drives the synthesis of ATP.

Recently the structure of nitrite reductase, a copper enzyme that converts NO_2^- to NO in the second step of a soil-denitrification process that eventually affords N_2, has been determined crystallographically. The information obtained from this structural analysis should stimulate research into the mechanistic details of the reaction. Similarly, X-ray structural results for enzymes such as hydrogenase and ceruloplasmin should promote the accumulation of mechanistic information about these interesting metalloenzymes in which atom- and group-transfer reactions play a central role. The nonheme diiron enzyme system methane monooxygenase represents another such frontier area, research into which is driven by a diversity of interests. Included are the use of MMOs in bioremediation of the environment, curiosity about how the hydroxylase enzyme might bind and activate methane, the importance of methane as a fuel and hydrocarbon resource, and interest in the details of the hydroxylation chemistry, especially in comparison with what is known about the related heme enzymes, the cytochromes P-450.

Another, almost untouched area is the role of metalloenzymes in sugar metabolism. Included are systems such as xylose isomerase, where a cobalt or magnesium center is responsible for converting a 2-hydroxyaldehyde into a 1-hydroxy-2-ketone, as well as other metalloisomerases. There are three major polymer classes in biology: proteins, nucleic acids, and carbohydrates. Given the activity of bioinorganic chemists in studying metal-promoted reactions in the first two classes, it is surprising that the third has until now received such scant attention. Metal-sugar chemistry would appear to hold considerable promise for both protein chemists and model builders.

13.8. Protein Tuning of the Active Sites

In Chapter 12 we saw how the same metal center could be embedded in a variety of different proteins to function in diverse ways, depending on the coordination number, choice of ligand that coordinated to the unit, local structural and dielectric changes, and substrate binding units contributed by the surrounding protein matrix. The tuning of these bioinorganic chips by the protein environment is similarly expected to emerge as an important feature of several of the systems already mentioned in this chapter, notably nitrogenase (FeMo cofactor), MMO (diiron center), hydrogenase (Fe_4S_4 cluster), urease (dinickel center), and the kinases and phosphatase enzymes (zinc center). The coupling of protein activities that results from embedding several different centers in a membrane constitutes another important frontier area involving active-site tuning. Perhaps the most challenging problem at the moment is to unravel the details of the oxygen-evolving complex (OEC) of photosystem II, in which four photoreduced manganese ions convert water into dioxygen. Present issues are the actual number of Mn ions involved in the catalytic center, their oxidation levels in the various photoreduced states, the O—O bond-forming state, the role of other redox-active components in the center, and the role of the protein environment in controlling and redox steps. Calcium and chloride ion participation, and the involvement of such cofactors as the reaction center chlorophyll, a putative tyrosyl radical, and the coupling of the reaction center chemistry with proton release, are additional features that reveal the intriguing complexity of this system. Challenging problems of this complexity require the collaborative interaction of a number of research groups bringing a wide range of expertise in order for significant progress to be made. Included are geneticists, photophysicists, theoretical chemists, biophysical chemists, inorganic chemists, biochemists, and synthetic model builders, all of whom contribute to the current excitement in bioinorganic chemistry research through their willingness to tackle such significant and interesting problems as the OEC.

Mention in the previous paragraph of the possible involvement of a tyrosyl radical in the OEC brings to mind the identification of this unit, as well as the imidazole radical, as an important component of a number of metalloenzymes. Included in the short but growing list are the small subunit of ribonucleotide reductase, discussed previously in Section 11.4.c, which has a stable tyrosyl radical that has been implicated in the enzyme mechanism; cytochrome c peroxidase, in which a tryptophan radical forms ~ 5 Å from the iron atom; and galactose oxidase, a copper enzyme having a cysteine-modified tyrosyl radical ligand. The last is one of an increasing number of systems in which an amino acid has been modified after translation of the protein. Such post-translational modification is yet another way that nature can tune the active site of a metalloenzyme, expanding the ligand donor and active-site amino-acid base to achieve a specific function. Understanding how tyrosine, tryptophan, and other radicals work in concert with, or help to modulate the properties of, a nearby metal ion constitutes a challenging and important task at the frontier of bioinorganic chemistry.

Study Problems

1. Suggest two additional frontier areas in bioinorganic chemistry.

Bibliography

Classic Work in Bioinorganic Chemistry
G. L. Eichhorn, ed. 1973. *Inorganic Biochemistry*, Volumes 1 and 2. Elsevier, Amsterdam.

Choice and Uptake of Metal Ions
M. J. Abrams and B. A. Murrer, 1993. "Metal Compounds in Diagnosis and Therapy." *Science* 261, 725–730.
M. Barinaga. 1992. "Giving Personal Magnetism a Whole New Meaning." *Science* 256, 967.
M. P. Blaustein. 1988. "Cellular Calcium: Nervous System." in B. E. C. Nordin, ed., *Calcium in Human Biology*. Springer-Verlag, London, 339–366.
K. Davis. 1993. "Cloning the Menkes Disease Gene." *Nature* 361, 98.
S. A. Hinton and D. Dean. 1990. "Biogenesis of Molybdenum Cofactors." *CRC Crit. Rev. Microbiol.* 17, 169–188.
K. J. Skinner. 1991. "The Chemistry of Learning and Memory." *C & E News* 69, 24–41.

Control and Utilization of Metal-Ion Concentrations
D. S. Bredt, P. M. Hwang, C. E. Glatt, C. Lowenstein, R. R. Reed, and S. H. Snyder. 1991. "Cloned and Expressed Nitric Oxide Synthase Structurally Resembles Cytochrome P-450 Reductase." *Nature* 351, 714–718.
H. L. Drake. 1988. "Biological Transport of Nickel," in J. R. Lancaster, Jr., ed., *The Bioinorganic Chemistry of Nickel*. VCH Publishers, Weinheim, 111–139.
C. A. Goode, C. T. Dinh, and M. C. Linder. 1989. "Mechanism of Copper Transport and Delivery in Mammals: Review and Recent Findings," in C. Kies, ed., *Copper Bioavailability and Metabolism*. Plenum Press, New York, 131–144.
E. D. Harris and S. S. Percival. 1989. "Copper Transport: Insights into a Ceruloplasmin-Based Delivery System," in C. Kies, ed., *Proceedings of the ACS Symposium on Copper Bioavailability and Metabolism*. Plenum Press, New York, 95–102.
E. Hubbard. 1989. "Metal Transport," in M. N. Hughes and R. K. Poole, eds., *Metals and Micro-Organisms*. Chapman and Hall, New York, 93–140.
T. V. O'Halloran. 1993. "Transition Metals in Control of Gene Expression." *Science* 261, 715–725.
S. Silver and P. Jasper. 1977. "Manganese Transport in Microorganisms," in E. D. Weinberg, ed., *Microorganisms and Minerals*. Marcel Dekker, New York, 105–149.
S. H. Snyder and D. S. Bredt. 1992. "Biological Roles of Nitric Oxide." *Sci. Amer.* 266, 68–77.
T. G. Traylor and V. S. Sharma. 1992. "Why NO?" *Biochemistry* 31, 2847–2849.

Metal Folding and Cross-Linking

R. J. Doyle. 1989. "How Cell Walls of Gram-Positive Bacteria Interact with Metal Ions," in T. J. Beveridge and R. J. Doyle, eds., *Metal Ions and Bacteria*. Wiley-Interscience, New York, 275–294.

F. G. Ferris, W. Shotyk, and W. S. Fyfe. 1989. "Mineral Formation and Decomposition by Microorganisms," in T. J. Beveridge and R. J. Doyle, eds., *Metal Ions and Bacteria*. Wiley Interscience, New York, 413–442.

A. M. Pyle and J. K. Barton. 1990. "Probing Nucleic Acids with Transition Metal Complexes," in S. J. Lippard, ed., *Progress in Inorganic Chemistry*, vol. 38. Wiley-Interscience, New York, 413–475.

E. D. Weinberg, ed. 1977. *Microorganisms and Minerals*. Marcel Dekker, New York.

Binding of Metal Ions to Biomolecules

H. M. Hassan and C. S. Moody. 1986. "Regulation of the Biosynthesis of Superoxide Dismutase in Procaryotes," in G. Rotilio, ed., *Superoxide and Superoxide Dismutase in Chemistry, Biology, and Medicine* Elsevier, Amsterdam, 274–279.

B. L. Iverson and R. A. Lerner. 1989. "Sequence-Specific Peptide Cleavage Catalyzed by an Antibody." *Science* 243, 1184–1188.

A.-F. Miller and W. H. Orme-Johnson. 1992. "The Dependence on Iron Availability of Allocation of Iron to Nitrogenase Components in *Klebsiella pneumoniae* and *Escherichia coli*." *J. Biol. Chem.* 267, 9398–9408.

R. N. Pau. 1989. "Nitrogenases Without Molybdenum." *TIBS* 14, 183–186.

S. Y. R. Pugh and I. Fridovich. 1986. "Introduction of Superoxide Dismutase in *Escherichia coli* B by Metal Chelating Agents," in G. Rotilio, ed., *Superoxide and Superoxide Dismutase in Chemistry, Biology, and Medicine*. Elsevier, Amsterdam, 280–286.

Electron-Transfer Proteins

J. Bolton, N. Mataga, and G. McLendon, eds. 1991. *Electron Transfer in Inorganic, Organic, and Biological Systems*, in ACS *Advances in Chemistry*, Volume 228. American Chemical Society, Washington, DC.

M. K. Johnson, R. B. King, D. M. Kurtz, Jr., C. Kutal, M. L. Norton, and R. A. Scott, eds. 1990. *Electron Transfer in Biology and the Solid State: Inorganic Compounds with Unusual Properties*, in ACS *Advances in Chemistry*, Volume 226. American Chemical Society, Washington, DC.

Substrate Binding and Activation

J. A. Piccirilli, J. S. Vyle, M. H. Caruthers, and T. R. Cech. 1993. "Metal Ion Catalysis in the *Tetrahymena* Ribozyme Reaction." *Nature* 361, 85–88.

A. M. Pyle. 1993. "Ribozymes: A Distinct Class of Metalloenzymes." *Science* 261, 709–714.

Atom- and Group-Transfer Chemistry

A. Boussac and A. W. Rutherford. 1988. "S-State Formation After Ca^{2+} Depletion in the Photosystem II Oxygen-Evolving Complex." *Chem. Scripta* 28A, 123–126.

G. C. Dismukes. 1988. "The Spectroscopically Derived Structure of the Manganese Site for Photosynthetic Water Oxidization and a Proposal for the Protein-Binding Sites for Calcium and Manganese." *Chem. Scripta* 28A, 99–104.

V. Förster and W. Junge. 1988. "Protolytic Reactions of the Photosynthetic Water Oxidase in the Absence and in the Presence of Added Ligands." *Chem. Scripta* 28A, 111–116.

N. Ito, S. E. V. Phillips, C. Stevens, Z. B. Ogel, M. J. McPherson, J. N. Keen, K. D. S. Yadav, and P. F. Knowles. 1991. "Novel Thioether Bond Revealed by a 1.7 Å Crystal Structure of Galactose Oxidase." *Nature* 350, 87–90.

G. Palmer. 1987. "Cytochrome Oxidase: A Perspective." *Pure and App. Chem.* 59, 749–758.

G. Renger. 1988. "On the Mechanism of Photosynthetic Water Oxidation to Diooxygen." *Chem. Scripta* 28A, 105–109.

B. E. Smith and R. R. Eady. 1992. "Metalloclusters of the Nitrogenases." *Eur. J. Biochem.* 205, 1–15.

Protein Tuning of the Active Sites

B. A. Barry and G. T. Babcock. 1988. "Characterization of the Tyrosine Radical Involved in Photosynthetic Oxygen Evolution." *Chem. Scripta* 28A, 117–122.

A. Boussac and A. W. Rutherford. 1988. "S-State Formation After Ca^{2+} Depletion in the Photosystem II Oxygen-Evolving Complex." *Chem. Scripta* 28A, 123–126.

G. C. Dismukes. 1988. "The Spectroscopically Derived Structure of the Manganese Site for Photosynthetic Water Oxidization and a Proposal for the Protein-Binding Sites for Calcium and Manganese." *Chem. Scripta* 28A, 99–104.

V. Förster and W. Junge. 1988. "Protolytic Reactions of the Photosynthetic Water Oxidase in the Absence and in the Presence of Added Ligands." *Chem. Scripta* 28A, 111–116.

N. Ito, S. E. V. Phillips, C. Stevens, Z. B. Ogel, M. J. McPherson, J. N. Keen, K. D. S. Yadav, and P. F. Knowles. 1991. "Novel Thioether Bond Revealed by a 1.7 Å Crystal Structure of Galactose Oxidase." *Nature* 350, 87–90.

G. Renger. 1988. "On the Mechanism of Photosynthetic Water Oxidation to Dioxygen." *Chem. Scripta* 28A, 105–109.

Index

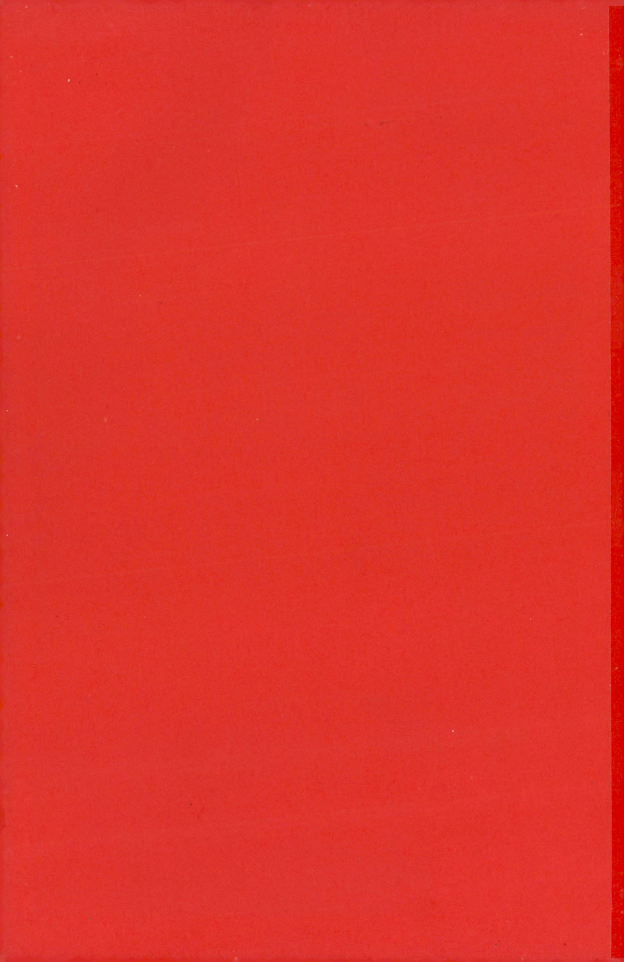